HYDROPONICS

HYDROPONICS

Ernesto Andres Roman
Joe Figel

Kruger Brentt
Publishers

2024

Kruger Brentt Publishers UK. LTD.
Company Number 9728962

Regd. Office: 68 St Margarets Road, Edgware, Middlesex HA8 9UU

ISBN: 978-1-78715-264-9

Disclaimer:

Every possible effort has been made to ensure that the information contained in this book is accurate at the time of going to press, and the publisher and author cannot accept responsibility for any errors or omissions, however caused. No responsibility for loss or damage occasioned to any person acting, or refraining from action, as a result of the material in this publication can be accepted by the editor, the publisher or the author. The Publisher is not associated with any product or vendor mentioned in the book. The contents of this work are intended to further general scientific research, understanding and discussion only. Readers should consult with a specialist where appropriate.

Every effort has been made to trace the owners of copyright material used in this book, if any. The author and the publisher will be grateful for any omission brought to their notice for acknowledgement in the future editions of the book.

For information on all our publications visit our website at http://krugerbrentt.com/

PREFACE

Hydroponics, once a niche agricultural technique, has rapidly transformed into a cornerstone methodology for plant biological research. This innovative approach, centered on cultivating plants in a soil-less environment while precisely controlling their nutrient intake, has ushered in a new era of understanding the complex intricacies of plant life. As our world grapples with escalating environmental challenges and an ever-growing demand for sustainable agriculture, the role of hydroponics in plant research becomes increasingly vital. In the ever-evolving realm of plant biological research, the practice of hydroponics has emerged as a standard methodology that continues to redefine the way we study and understand plant life.

The significance of hydroponics in modern agriculture and plant cultivation cannot be overstated. This innovative method of growing plants without soil, utilizing nutrient-rich water solutions, has far-reaching implications and benefits that contribute to various aspects of food production, environmental sustainability, and scientific research.One of the most prominent advantages of hydroponics is its exceptional resource efficiency. By minimizing water consumption compared to traditional soil-based farming, it addresses the growing concern of water scarcity. Hydroponic systems can reduce water usage by up to 90%, making it a critical solution for regions struggling with limited water resources. Furthermore, its controlled environment and optimized nutrient delivery result in faster growth rates and higher crop yields, which are essential for increasing food production to feed a growing global population.

The soil-less nature of hydroponics eliminates the risk of soil-borne diseases and pests, reducing the need for chemical pesticides and herbicides. This enhances food safety by reducing the potential for harmful contaminants in crops. Additionally, hydroponics enables the cultivation of crops in urban areas and other locations with limited arable land. Vertical and stacked cultivation methods maximize space, making it an ideal solution for urban agriculture and reducing the need for large expanses of farmland. Hydroponic systems offer flexibility in nutrient formulations, enabling growers to tailor nutrient solutions to the specific requirements of different plant species, growth stages, and environmental conditions. This precision control allows for experimentation and research in plant biology, providing valuable insights into plant responses to various factors in a controlled and repeatable manner.

This book represents a comprehensive exploration of the vital role hydroponics plays in advancing our knowledge of the intricate mechanisms governing plant growth, development, and responses to various environmental stimuli. Hydroponics, with its soil-less approach and controlled nutrient delivery, not only offers a precise and adaptable platform for experimentation but also paves the way for sustainable agriculture solutions in an increasingly resource-constrained world. This preeminent technique has opened new avenues for investigating plant physiology, genetics, and biochemistry, making it indispensable for researchers, educators, and students in the field of plant biology. The chapters that follow delve into the core principles of hydroponic systems, nutrient solutions, plant growth, and experimental design, all while emphasizing the pivotal role of hydroponics in pushing the boundaries of plant biology. Additionally, the book underscores the significance of sustainable and resource-efficient plant research, reinforcing the critical connection between hydroponics and our collective responsibility to safeguard the environment.

The book opens by establishing a strong foundation in hydroponics, explaining its core principles, techniques, and advantages in plant research. The book provides insights into formulating and optimizing nutrient solutions for various plant species, emphasizing the importance of nutrient management in achieving research objectives. The book offers a wealth of knowledge about how this technique enables researchers to investigate and manipulate the factors influencing plant growth, from root development to flowering and fruiting. The book showcases real-world applications of hydroponics in plant biology. Through case studies, readers will see how hydroponics has been instrumental in advancing scientific knowledge and practical solutions in agriculture and environmental conservation. The book is a valuable resource for researchers, scholars, and students, aspiring Hydroponic Gardeners, commercial hydroponic farmers, and sources of inspiration for a variety of readers interested in the field of hydroponics.

Lastly, we would like to express our sincere gratitude to the experts, researchers, and professionals who have contributed their knowledge and insights to the creation of this book. In spite of the best efforts, it is possible that some errors may have occurred into the compilation and editing of the book. Further queries, constructive suggestions and criticisms for the improvement of the book are always welcome and shall be thankfully acknowledged.

Ernesto Andres Roman

Joe Figel

CONTENTS

CHAPTER-1

HYDROPONIC SYSTEMS: NUTRIENT SOLUTIONS

INTRODUCTION: AN OVERVIEW

Hydroponic systems represent an innovative and sustainable approach to modern agriculture. Unlike traditional soil-based farming, hydroponics relies on a soilless cultivation method that uses nutrient-rich water solutions to deliver essential elements directly to plant roots. This method allows for greater control over growing conditions, efficient resource use, and year-round cultivation, making it an increasingly popular choice for both commercial growers and home gardeners. In this paragraph, we'll delve into the various aspects of hydroponic systems, from their benefits to different techniques and the plants they can support.

Hydroponic crop production has significantly increased in recent years worldwide, as it allows a more efficient use of water and fertilizers, as well as a better control of climate and pest factors. Furthermore, hydroponic production increases crop quality and productivity, which results in higher competitiveness and economic incomes.

Among factors affecting hydroponic production systems, the nutrient solution is considered to be one of the most important determining factors of crop yield and quality. This chapter aims to explain aspects related to plant nutrition and its effects on production of hydroponic crops, considering basic aspects such as nutrient solutions and their development through the years; components of nutrient solutions (macro and micronutrients), taking into account criteria of nutrimental essentiality in higher plants and their classification, as well as a brief description of their functions in plants; we define the concept of benefic element and its classification, and cite some examples of their addition to nutrient solutions. The concept of pH of the nutrient solution is also defined, as well as its effect on nutrimental availability; osmotic potential of the nutrient solution and its relationship with electric conductivity are discussed, besides their used units and their equivalences, and the influence of both factors on the nutrient uptake in plants; we highlight the importance of oxygenation in the nutrient solution; climate factors affecting nutrient solutions behaviour are also reported, emphasizing on temperature; formulation and preparation of nutrient solutions considering different fertilizer sources and water quality are described as well; finally, we raise topics related to the management of nutrient solutions depending on the species nutrimental needs and on the hydroponic system used, including flow diagrams and figures that facilitate readers comprehension of concepts and principles. Therefore, this chapter aims to be a practical guide to those interested in hydroponic crops, with a strong theoretical support.

HYDROPONIC NUTRIENT SOLUTIONS

Hydroponic nutrient solutions are the lifeblood of hydroponic systems, as they provide essential elements for plant growth in the absence of soil. These solutions are carefully formulated mixtures of water and mineral nutrients, including macronutrients like nitrogen, phosphorus, and potassium, as well as micronutrients like iron, magnesium, and calcium. By precisely controlling the composition and pH levels of these nutrient solutions, hydroponic growers can optimize plant health and yield. Selecting the right nutrient solution and managing its concentration and balance is a fundamental aspect of successful hydroponic cultivation.

A nutrient solution for hydroponic systems is an aqueous solution containing mainly inorganics ions from soluble salts of essential elements for higher plants. Eventually, some organic compounds such as iron chelates may be present (Steiner, 1968). An essential element has a clear physiological role and its absence prevents the complete plant life cycle (Taiz & Zeiger, 1998). Currently 17 elements are considered essential for most plants, these are carbon, hydrogen, oxygen, nitrogen, phosphorus, potassium, calcium, magnesium, sulphur, iron, copper, zinc, manganese, molybdenum, boron, chlorine and nickel (Salisbury& Ross, 1994). With the exception of carbon (C) and oxygen (O), which are supplied from the atmosphere, the essential elements are obtained from the growth medium. Other elements such as sodium, silicon, vanadium, selenium, cobalt, aluminum and iodine among others, are considered beneficial because some of them can stimulate the growth, or can compensate the toxic effects of other elements, or may replace essential nutrients in a less specific role (Trejo-Téllez et al., 2007). The most basic nutrient solutions consider in its composition only nitrogen, phosphorus, potassium, calcium, magnesium and sulphur; and they are supplemented with micronutrients.

The nutrient composition determines electrical conductivity and osmotic potential of the solution. Moreover, there are other parameters that define a nutrient solution as discussed below in detail.

pH of the Nutrient Solution

The pH is a parameter that measures the acidity or alkalinity of a solution. This value indicates the relationship between the concentration of free ions H^+ and OH^- present in a solution and ranges between 0 and 14.

In soil, the Troug diagram illustrates the pH effect on the availability of nutrients to plants (Fig. 1.1). Similarly, changing the pH of a nutrient solution affects its composition, elemental speciation and bioavailability. The term "speciation" indicates the distribution of elements among their various chemical and physical forms like: free ions, soluble complexes, chelates, ion pairs, solid and gaseous phases and different oxidation states (De Rijck & Schrevens, 1998a).

An important feature of the nutrient solutions is that they must contain the ions in solution and in chemical forms that can be absorbed by plants, so in hydroponic systems the plant productivity is closely related with to nutrient uptake and the pH regulation (Marschner, 1995). Each nutrient shows differential responses to changes in pH of the nutrient solution as described below.

In the nutrient solution, NH_3 only forms a complex with H^+. For a pH range between 2 and +7, NH_3 is completely present as NH_{+4} (Fig. 1.2). Increasing the pH above 7 the concentration of $+NH_4$ decreases, while the concentration of NH_3 augments (De Rijck & Schrevens, 1999).

Tyson et al. (2007) in a study to determine the nitrification rate response in a perlite trickling biofilter (root growth medium) exposed to hydroponic nutrient solution, varying NO_{3-} concentrations and two pH levels (6.5 and 8.5), founded that metrication was signicantly impacted by water pH. The increased ammonia oxidation rate (1.75) compared to nitrite oxidation rate (1.3) at pH 8.5 resulted in accumulation of NO_{2-} to levels near those harmful to plants (observed peak of 4.2 mg L_{-1} NO_{2-}). The potential for increased levels of un-ionized ammonia, which reduced plant nutrient uptake from micronutrient precipitation, are additional problems associated with pH 8.5.

Phosphorus is an element which occurs in forms that are strongly dependent on environment pH. In the root zone this element can be found as PO_4^{3-}, H_2PO_{4-}, and H_2PO_{-4} ions; the last two ions are the main forms of P taken by plants. On inert substrates, the largest amount of P available in a nutrient solution is presented when its pH is slightly acidic (pH 5). In alkaline and highly acidic solutions the concentration of P decreases in a-significant way (Dyśko et al., 2008). Namely, with pH 5, 100% of P is present as H_2PO_4 ; this-2 form converts into HPO_4^{-2} at pH 7.3 (pKa2), reaching 100% at pH 10. The pH range that-2dominates the ion $H_2PO_4^{-2}$ on HPO_{4-} is between 5 and 6 (De Rijck & Schrevens, 1997). The pH-dependent speciation of P is showed in Fig. 1.3. Potassium is almost completely present as a free ion in a nutrient solution with pH values-2 from 2 to 9; only small amounts of K^+ can form a soluble complex with SO_4^{-2} or can be bound to Cl^- (De Rijck & Schrevens, 1998a). Like potassium, calcium and magnesium are available to plants in a wide range of pH; however, the presence of other ions interferes in their availability due to the formation of compounds with different grade of solubility.

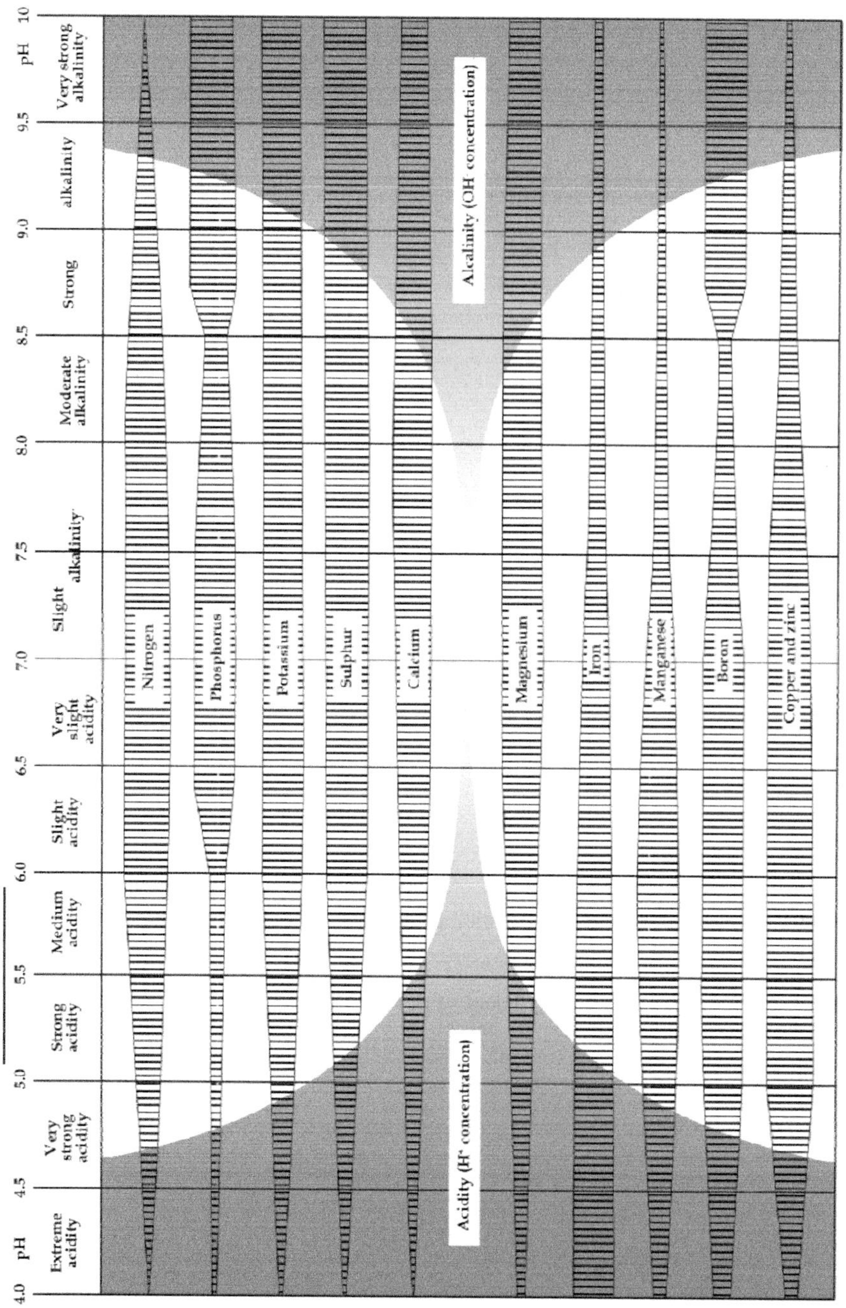

Fig. 1.1. Troug diagram of nutrient availability. Each nutrient is represented with a band; the thickness is proportional to the availability.

As water naturally contains HCO3 , this anion turns into CO3 when the pH is higher than 8.3 or to H_2CO_3 when it- 2-is less than 3.5; the H_2CO_3 is in chemical equilibrium with the carbon dioxide in the atmosphere. Thus at a pH above 8.3, Ca^{2+} and Mg^{2+} ions easily precipitate as carbonates (Ayers& Westcot, 1987). Also, as mentioned above, when the pH of the nutrient solution increases,-2 the HPO_4^{-2} ion predominates, which precipitates with Ca^{2+} when the product of the concentration of these ions is greater than 2.2, expressed in mol m^{-3} (Steiner, 1984). Sulphate also forms relatively strong complexes with Ca^{2+} and Mg^{2+} (De Rijck & Schrevens, 1998b). As-2-pH increases from 2 to 9, the amount of SO4 , forming soluble complexes with Mg^{2+} and with K^+ as KSO_{4-} increases (De Rijck & Schrevens, 1999).

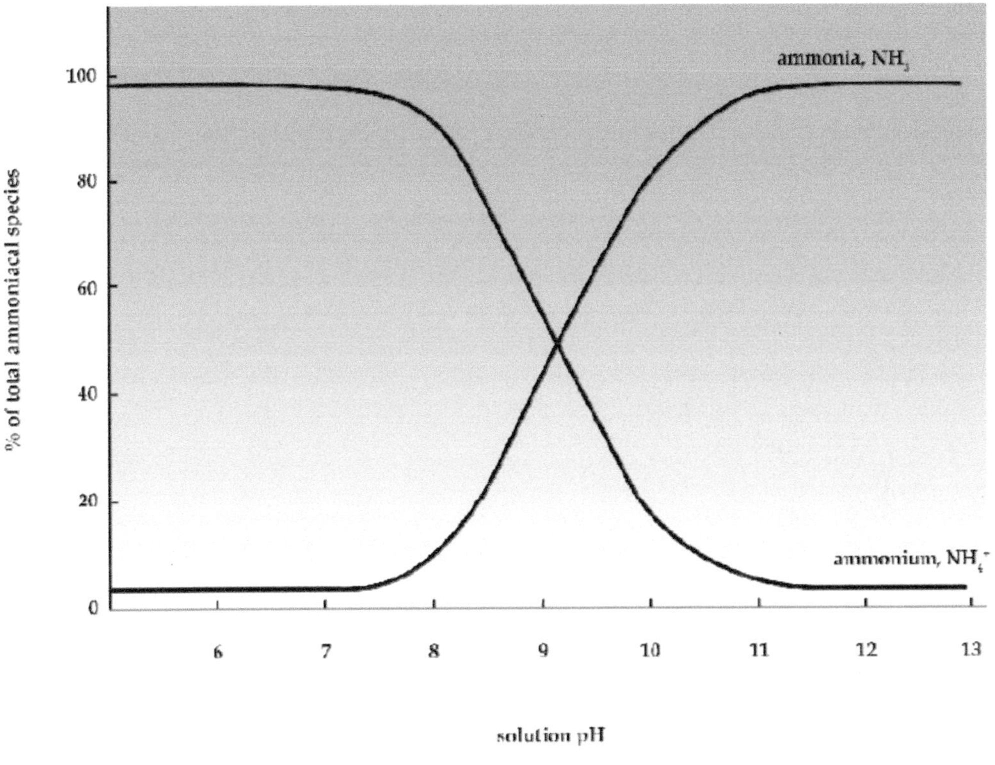

Fig. 1.2. Ammoniacal speciation in function of pH.

Iron, copper, zinc, boron, and manganese, become unavailable at pH higher than 6.5 (Timmons et al., 2002; Tyson, 2007). In *Triticum aestivum*, the manganese precipitation on root surfaces was correlated with a plant-induced rise in pH of culture above 5.5 (Macfie &Taylor, 1989). Boron is mainly uptaken by plants as boric acid, which is not dissociated until pH is close to 7; to greater pH values, boric acid accepts hydroxide ions to form anionic species (Tariq & Mott, 2007) (Fig. 1.4).

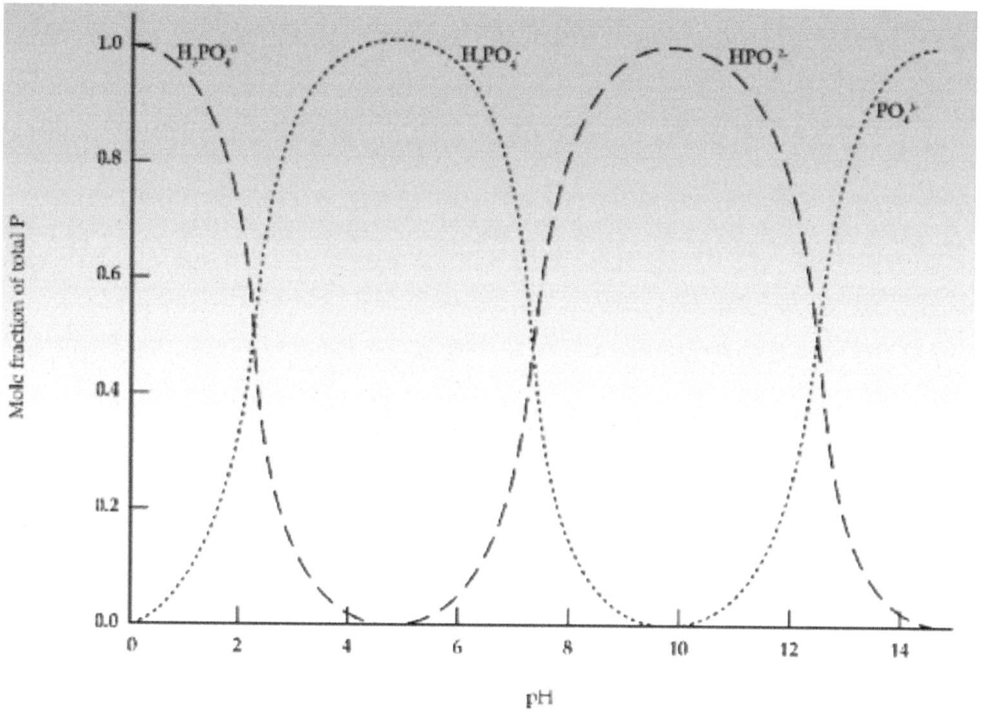

Fig. 1.3. Speciation of P depending on pH.

Therefore, nutrient availability for plant uptake at pH above 7 may be restricted due to precipitation of Fe^{2+}, Mn^{2+}, $PO3^{-4}$, Ca^{2+} and Mg^{2+} to insoluble and unavailable salts (Resh, 2004). The proper pH values of nutrient solution for the development of crops, lies between5.5 and 6.5.

ELECTRICAL CONDUCTIVITY OF THE NUTRIENT SOLUTION

The total ionic concentration of a nutrient solution determines the growth, development and production of plants (Steiner, 1961). The total amount of ions of dissolved salts in the nutrient solution exerts a force called osmotic pressure (OP), which is a colligative property of the nutrient solutions and it is clearly dependent of the amount of dissolved solutes (Landowne, 2006). Also, the terms solute potential or osmotic potential are widely used in nutrient solution, which represent the effect of dissolved solutes on water potential; solutes reduce the free energy of water by diluting the water (Taiz & Zeiger, 1998). Thus, the terms osmotic pressure and osmotic potential can be used interchangeably, still important considering the units that are used, commonly atm, bar and MPa (Sandoval et al., 2007).

An indirect way to estimate the osmotic pressure of the nutrient solution is the electrical conductivity (EC), an index of salt concentration that defines the total amount of salts in a solution. Hence, EC of the nutrient solution is a good indicator of the amount of available

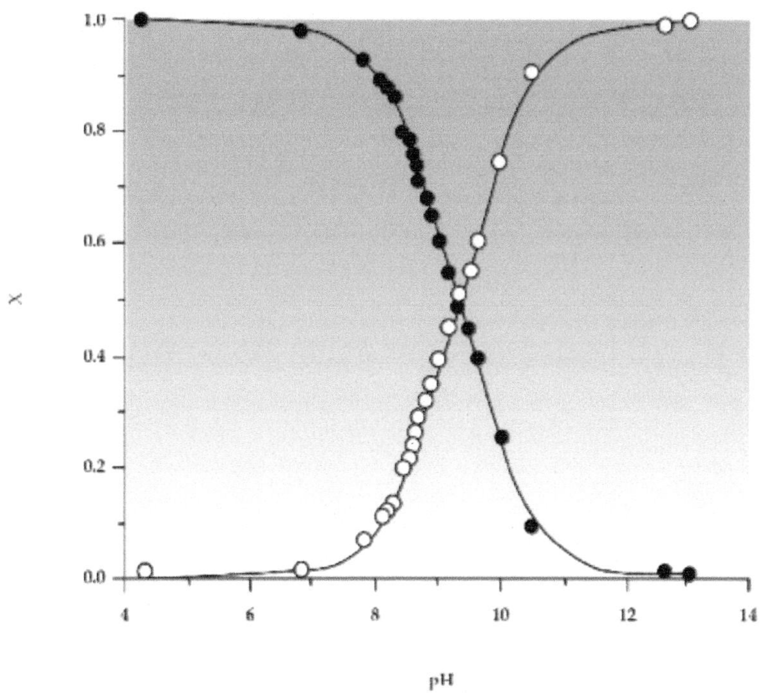

Fig. 1.4. Transformation of boric acid (black circles) and anion forms of boric acid (white circles) as a function of pH (Bishop et al., 2004).

ions to the plants in the root zone (Nemali & van Iersel, 2004). Estimation of the osmotic pressure of a nutrient solution from EC can be done by using the following empirical relations (Sandoval, 2007):

$$OP\ (atm) = 0.36\ X\ EC\ (in\ dS\ m^{-1}\ at\ 25\ ^{\circ}C)$$

$$OP\ (bar) =-0.36\ X\ EC\ (in\ dS\ m^{-1}\ at\ 25\ ^{\circ}C)$$

$$OP\ (MPa) = OP\ (bars)\ X\ 0.1$$

The ions associated with EC are Ca^{2+}, Mg^{2+}, K^{+}, Na^{+}, H^{+}, NO_3 , SO_4, Cl , HCO_3, $OH^{-\ 2}$ (United States Department of Agriculture [USDA], 2001). The supply of micronutrients, namely Fe, Cu, Zn, Mn, B, Mo, and Ni, are very small in ratio to the others elements (macronutrients), so it has no a significant effect on EC (Sonneveld & Voogt, 2009).

The ideal EC is specific for each crop and dependent on environmental conditions (Sonneveld & Voogt, 2009); however, the EC values for hydroponic systems range from 1.5 to 2.5 ds m^{-1}. Higher EC hinders nutrient uptake by increasing osmotic pressure, whereas lower EC may severely affect plant health and yield (Samarakoon et al., 2006). The decrease in water uptake is strongly and linearly correlated to EC (Dalton et al., 1997). Table 1 shows the classification of crops in function of salinity tolerance.

As noted in Table 1.1, some crops can grow with high levels of EC and even a proper management of EC of the nutrient solution can provide and effective tool to improve vegetable quality (Gruda, 2009). In particular, parameters of fruit quality such as soluble solids content, titratable acidity and dry matter augmented by increasing EC level of nutrient solution from 2 to 10 dS m^{-1}. As a consequence, deep sea water (DSW) is being used for nutrient solution due to its high amount of Na^+, Mg^{2+}, K^+ and Ca^{2+} (Chadirin et al., 2007).

Table 1.1. Threshold EC for salinity groups and example of crops (Jensen, 1980; Tanji, 1990).

Salinity Group	Threshold EC, dS m^{-1}	Example of Crops
Sensitive	1.4	lettuce, carrot, strawberry, onion
Moderately sensitive	3.0	broccoli, cabbage, tomato, cucumber, radish, pepper
Moderately tolerant	6.0	soybean, ryegrass
Tolerant	10.0	bermuda-grass, sugarbeet, cotton

NUTRIENT SOLUTION AND ITS COMPOSITION

As previously stated, nutrient solutions usually contain six essential nutrients: N, P, S, K, Ca and Mg. Thereby Steiner created the concept of ionic mutual ratio which is based on the mutual ratio of anions: NO_{3^-}, $H_2PO_{4^-}$ and SO_{42^-}, and the mutual ratio of cations K+, Ca^{2+}, Mg^{2+}. Such a relationship is not just about the total amount of each ion in the solution, but in the quantitative relationship that keep the ions together; if improper relationship between them take place, plan performance can be negatively affected(Steiner, 1961, 1968).

In this way, the ionic balance constraint makes it impossible to supply one ion without introducing a counter ion. A change in the concentration of one ion must be accompanied by either a corresponding change for an ion of the opposite charge, a complementary change for other ions of the same charge, or both (Hewitt, 1966).

When a nutrient solution is applied continuously, plants can uptake ions at very low concentrations. So, it has been reported than a high proportion of the nutrients are not used by plants or their uptake does not impact the production. For example, it was determined that in anthurium, 60% of nutrients are lost in the leachate (Dufour & Guérin, 2005); but in closed systems, however, the loss of nutrients from the root environment is brought to a minimum (Voogt, 2002). Also it has been shown that the concentration of nutrient solution can be reduced by 50% without any adverse effect on biomass and quality in gerbera (Zheng et al., 2005) and geranium (Rouphael et al., 2008). Accordingly, Siddiqi et al. (1998) reported no adverse effect on growth, fruit yield and fruit quality in tomato when reduction of macronutrient concentrations to 50% of the control level as well as cessation of replenishment of the feed solution for 16 days after 7 months of growth at control levels were applied. However, it is expected that in particular situations, too low concentrations do not cover the minimum demand of certain nutrients.

On the other hand, high concentrated nutrient solutions lead to excessive nutrient uptake and therefore toxic effects may be expected. Conversely, there are evidences of positive effects of high concentrations of nutrient solution. In salvia, the increase of Hoagland concentration at 200% caused that plants flowered 8 days previous to the plants at low concentrations, increasing total dry weight and leaf area (Kang & van Iersel, 2004). Likewise, high levels of K^+in the nutrient solution (14.2 meq L^{-1} vs 3.4 meq L^{-1}) increased fruit dry matter, total soluble solids content and lycopene concentration of tomato (Fanasca et al., 2006).

The explanation of these apparent controversial responses is the existence of optimal concentrations of certain nutrients in a solution for a culture under special environmental conditions as well as their relative proportions and not their absolute concentrations as determining factors (Juárez et al., 2006). In order to prevent contradictory observations, Dufour & Guérin (2005) recommend: a) to monitor the availability of nutrients through changes in the ionic composition of the substrate by analysis of percolate, and b) to asses plant nutrient uptake by nutrient content analysis in leaves. Moreover, Voogt (2002) indicates that the nutrient solution composition must reflect the uptake ratios of individual elements by the crop and as the demand between species differs, the basic composition of a nutrient solution is specific for each crop. It must also be taken into account that the uptake differs between elements and the system used. For instance, in open-systems with free drainage, much of the nutrient solution is lost by leachate.

There are several formulations of nutrient solutions. Nevertheless, most of them are empirically based. Table 1.2 comprises some of them.

Table 1.2. Concentration ranges of essential mineral elements according to various authors (adapted from Cooper, 1988; Steiner, 1984; Windsor & Schwarz, 1990).

Nutrient	Hoagland & Arnon (1938)	Hewitt (1966)	Cooper (1979)	Steiner (1984)
	mg L^{-1}			
N	210	168	200-236	168
P	31	41	60	31
K	234	156	300	273
Ca	160	160	170-185	180
Mg	34	36	50	48

Nutrient	Hoagland & Arnon (1938)	Hewitt (1966)	Cooper (1979)	Steiner (1984)
S	64	48	68	336
Fe	2.5	2.8	12	2-4
Cu	0.02	0.064	0.1	0.02
Zn	0.05	0.065	0.1	0.11
Mn	0.5	0.54	2.0	0.62
B	0.5	0.54	0.3	0.44
Mo	0.01	0.04	0.2	Not considered

The Temperature of The Nutrient Solution

The temperature of the nutrient solution influences the uptake of water and nutrients differentially by the crop.

Two nutrient solution temperatures (cold and warm solution, 10 and 22°C, respectively) were evaluated during two flowering events of rose plants (*Rosa × hybrida* cv. Grand Gala).

- Generally, cold solution increased NO_3 uptake and thin-white roots production, but decreased water uptake. Nutrient solution temperature also had an effect on the photosynthetic apparatus. In general terms, the effective quantum yield and the fraction of open PSII reaction centres were higher in rose plants grown at cold solution (Calatayud et al., 2008).

In spinach seedlings, three temperatures of irrigation water (24, 26 and 28°C) were evaluated during 8 weeks. Leaf length, leaf number and total fresh and dry biomass weights per plant were higher in plants grown at elevated temperatures, with optimum growth being recorded at 28°C (Nxawe et al., 2009).

In tomato plants, rates of water and nutrients uptake by roots (which varied depending on solar radiation) were studied. High solution temperature (35°C) produced effects in the short and long-term. In the short-term, water and nutrients uptake were activated through a decrease in water viscosity, and membrane transport was affected. In the long term, oxygen solubility was reduced, while enzymatic oxidization of phenolic compounds in root epidermal and cortex tissues were stimulated, but nutrient concentration in root xylem sap diminished, and the root xylem sap concentration of N, K, Ca became lower than those in the nutrient solution (Falah et al., 2010).

Graves (1983) observed that at temperatures below 22°C the dissolved oxygen in the nutrient solution is sufficient to cover the demand of this element in tomato plants. Nevertheless, the requirement diminished as a consequence of a reduction in a number of physiological processes, including respiration, which further impacts plant growth. Conversely, temperatures over 22°C, oxygen demand is not covered by the nutrient solution as higher temperatures increase the diffusion of this gas. At high temperatures of the nutrient solution an increased vegetative growth to a greater extent than desirable is observed, which reduces fructification.

To assess the importance of temperature on the solubility of oxygen, Table 1.3 depicts data for temperatures that are usually filed within greenhouses, so that temperature has a direct relationship to the amount of oxygen consumed by the plant and reverse relationship with dissolved oxygen from the nutrient solution.

Table 1.3. Solubility of oxygen in water pure at various temperatures at 760 mm Hg of atmospheric pressure.

Temperature, °C	Oxygen solubility, mg L^{-1} of pure water
10	11.29
15	10.08
20	9.09
25	8.26
30	7.56
35	6.95
40	6.41
45	5.93

Management of Nutrient Solution

Soilless cultivation allows a more accurate control of environmental conditions that offers possibilities for increasing production and improving quality of crops. In particular, in the nutrient solution parameters such as temperature, pH, electrical conductivity, oxygen content, among others can be manipulated. If these parameters are not controlled properly and in timing, advantage can be translated into disadvantages. Then, several ways to control some parameters of the nutrient solution are to be reviewed in the following.

pH Regulation

As mentioned above, the pH value determines the nutrient availability for plants. Accordingly, its adjustment must be done daily due to the lower buffering capacity of soilless systems (Urrestarazu, 2004).

The changes in the pH of a nutrient solution depending on the difference in the magnitude of nutrient uptake by plants, in terms of the balance of anions over cations. When the anions are up taken in higher concentrations than cations, for example nitrate, the plant excretes-OH^- or HCO3anions, to balance the electrical charges inside, which produces increasing in the pH value. This process is called physiological alkalinity (Marschner, 1995).

Hence, incorporation of ammonium as N source in the nutrient solution regulates the pH and therefore nutrient availability is ensured. Breteler & Smit (1974) reported that ammonium depressed the pH of nutrient solution even in the presence of nitrate. In rose plants, the addition of ammonium in a nutrient solution containing nitrate produced a total nitrogen uptake increase during shoot elongation; and an increase in P concentration in the roots (Lorenzo et al., 2000). The proportion of total nitrogen is added to the nutrient solution as ammonium is dependent on the crop. On other hand, the chemical adjustment is widely used, namely the addition of acids to reduce the pH value. The pH is closely related to the concentration of HCO_{3-} and $CO3^{2-.}$, when an acid is applied, the CO3ion is transformed to HCO_3 , and then HCO_3 is converted2-into H_2CO_3. Carbonic acid is partially dissociated in H_2O and CO_2 (De Rijck & Schrevens, 1997). Regulation of pH is normally carried out by using nitric, sulphuric or phosphoric acid, and such acids can be use deither individually or combined.

Management of Electrical Conductivity (EC)

Electrical conductivity (EC) is modified by plants as they absorb nutrients and water from the nutrient solution. Therefore, a decrease in the concentration of some ions is and an increase in the concentration of others is observed simultaneously, both in close and open systems. For example, in a closed hydroponic system with a rose crop, the composition of the nutrient solution in the tank was measured. It was observed that the concentration of Fe decreased very fast, while that of Ca^{2+}, Mg^{2+} and Cl^- increased; moreover, concentrations of-2K^+, Ca^{2+} and SO_4 did not reach critical levels (Lykas et al., 2001). Instead, in an open system with recirculation of nutrient solution, an increase in the EC value due to the accumulation of high levels of some ions like bicarbonates, sulphates and chlorides is observed (Zekki et al. 1996). So, the recycling of nutrient solution represents a point of discussion. Moreover, the substrates can retain ions and consequently the EC increases. To reduce the salt accumulation in substrates, the controlled leaching with water of good quality is an alternative (Ansorena, 2004). The use of mulching with polyethylene or polypropylene sheet reduces the water consumption, increases the calculated water use efficiency and decreases the EC of the substrate; so the mulching is an alternative to control of EC too (Farina et al., 2003).

On the one hand, positive evidences of nutrient solutions reuse are reported, which necessarily involves regulation of the EC. Therefore, recycling and reuse of solutions is a trend in searching for sustainable agricultural production systems (Andriolo et al., 2006). Brun et al. (2001) reported recycling systems based on EC control, consisting of adding a water complement to the drainage to decrease the EC and a complement nutrient solution to obtain the desired EC. Carmassi et al. (2003) developed a simple model for the changes in ion concentration and EC of recirculating nutrient solution in closed-loop soilless culture on the basis of balance equation for nutrient uptake by hydroponically-grown plants. In this model, the crop evapotranspiration is compensated by refilling the mixing tank with complete nutrient solution.

Recently, an *in situ* optimal control method of nutrient solution composition has been proposed. Instead of modelling the correlations between greenhouse vegetable growth and nutrient solution, this method is based on Q-learning searches for optimal control policy through systematic interaction with the environment (Chen et al., 2011).

Even though, Bugbee (2004) indicates that the monitoring ions in solution is not always necessary. In fact, the rapid depletion of some nutrients often causes people to add toxic amounts of nutrients to the solution. Besides, it has been demonstrated the existence of a wide cultivable microbial community in the nutrient solution before recycling and recirculation, which supports the necessity of disinfecting nutrient solutions used in soilless cultivation systems, during the recycling process, in order to ensure crop sanitation and avoiding plant disease spreading (Calmin et al., 2008).

Temperature Control

The temperature of the nutrient solution has a direct relation to the amount of oxygen consumed by plants, and an inverse relation to the oxygen dissolved in it, as it was previously indicated. Temperature also affects solubility of fertilizer and uptake capacity of roots, being evident the importance of controlling this variable especially in extreme weathers. Each plant species has a minimum, optimum, and maximum temperature for growth, which requires the implementation of heating or cooling systems for balancing the nutrient solution temperature.

The underground water pipe system for energy-saving control of nutrient solution temperature consists of a large-sized pipe filled with water under the ground, and a unit for circulating the nutrient solution between the cultivation bed and the underground water pipe. The temperature condition in the underground water pipe 1.5 m below the ground surface is stable as compared to that in greenhouses which excessively high temperatures in summer and low in winter. During the circulation, heat can be exchanged between the nutrient solution and the water stored in the underground water pipe. Furthermore, this circulation warmed the nutrient solution excessively chilled in cold winter nights (Hidaka et al., 2008).

Nam et al. (1996) evaluated the cooling capacities of three different systems, which used either polyethylene or stainless tube in the solution tank, or a counter flow type with double pipes, having 41, 70 or 81% of cooling load in a hydroponic greenhouse, respectively.

Villela et al. (2004) evaluated the cooling of the nutrient solution at about 12°C by using a heating exchange device on the productivity of two varieties of strawberry. The cooling of the nutrient solution conferred better productivity of Sweet Charlie variety; whilst it didn't cause any effects over the Campinas variety.

As stated above, deep sea water (DSW) can be used in the preparation of nutrient solutions due to its nutrient content. Likewise, it is one alternative for nutrient cooling systems in hot season due to its low temperature. Cold DSW pumped inside pipe through cultivation bed might decrease temperature of nutrient solution by heat exchange between nutrient solution and DSW. For environment reason, after being used for cooling system, DSW that contained abundant nutrient can be used as nutrient supplement for tomato plant by diluting into standard nutrient solution. It is suggested that DSW might increase fruit quality because of its enrichment of nutrient solution (Chadirin et al., 2006).

Oxygenation of Nutrient Solution

The consumption of O_2 increases when the temperature of nutrient solution increases too. Consequently, it produces an increase in the relative concentration of CO_2 in the root environment if the root aeration is not adequate (Morard & Silvestre, 1996).

The concentration of oxygen in the nutrient solution also depends on crop demand, being higher when the photosynthetic activity increases (Papadopoulous et al., 1999). A decrease bellow 3 or 4 mg L^{-1} of dissolved oxygen, inhibits root growth and produces changes to a brown color, which can be considered as the first symptom of the oxygen lack (Gislerød & Kempton, 1983).

Nonetheless, substrates under long cultivation periods usually present increase of organic matter content and microorganism activity, which could lead to an increase of the competition for oxygen in the root environment. Yet, roots are densely matted within the substrate, which alters diffusion and supply of oxygen (Bonachela et al., 2010).

The supply of pure, pressurized oxygen gas to the nutrient solution is an oxygen-enriched method often used for research purposes, and it is called oxyfertigation (Chun & Takakura, 1994).

Bonachela et al. (2010) evaluated the response to oxygen enrichment of nutrient solution in of autumn-winter sweet pepper and spring melon crops grown on rockwool slabs and perlite grow-bags, compared to non-enriched crops. The pressurized oxygen gas was dissolved in the nutrient solution during each irrigation with a gas injector within the irrigation pipe. The use of inexpensive systems of substrate oxygen enrichment should be restricted to rockwool substrates and to crop periods when a high oxygen demand coincides with low oxygen availability, such as the period from melon flowering phase.

The supply of potassium peroxide as an oxygen generator on vegetable crops grown in commercial substrates once a week was evaluated in sweet pepper, melon and cucumber. Results indicated that the application of potassium peroxide at a concentration of 1 g L^{-1} is the best fraction to use in soilless culture. The treatment with potassium peroxide increases the yield of sweet pepper and melon in 20 and 15% respectively, in comparison to the control, whereas there was no significant difference in cucumber yield (Urrestarazu & Mazuela, 2005).

PREPARATION OF NUTRIENT SOLUTION

Nutrient Solution Design

Hansen (1978) indicates that the addition of plant nutrients to hydroponic systems may be performed according to the plant nutrient requirement. Application of nutrients may be performed according to analyses of a specific crop stage that may describe the consumption of the various typical nutrients of the particular crop or by means of analyses of the total plant needs quantitatively adjusted to the rate of growth and the amounts of water supplied. Thus, the composition and concentration of the nutrient solution are dependent on culture system, crop development stage, and environmental conditions (Coic, 1973; Steiner, 1973).

Likewise, Steiner (1968) proposed that in soilless cultures any ionic ratio and any total concentration of ions can be given, as precipitation limits for certain combinations of ions are considered. Thus, the selection the concentration of a nutrient solution should be such that water and total ions are absorbed by the plant in the same proportion in which those are present in the solution.

Steiner (1961) developed a method to calculate a formula for the composition of a nutrient solution, which satisfies certain requirements. Later on he evaluated five different ratios of NO_3:-anions (NO_{3-}+H_2PO_{4-}+SO_4^{-2}) and three of K+:cations (K^++Ca^++2Mga2+), combining also the two groups, resulting in a full factorial design (Fig. 1.5); all solutions had the same osmotic pressure and pH alue. In this system, the relative concentration of K^+ increases at the expense of a2+ and Mg^{2+} concentrations. Further more, the ratio H_2PO_4-SO4^{2-} between Ca^{2+} and Mg^{2+} is constant, while the changes in the NO_3 concentration-2- -are produced at expense of the H_2PO_4 and SO_4 - concentrations (Steiner, 1966).

On the other hand, Van Labeke et al. (1995) studied *Eustoma grandiflorum* responses to different nutrient solutions differing in ion ratios using an experimental as a {3.1} simplex centroid design, one in the cation factor-space and the other in the anion factor-space, which is depicted in Fig. 1.6. Then, De Rijck & Schrevens (1998c) investigated the effects of the mineral composition of the nutrient solution and the moisture content of the substrate on the mineral content of hydroponically grown tomato fruits, using "design and analysis of mixture systems", a {3.1} simplex lattice design extended with the overall centroid set-up in the cation factor-space (K^+, Ca^{2+} and Mg^{2+}) of the nutrient solution. For each nutritional composition two moisture contents (40 and 80% of volume) of the substrates were investigated. After this short sample illustrates some aspects to be considered in the preparation of nutrient solutions.

Water Quality

According to Tognoni et al. (1998), water quality in hydroponics can be limited to the concentrations of specific ions and phytotoxic substances relevant for plant nutrition as well as the presence of organisms and substances that can clog the irrigation systems.

Regarding the presence of organisms both in water for preparing nutrient solution and in recirculating nutrient solution, its control can be achieved by heat treatment, UV radiation and membrane filtration. However, cheaper chemical treatments as sodium hypochlorite, chlorine dioxide and copper silver ionization may partly solve the pathogen problem; with the disadvantages that introduce a potential accumulation of other elements in closed systems (van Os, 2010).

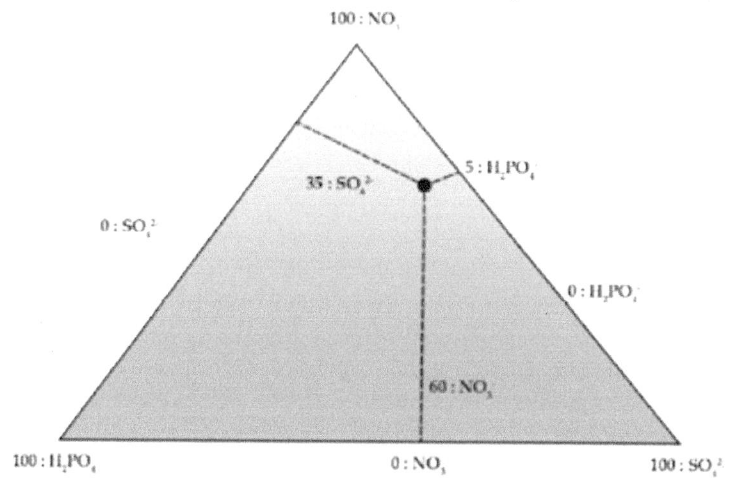

a

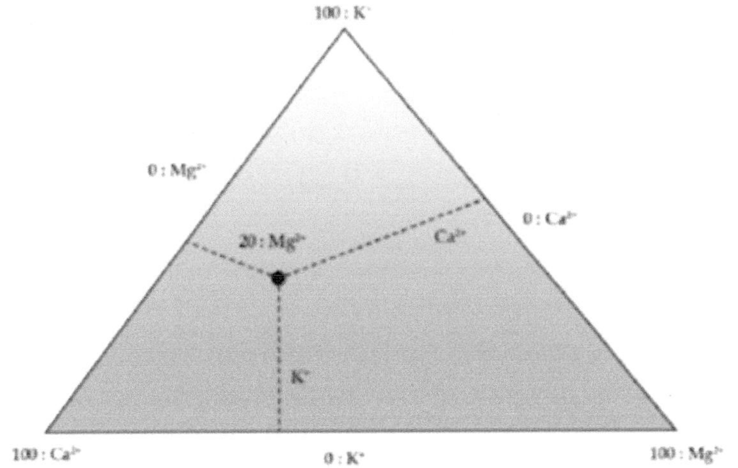

b

Fig. 1.5. Proposed ratios NO_3 :anions (NO_3 +H_2PO_4 +SO4) (a) and K :cations (K+Ca+Mg)(b) by Steiner (1966).

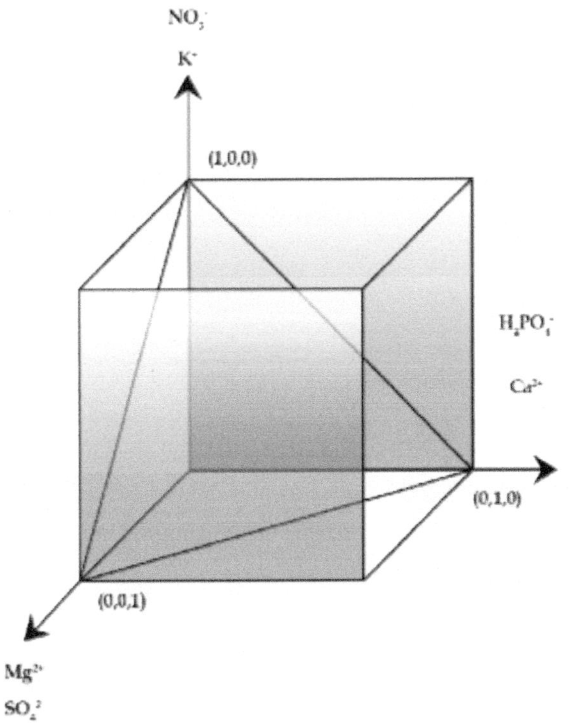

Fig. 1.6. Experimental {3.1} simplex centroid design used by Van Lecke et al. (2006).

It is necessary to carry out a chemical analysis of water to be used in the nutrient solution. Knowing the kind and concentration of ions allows identifying on the one hand, those that are needed in the nutrient solution and therefore can be subtracted from the original formulation; and on the other hand, to take decisions about ions not needed in the nutrient solution.

As previously mentioned, DSW can be used for preparing nutrient solutions. Chadirin et al. (2007) reported that plants treated with circulated water having an EC of 20 dS m^{-1} produced tomatoes with highest soluble solids, %8.0 Brix or increased yield in %30 in comparison to the control. Nevertheless, Pardossi et al. (2008) showed that nutrient solution with high EC (9 dS m^{-1}) had no important effects on both crop yield and fruit quality.

Fertilizer Source for Nutrient Solution

Table 1.4 has a list of commonly used fertilizers and acids in hydroponics, as well as some characteristics of interest for plant nutrition applications.

Table 1.4. Fertilizers containing macronutrients that are commonly used in the preparation of nutrient solutions.

Fertilizers	Formula	Nutrient percentage	Solubility, g L^{-1} at 20 ^{o}C
Calcium nitrate	$Ca(NO_3)2\ 5H_2O$	N: 15.5; Ca: 19	1290
Potassium nitrate	KNO_3	N: 13; K:38	316
Magnesium nitrate	$Mg(NO_3)2\ 6H_2O$	N: 11; Mg:9	760
Ammonium nitrate	$NH4NO_3$	N:35	1920
Monopotassium phosphate	KH_2PO_4	P: 23; K: 28	226
Monoammonium phosphate	$NH4H_2PO_4$	N; 12; P: 60	365
Potassium sulphate	K2SO4	K: 45; S: 18	111
Magnesium sulphate	$MgSO_4\ 7H_2O$	Mg: 10; S: 13	335
Ammonium sulphate	(NH4)2SO4	N: 21; S: 24	754
Potassium chloride	KCl	**K: 60; Cl: 48**	330

PROSPECTS: AT A GLANCE

- The fundamental component in hydroponic system is represented by the nutrient solution. The control of nutrient solution concentration, referred as electrical conductivity or osmotic pressure, allows the culture of a great diversity of species. Moreover, the accurate control of nutrient supply to the plant represents the main advantage of soilless culture. Additionally, the regulation of pH, root temperature among others factors, leads to increased yield and quality.

- Hydroponics is a versatile technology; appropriate for both village and backyard production systems to high-tech space stations. Hydroponic technology can be an efficient mean for food production from extreme environmental ecosystems such as deserts, mountainous regions, or arctic communities. In highly populated areas, hydroponics can provide locally grown high-value crops such as leafy vegetables or cut flowers.
- The future use of controlled environment agriculture and hydroponics must be cost- competitive with those of open field agriculture. Therefore, associated technologies such as artificial lighting, plastics, and new cultivars with better biotic and abiotic resistance will increase crop yields and reduce unit costs of production.
- Prospects for hydroponics may improve if governments design public policies supporting subsidies for such production systems. Besides economic benefits, hydroponics implies conservation of water, cogeneration of energy, income-producing employment for, reducing the impact on welfare rolls and improving the quality of life.
- Nowadays, development and use of hydroponics has enhanced the economic well-being of many communities both in developing and developed countries.

REFERENCES

Andriolo, J. L.; Godoi, R. S.; Cogo, C. M.; Bortolotto, O. C.; Luz, G. L. & Madaloz, J. C. (2006).

Growth and Development of Lettuce Plants at High NH4 :NO_3 Ratios in the + - Nutrient Solution. *Horticultura Brasileira*, Vol.24, No.3, (Jul-Set 2006), pp. 352-355, ISSN 0102-0811

Ansorena, J. (1994). *Sustratos. Propiedades y Caracterización*. Mundi-Prensa, ISBN 978-84-7114- 481-2, Madrid, España

Ayers, C. J. & Westcot, D. W. (1987). *La Calidad del Agua en la Agricultura*. FAO. Serie Riego y Drenaje No. 29. Roma, Italia.

Bishop, M.M; Shahid, N.; Yang, J. & Barron, A. (2004) Determination of the Mode and Efficacy of the Cross-Linking of Guar by Borate Using MAS 11B NMR of Borate Cross-Linked Guar in Combination with Solution 11B NMR of Model Systems, *Dalton Transactions*, Vol.2004,No. 17, pp. 2621–2634, ISSN 0300-9246

Bonachela, S.; Acuña, R. A.; Magan, J. J. & Malfa, O. (2010). Oxygen Enrichment of Nutrient Solution of Substrate-Grown Vegetable Crops under Mediterranean Greenhouse Conditions: Oxygen Content Dynamics and Crop Response. *Spanish Journal of Agricultural Research*, Vol.8, No.4, (Dec 2010), pp. 1231-1241, ISSN: 1695-971-X

Breteler, H. & Smith, A. L. (1974). Effect of Ammonium Nutrition on Uptake and Metabolism of Nitrate in Wheat. *Netherlands Journal of Agricultural Science*, Vol.22, No.1, (Jan 1974), pp. 73–81, ISSN 0028-2928.

Brun, R.; Settembrino, A. & Couve, C. (2001). Recycling of Nutrient Solutions for Rose (*Rosa hybrida*) in soilless culture. *Acta Horticulturae,* Vol.554, No.1, (Jun 2001), pp. 183- 192..ISSN 0567-7572

Bugbee, B. (2004). Nutrient Management in Recirculating Hydroponic Culture, *Acta Horticulturae*, Vol. 648, No.1, (Feb 2004), pp. 99–112, ISSN 0567-7572

Calatayud, A.; Gorbe, E.; Roca D.; & Martínez P. F. (2008). Effect of Two Nutrient Solution + Temperatures on Nitrate Uptake, Nitrate Reductase Activity, NH4 Concentration and Chlorophyll a Fluorescence in Rose Plants. *Environmental and Experimental Botany*, Vol.64, No.1, (September 2008), pp. 65-74, ISSN 0098-8472

Calmin, G.; Dennler, G.; Belbahri, L.; Wigger, A. & Lefort F. (2008). Molecular Identification of Microbial Communities in the Recycled Nutrient Solution of a Tomato Glasshouse Soil-Less Culture. *The Open Horticulture Journal*, Vol.1, No.1, (Jan 2008), pp. 7-14, ISSN: 18748406

Carmassi, G.; Incrocci, L.; Malorgio, M.; Tognoni, F. & Pardossi, A. (2003). A Simple Model for Salt Accumulation in Closed-Loop Hydroponics. *Acta Horticulturae,* Vol.614, No.1, (September, 2003), pp. 149-154, ISSN 0567-7572

Coic, Y (1973) Les Problèmes de Composition et de Concentration des Solutions Nutritives en Culture Sans Sol, *Proceedings of IWOSC 1973 3rd International Congress on Soilless Culture*, pp. 158-164, Sassari, Italy, May 7-12, 1973.

Cooper, A. (1988). "1. The system. 2. Operation of the system". In: *The ABC of NFT. Nutrient Film Technique*, 3-123, Grower Books (ed.), ISBN 0901361224, London, England.

Charidin, Y.; Suhardiyano, H. & Matsuoka, T. 2006. Application of Deep Sea Water for Nutrient Cooling System in Hydroponic Culture, *Proceedings of APAARI the International Symposium on Sustainable Agriculture in Asia*, pp. 1-4, Bogor, Indonesia, September 18-21, 2006.

Chadirin, Y.; Matsuoka, T.; Suhardiyanto, H. & Susila. A. D. (2007). Application of Deep Sea Water (DSW) for Nutrient Supplement in Hydroponics Cultivation of Tomato: Effect of supplemented DSW at Different EC Levels on Fruit Properties. *Bulletin Agronomy*, Vol.35, No.2, pp. 118 – 126, ISSN 216-3403

Chun, C. & Takakuta, T. (1994). Rate of Root Respiration of Lettuce under Various Dissolved Oxygen Concentrations in Hydroponics. *Environment Control in Biology*, Vol.32, No.2, (Apr 1994), pp. 125-135, ISSN 1883-0986

Dalton, F. N.; Maggio, A. & Piccinni, G. (1997). Effect of Root Temperature on Plant Response Functions for Tomato: Comparison of Static and Dynamic Salinity Stress Indices. *Plant and Soil,* Vol. 192, No.2, (May 1997), pp. 307-319, ISSN 0032-079X

De Rijck G. & Schrevens E. (1997) pH Influenced by the Elemental Composition of Nutrient Solutions. *Journal of Plant Nutrition*, Vol.20, No.7-8, (Jul 1997) 911-923. ISSN. 0190- 4167

De Rijck G.; Schrevens E. (1998a) Cationic Speciation in Nutrient Solutions as a Function of pH. *Journal of Plant Nutrition*, Vol.21, No.5 (May 1998), pp. 861-870, ISSN. 0190-4167

De Rijck, G. & Schrevens, E. (1998b). Elemental bioavailability in nutrient solutions in relation to complexation reactions. *Journal of Plant Nutrition*, Vol.21, No.10, (Oct 1998), pp. 2103-2113, ISSN 0190-4167.

De Rijck G. & Schrevens E (1998c) Comparison of the Mineral Composition of Twelve Standard Nutrient Solutions. *Journal of Plant Nutrition*, Vol.21, No.10, (Oct 1998), pp. 2115-2125. ISSN 0190-4167

De Rijck G. & Schrevens, E. (1999) Anion Speciation in Nutrient Solutions as a Function of pH *Journal of Plant Nutrition*, Vol.22, No.2, (Feb 1999), pp. 269-279. ISSN 0190-4167

Dufour, L. & Guérin, V. (2005). Nutrient Solution Effects on the Development and Yield of Anthurium andreanum Lind. in Tropical Soilless Conditions. *Scientia Horticulturae*, Vol.105, No.2, (Jun 2005), pp. 269-282, ISSN 0304-4238

Dyśko, J. ; Kaniszewski, S. & Kowalczyk, W. (2008). The Effect of Nutrient Solution pH on Phosphorus Availability in Soilless Culture of Tomato. *Journal of Elementology*, Vol. 13, No.2, (Jun 2008), pp. 189-198, ISSN 1644-2296

Falah, M. A. F.; Wajima, T.; Yasutake, D.; Sago, Y. & Kitano, M. (2010). Responses of Root Uptake to High Temperature of Tomato Plants (*Lycopersicon esculentum* Mill.) in Soil-less Culture. *Journal of Agricultural Technology*, Vol.6, No.3, (Jul 2010), pp. 543- 558, ISSN 1686-9141

Fanasca, S.; Colla, G.; Maiani, G.; Venneria, E.; Rouphael, Y.; Azzini, E. & Saccardo, F. (2006). Changes in Antioxidant Content of Tomato Fruits in Response to Cultivar and Nutrient Solution Composition. *Journal of Agricultural and Food Chemistry,* Vol.54, No. 12, (Jun 2006), pp. 4319-4325, ISSN 0021-8561

Farina, E.; Allera, C.; Paterniani, T. & Palagi, M. (2003). Mulching as a Technique to Reduce Salt Accumulation in Soilless Culture. *Acta Horticulturae,* Vol.609, No.1, (May 2003), pp. 459-466, ISSN 0567-7572

Chen, F.; He, H. & Tang, Y. (2011). In-situ Optimal Control of Nutrient Solution for Soilless Cultivation, *Proceedings of ICACC 2011 3rd International Conference on Advanced Computer Control*, pp. 412-416, Harbin, China, Jan 18-20, 2011.

Gislerød, H. R. & Adams, P. (1983). Diurnal Variations in the Oxygen Content and Requirement of Recirculating Nutrient Solutions and in the Uptake of Water and

Potassium by Cucumber and Tomato Plants. *Scientia Horticultura*, Vol.21, No.4, (Dec 1983), pp. 311–321, ISSN 0304-4238

Graves, C. J. (1983). The Nutrient Film Technique. *Horticultural Reviews*, Vol.5, No.1, (Jan 1983), pp. 1-44, ISSN 978-0-470-38642-2

Gruda, N. (2009). Do Soilles Culture Systems have an Influence on Product Quality of Vegetables? *Journal of Applied Botany and Food Quality*, Vol.82, No.2, pp. 141-147, ISNN 1613-9216

Hansen, M. (1978). Plant Specific Nutrition and Preparation of Nutrient Solutions. *Acta Horticulturae,* Vol.82, No.1, (Apr 1978), pp. 109-112, ISSN 0567-7572

Hewitt, E. J. (1996). *Sand and Water Culture Methods Used in the Study of Plant Nutrition.* Technical Communication No. 22. Commonwealth Bureau of Horticulture and Plantation Crops, East Malling, Maidstone, Kent, England

Hidaka, K.; Kitano, M.; Sago, Y.; Yasutake, D.; Miyauchi, K.; Affan, M. F. F.; Ochi, M. & Imai,

S. (2008). Energy-Saving Temperature Control of Nutrient Solution in Soil-Less Culture Using an Underground Water Pipe. *Acta Horticulturae,* Vol.797. No.1, (Sep 2008), pp. 185-191, ISSN 0567-7572

Jensen, M. H. & Collins, W. L. (1985). Hydroponic Vegetable Production. *Horticultural Reviews*, Vol.7, pp. 483- 559, ISSN 9780870554926

Juárez H., M. J.; Baca C., G. A.; Aceves N., L. A.; Sánchez G., P.; Tirado T., J. L.; Sahagún C.,

J. & Colinas D. L., M. T. (2006). Propuesta para la Formulación de Soluciones Nutritivas en Estudios de Nutrición Vegetal. *Interciencia*, Vol.31, No.4, (Apr 2006), ISSN 0378-1844

Kang, J. G. & van Iersel, M. W. (2004). Nutrient Solution Concentration Affects Shoot: Root Ratio, Leaf Area Ratio, and Growth of Subirrigated Salvia (*Salvia splendens*). *HortScience*, Vol.39, No.1, (Feb 2004), pp. 49-54, ISSN 0018-5345

Landowne, D. (2006). *Cell Physiology*, McGraw-Hill Medical Publishing Division, ISBN 0071464743, Miami, FL., U. S. A.

Lorenzo, H.; Cid, M. C; Siverio, J. M. & Caballero, M. (2000). Influence of Additional Ammonium Supply on Some Nutritional Aspects in Hydroponic Rose Plants. *The Journal of Agricultural Science*, Vol.134, No.4, (Sep, 2000), pp. 421-425, ISSN 0021- 8596

Lykas, C. H.; Giaglaras, P. & Kittas, C. (2001). Nutrient Solution Management Recirculating Soilless Culture of Rose in Mild Winter Climates. *Acta Horticulturae,* Vol.559, No.1, (Oct 2001), pp. 543-548, ISSN 0567-7572

Macfie, S. M. & Taylor, G. J. (1989). The Effects of pH and Ammonium on the Distribution of Manganese in *Triticum aestivum* Grown in Solution Culture. *Canadian Journal of Botany*, Vol.67, No.11, (Nov 1989), pp. 3394-3400 ISSN 0008-4026

Marschner, H. (1995). *Mineral Nutrition of Higher Plants*, Academic Press, ISBN 0-12-473542-8, New York, U. S. A.

Morard, P. & Silvester, J. (1996). Plant Injury Due to Oxygen Deficiency in the Root Environment of Soilless Culture: A Review. *Plant and Soil*, Vol.184, No.2, pp. 243- 254, ISBN 0032-079X

Nam, S. W.; Kim, M. K. & Son, J. E. (1996). Nutrient Solution Cooling and its Effect on Temperature of Leaf Lettuce in Hydroponic System. *Acta Horticulturae,* Vol.440, No. 1, (Dec 1996), pp.: 227-32. ISSN 0567-7572

Nxawe, S.; Laubscher, C. P. & Ndakidemi, P. A. (2009). Effect of Regulated Irrigation Water Temperature on Hydroponics Production of Spinach (*Spinacia oleracea L*). *African Journal of Agricultural Research*, Vol.4, No.12, (December, 2009), pp. 1442-1446, ISSN 1991- 637X

Nemali, K. S. & van Iersel, M. W. (2004). Light Intensity and Fertilizer Concentration: I. Estimating Optimal Fertilizer Concentration from Water-Use Efficiency of Wax Begonia. *HortScience*, Vol.39, No.6, (Oct 2004), pp. 1287-1292. ISSN 0018-5345

Papadopoulous, A. P.; Hao., X.; Tu, J. C. & Zheng, J. (1999). Tomato Production in Open or Closed Rockwool Culture Systems with NFT or Rockwool Nutrient Feedings. *Acta Horticulturae,* Vol.481. No.1, (Jan 1999), pp. 89-96, ISSN 0567-7572

Pardossi, A.; Incrocci, L.; Massa, D.; Carmassi, G. & Maggini, R. (2009). The Influence of Fertigation Strategies on Water and Nutrient Efficiency of Tomato Grown in Closed Soilless Culture with Saline Water. *Acta Horticulturae*, Vol.807, No.2, (Jan 2009), pp. 445-450, ISSN 0567-7572

Resh, H. M. (2004). *Hydroponic Food Production*, Newconcept Press, Inc., ISBN-10: 093123199X, Mahwah, NJ., U. S. A.

Rouphael, Y.; Colla, G., (2009). The Influence of Drip Irrigation or Subirrigation on Zucchini Squash Grown in Closed-Loop Substrate Culture with High and Low Nutrient Solution Concentrations. *HortScience*, Vol.44, No.2, (Apr 2009), pp. 306-311, ISSN 0018-5345

Salisbury, F. B. & Ross, C. W. (1992). *Plant Physiology.* Wadsworth Publishing Company, ISBN 0-534-15162-0, California, U. S. A.

Samarakoon, U.C.; Weerasinghe, P. A. & Weerakkody, A. P. (2006). Effect of Electrical Conductivity [EC] of the Nutrient Solution on Nutrient Uptake, Growth and Yield of Leaf Lettuce (*Lactuca sativa* L.) in Stationary Culture. *Tropical Agricultural Research*, Vol.18, No. 1, (Jan 2006), pp. 13-21 ISSN 1016.1422

Sandoval V., M.; Sánchez G., P. & Alcántar G., G. (2007). Principios de la Hidroponía y del Fertirriego, In: *Nutrición de Cultivos*, G. Alcántar G & L. I. Trejo-Téllez, L. I. (Eds.), 374-438, Mundi-Prensa, ISBN 978-968-7462-48-6, México, D. F., México.

Siddiqi, M. Y.; Kronzucher, H. J.; Britto, D. T. & Glass, A. D. M. (1998). Growth of a Tomato Crop at Reduced Nutrient Concentrations as a Strategy to Limit Eutrophication. *Journal of Plant Nutrition*, Vol.21, No.9, (Sep 1998), pp. 1879-1895. ISSN 0190-4167

Steiner, A. A. (1961). A Universal Method for Preparing Nutrient Solutions of a Certain Desired Composition. *Plant and Soil*, Vol.15, No.2, (October, 1961), pp. 134-154, ISBN 0032-079X

Steiner, A. A. (1966). The Influence of Chemical Composition of a Nutrient Solution on the Production of Tomato Plants. *Plant and Soil*, Vol.24, No.3, (June1966), pp. 454-466, ISBN 0032-079X

Steiner, A.A. (1968). Soilless Culture, *Proceedings of the IPI 1968 6th Colloquium of the Internacional Potash Institute*, pp: 324-341, Florence, Italy

Steiner, A. A. (1973). The Selective Capacity of Tomato Plants for Ions in a Nutrient Solution, *Proceedings of IWOSC 1973 3rd International Congress on Soilless Culture*, pp. 43-53, Sassari, Italy, May 7-12, 1973

Steiner, A. A. (1984). The Universal Nutrient Solution, *Proceedings of IWOSC 1984 6th International Congress on Soilless Culture*, pp. 633-650, ISSN 9070976048, Wageningen, The Netherlands, Apr 29-May 5, 1984

Sonneveld, C. & Voogt, W. (2009). *Plant Nutrition of Greenhouse Crops*, Springer, ISBN 9048125316, New York, U. S. A.

Taiz, L. & Zeiger, E. (1998). *Plant Physiology.* Sinauer Associates, Inc. Publishers. Sunderland, ISBN : 0878938311, Massachusetts, U. S. A.

Tanji, K. K. (1990). *Agricultural salinity assessment and management.* American Society of Civil Engineers, ISBN-10: 0872627624, New York, U. S. A.

Tariq, M. & Mott, C. J. B. (2007). The Significance of Boron in Plant Nutrition an environment-A Review. *Journal of Agronomy,* Vol.6, No.1, (Jan 2007), pp. 1-10, ISSN 1812-5379

Timmons, M. B; Ebeling, J. M.; Wheaton, F. W.; Summerfelt, S. T. & Vinci, B. J. (2002). *Recirculating aquaculture systems*. Cayuga Aqua Ventures, ISBN 0-9712646-1-9, Ithaca, NY.

Tognoni, F.; Pardossi, A. & Serra, G. (1998). Water Pollution and the Greenhouse Environmental Costs. *Acta Horticulturae*, Vol.458, No.1, (Apr 1998), pp. 385–394, ISSN 0567-7572

Trejo-Téllez, L. I.; Gómez-Merino, F. C. & Alcántar G., G. (2007). Elementos Benéficos, In: *Nutrición de Cultivos*, G. Alcántar G & L. I. Trejo-Téllez, L. I. (Eds.), 50-91, Mundi- Prensa, ISBN 978-968-7462-48-6, México, D. F., México

Tyson, R. V. (2007). Reconciling pH for Ammonia Biofiltration in a Cucumber/ Tilapia Aquaponics System Using a Perlite Medium. *Journal of Plant Nutrition*, Vol.30, No.6, (Jun 2007), pp. 901–913, ISSN 0190-4167

Tyson, R. V.; Simonne, E. H.;Davis, M.; Lamb, E. M.; White, J. M. & cation Rate in Perlite Medium.⇒Treadwell, D. D. (2007). Effect of Nutrient Solution, Nitrate-Nitrogen Concentration, and pH on Nitri *Journal of Plant Nutrition*, Vol.30, No.6, (Jun, 2007), pp. 901-913, ISSN 0190-4167

Urrestarazu, M. (2004). *Tratado de Cultivo sin Suelo.* Mundi-prensa, ISBN 84-8476-139-8, Madrid, España

Urrestarazu, M. & Mazuela, P. C. (2005). Effect of Slow-Release Oxygen Supply by Fertigation on Horticultural Crops under Soilless Culture. *Scientia Horticulturae*, Vol. 106, No.4, (November 2005), pp. 484-490, ISSN 0304-4238

USDA. (2001). *Soil Quality Test Kit Guide.* Natural Resources Conservation Service. Washington, D. C., U. S. A.

Van Labeke, M. C.; Dambre, P.; Schrevens. E. & De Rijck G. (1995) Optimisation of the Nutrient Solution for *Eustoma glandiflorum* in Soilless Culture. *Acta Horticulturae*, Vol.401, No.1, (Oct 1995), pp. 401-408, ISSN 0567-7572

Van Os, E. A. (2010). Disease Management in Soilless Culture Systems. *Acta Horticulturae*, Vol.883, No.1, (Nov 2010), pp. 385-393, ISSN 0567-7572

Villela, J.; Luiz, V. E.; Araujo, J. A. C. de & Factor, T. L. (2004). Nutrient Solution Cooling Evaluation for Hydroponic Cultivation of Strawberry Plant. *Engenharia Agrícola*, Vol.24, No.2, (May-Aug 2004), pp. 338-346, ISSN 0100-6916

Voogt, W. (2002). Potassium management of vegetables under intensive growth conditions, In: *Potassium for Sustainable Crop Production.* N. S. Pasricha & S. K. Bansal SK (eds.), 347-362, International Potash Institute, Bern, Switzerland.

Windsor, G. & Schwarz, M. (1990). *Soilless Culture for Horticultural Crop Production.* FAO, Plant Production and Protection. Paper 101. Roma, Italia.

Zekki, H.; Gauthier, L. & Gosselin A. (1996). Growth, Productivity and Mineral Composition of Hydroponically Cultivated Greenhouse Tomatoes, With or Without Nutrient Slution Recycling. *Journal of the American Society for Horticultural Science*, Vol.121, No.6, (Nov 1996), pp. 1082-1088, ISSN 0003-1062

Zheng, Y.; Graham, T. H.; Richard, S. & Dixon, M., (2005). Can Low Nutrient strategies be Used for Pot Gerbera Production in Closed-Loop Subirrigation? *Acta Horticulturae,* Vol.691, No.1, (October 2005), pp. 365-372. ISSN 0567-7572

CHAPTER-2

PARAMETERS REQUIRED FOR *IN VITRO* HYDROPONIC PEA PLANTLET FLOWERING AND FRUITING

INTRODUCTION: AN OVERVIEW

Achieving successful flowering and fruiting in pea plantlets cultivated in an *in vitro* hydroponic setting is a subject of significant interest in modern agricultural research. The controlled environment of *in vitro* hydroponics offers researchers a unique opportunity to manipulate various parameters to optimize the growth and reproductive phases of these plants. In this context, understanding the essential parameters and factors necessary for inducing flowering and fruit development in pea plantlets is crucial.

Flowering *in vitro* has been reported infrequently in tissue culture and the subsequent occurrence of fruiting structures from these flowers is rare (Al-Juboory et al., 1991; Bodhipadma & Leung, 2003; Dickens & Van Staden 1985; 1988; Franklin et al., 2000; Ishioka et al., 1991; Lee et al., 1991; Pasqua et al., 1991; Rastogi & Sawhney 1986; Tisserat & Galletta 1993; 1995). Fruits are complex organs composed of unique tissues that are a source of many important food products, nutrients and phytochemicals. The biosynthesis of phytochemicals common in fruits by cultured vegetative cells and tissues is difficult and usually not achievable; when achieved they usually occur at lesser yields than in vivo derived fruits (Tisserat et al., 1989a; 1989b). Unfortunately, fruit tissues and organs are difficult to establish, maintain and proliferate *in vitro* as such, mainly because they fail to retain their unique tissue and organ integrity within a sterile environment and often generate into a undifferentiated mass (*i.e.* callus) with an altered biochemical metabolism compared to that obtained from the original fruit tissues (Hong et al., 1989; Tisserat et al., 1989a; 1989b). Nevertheless, development of sterile fruit production systems would be useful in order to study the reproductive processes, provide a source of important secondary natural products *in vitro*, provide sterile produce for at-risk populations with weakened immune systems, and aid in breeding projects (Bodhipadma & Leung, 2003; Butterweck, 1995; Kamps, 2004; Ochatt et al., 2002; Pryke & Taylor, 1995).

According to the Centers for Disease Control and Prevention (CDC) each year 76 million people in the USA get food sickness, of these 325,000 are hospitalized and 5,000 die (Anon., 2004). One in every 6 Americans becomes a victim to a serious food poisoning per year (CDC, 2011) The majority of these victims have a weakened immune system and cannot effectively fight infections normally (U.S. Department of Agriculture-Food Safety and Inspection Service, 2006; Hayes et al., 2006). Peoples with high risk to food borne infections include: young children, pregnant women, older adults, and persons with weakened immune systems, including those with HIV/AIDS infection, cancer, diabetes, kidney disease, and transplant patients, or those individuals undergoing chemotherapy (U.S. Department of Agriculture-Food Safety and Inspection Service, 2006; CDC, 2011; Hayes et al., 2006; Kamps, 2004). An analysis of food-poisoning outbreaks from 1990-2003 revealed that contaminated produce is responsible for the greatest number of individual food-borne illnesses (Table 2.1) (Anon., 2004).

Table 2.1. Analysis of monitored food-poisoning outbreaks and number of illnesses from 1990-2003 (Anon. 2004).

Food Source	Outbreaks No. (%)	Illnesses No. (%)	Common Food Pathogens
Seafood	720 (26%)	8,044 (10%)	*Vibrio parahaemolyticus*
Produce	428 (16%)	23,857 (29%)	*Salmonella*
Poultry	355 (13%)	11,898 (14%)	*Clostridium perfringens*
Beef	338 (12%)	10795 (13%)	*C. perfringens, Escherichia coli*
Eggs	306 (11%)	10449 (13%)	*Salmonella*
Multi-ingredient foods	591 (22%)	17,728 (21%)	Multiple, unknown
Totals:	2,738	82,771	

Other studies suggest that of the 200,000 to 800,000 cases of food poisoning occurring daily in the USA over one-third to as much as one-half are due to produce contaminates, with≈90% of that being bacterial (U.S. Department of Agriculture-Food Safety and Inspection Service, 2006; Kamps, 2004; Wagner, 2011). In addition, food poisoning is also caused by heavy metals, fungus, viruses, chemicals (e.g., pesticides), and parasites. All of these contaminates are commonly found on produce, and still produce food poisoning is usually overlooked as the source of the food borne illness (CDC, 2010; Kamps, 2004). Obviously, growing sterile produce would have a niche market for at-risk populations susceptible to food poisoning. One can speculate that the high cost of sterile produce would be acceptable to at-risk populations in certain situations.

Elucidation of the wide range of parameters responsible for flowering and fruiting from cultured plantlets on demand in a single study has not been conducted. A sterile hydroponics culture system, termed the automated plant culture system (APCS) has been employed to obtain reproductive activity (i.e. flowering and fruiting) from cucumber and pepper plantlets (Tisserat & Galletta 1993; 1995). The APCS mimics the hydroponic system by employing a large plant growth culture vessel and immersing and draining the plant periodically with nutrient medium kept in a larger medium reservoir (e.g. 1000-ml aliquot) (Fig.2.1). The in vitro hydroponic system differs from traditional agar tissue culture vessel (e.g. 25-ml aliquot) in a number of characteristics: 1) plant cultures in the APCS remain intact and stationary for the duration of their growth once established whereas plantlets grown in agar-based cultures must be repeatedly and frequently removed and replanted to obtain fresh medium, 2) the APCS provides considerably more nutrient medium (≈ 40 x) for culture growth (stored in a separate reservoir) than the agar media provided for plantlets grown in culture tubes, 3) plantlets grown in the APCS can achieve considerably higher growth rates and physical sizes than plantlets grown in agar grown cultures, and 4) flowering and fruiting is more readily achievable in plantlets grown in the APCS than in agar cultures tubes.

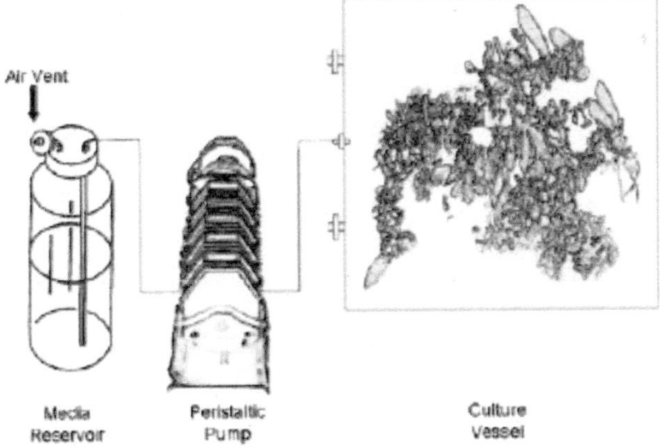

Fig. 2.1. Diagrammatic representation of automated plant culture system.

The very fact that plantlets grown in the APCS can obtain larger sizes than plantlets grown in smaller culture vessels may in itself be the prerequisite to obtain subsequent reproductive activities *in vitro* compared to plantlets grown in the tissue culture vessel (Tisserat & Galletta, 1993; 1995). Further, intermittent forced ventilation of the culture vessel with sterile air greatly aided in the formation of flowers and fruits from cucumbers in the APCS (Tisserat & Galletta, 1993).

Flowering is considered a complex, morphological event regulated by a combination of genetic, hormonal, nutritional and environmental factors. In prior studies, the production of fruits from flowers *in vitro* was achieved with a great degree of cultural manipulation (Al- Juboory et al., 1991; Dickens & Van Staden 1985; 1988; Ishioka et al., 1991; Pasqua et al., 1991; Rastogi & Sawhney 1986; Tisserat & Galletta 1993; 1995). This study will investigate a number of the parameters which have been identified as considered important factors to promote flowering and fruiting reports. Specifically, these factors included: cultivar types, vessel size, medium volume, inorganic salt concentrations, sucrose levels, medium pH, growth regulators, and photoperiod. The influence of cv. type and plant density on sterile pea reproductive activity was also investigated. These factors are addressed using peas (*Pisum sativum* L.) as the bioassay species. Peas are identified as a species that quickly flowers and fruits *in vitro* and therefore would be an ideal plant to study reproductive activities *in vitro* (Franklin et al., 2000). An understanding of how these factors affect flowering and fruiting is important to achieve maximum fruiting yields *in vitro.*

MATERIALS AND METHODS

Cultures and Media

Pea (cv. 'Oregon Sugar Pod II') plantlets were obtained from one-week-old sterilely germinated seeds. At this time these seedlings were usually one to 2-cm in length and possessed 2 to 3 leafs and a rudimentary tap root. BM contained MS salts and the following in mg / liter: thiamine•HCl, 0.5; myo-inositol, 100; sucrose, 30,000; and agar, 10,000 (Sigma Chemical Company, St. Louis, MO). The pH was adjusted to 5.7 ± 0.1 with 0.1 N HCl or NaOH before the addition of agar. Liquid BM pH value was adjusted to 5.0 ± 0.1. BM was dispensed in 25-ml aliquots into culture tubes (150 mm H x 25 mm diam.; 55 mm^3 cap.) and baby food containers (76 mm H x 60 mm diam.; 143 mm^3 cap.) (Sigma Chemical Company); and -50ml aliquots into Magenta (GA 7) polycarbonate containers (70 mm L x 70 mm W x 100 mm H; 365 mm^3 cap.) (Sigma Chemical Company), one-pint Mason jar (130 mm H x 78 mm diam.; 462 mm^3 cap.) (Kerr, Lancaster, PA), 38 x 250 mm culture tubes (250 mm H x 38 mm diam.; 270 mm^3 cap.) (Bellco Glass Inc., Vineland, NJ) and one-quart Mason jar (180 mm H x 91 mm diam.; 925 mm^3 cap.) and -100ml aliquots were dispensed into ½-gallon Mason jars(250 mm H x 107 mm diam.; 1,850 mm^3 cap.). Vessels were capped with polypropylene lids. The APCS employed a reservoir of a ½-gallon Mason jar containing one-liter liquid BM. All media was sterilized for 15 min at 1.05 kg / cm^2 and 121 °C.

Automated Plant Culture System

To determine the influence of vessel type on pea reproductive activity, a single plantlet was cultured in culture tubes, baby food jars, Magenta containers, 1-pint

Mason jars, 1-quart Mason jars, 1-quart Mason jar with APCS, 6-liter Bio-Safe container (379 mm L x 170 mm W x 178 mm H; 6000 mm^3 cap.) (Nalgene Co., Rochester, NY) with APCS and a -16.4liter mega- vessel with APCS. The procedure to construct the APCS has been outlined in detail elsewhere (Tisserat 1996). The APCS vessels employed were a mega-vessel consisting of two interlocking polycarbonate pans (325 mm L x 265 mm W x 600 mm H; -16,400mm^3 cap.) (Cambro, Huntington Beach, CA), a polycarbonate Bio-Safe container, or a ½-gallon Mason jar. Cultures were soaked 4 times daily for 5 minutes and then the medium was evacuated. The APCS (i.e., in vitro hydroponics system) consisted of two digital programmable timers, a peristaltic pump fitted with "easy-load" pump head (Model L/S16-Masterflex—Cole Parmer, Chicago, IL), a culture chamber, silicone tubing, and a single medium reservoir (Fig 2.1). A variety of culture chambers could be employed with this pumping system. Digital timers (Intermatic Inc., Chicago, IL) controlled the operation of the pumps forward and reverse flow in order to fill and drain the culture vessels. Culture vessels were fitted with bacterial hydrophobic air vents to accommodation ventilation and polypropylene spigots were accommodate media filling and removal. In some cases, a layer of glass gravel, 50 mm in depth was added to culture vessels to mimic a loose "soil-type" environment for plant roots. APCS culture vessels were ventilated with 30-min air exchanges at 10 applications per day via air vents. Air provided to the culture systems was from compressed air, pretreated through a charcoal filter and regulated to a 300 ml/min flow rate. Digital timers were controlled ventilation treatments, when applied.

Experiments

The influence of medium volume on vegetative and reproductive growth was addressed by culturing a single plantlet within ½-gallon Mason jar with an APCS and employing the medium reservoir volumes of 150, 250, 350, 500, or 1000 ml. In subsequent experiments, a test of medium volume employing 1000, 1500, 2000, and 2500-ml medium reservoir volumes was conducted. A comparison was also made employing an APCS and either ½- or full-strength MS salts using a 1000-ml reservoir volume. The influence of various concentrations of sucrose on pea growth responses was conducted by employing a ½-gallon Mason jar with an APCS. Sucrose levels were tested at 0, 0.5, 1, 1.5, 3, 5, 7.5, or 10 % levels. The influence of pH on the growth of pea seedlings was tested at 4, 5, 5.7, 6, 7, or 8 levels in 1-quart Mason jars containing 50-ml BM. To test the effect of growth regulators on pea seedlings vegetative and reproductive growth, a single plantlet was grown on 0, 0.01, 0.1 and 1.0 mg / l BA, NAA, or GA_3 in -1quart Mason jars containing -50ml BM. The influence of co-culture of more than one plantlet per vessel on the growth responses of pea was tested by planting 3 ,1 or 5 plantlets in 38 x 250 mm culture tubes, -1pint jar, -1quart jar, ½-gallon jar and ½-gallon jar with an APCS. The influence of pea cultivar on the growth responses was tested by culturing plantlets in the APCS of the following cvs.: 'Bush Snappy', 'Oregon Sugar Pod II', 'Super Snappy' and 'Wando'. Photoperiods effects were tested at 8, 12, 16, and 24 h on a single plantlet grown in a 1-quart Mason jar containing 50-ml BM.

Ten replicates/treatment were planted and experiments were repeated at least three times. Following eight weeks incubation, data on culture fresh weight, leaf number, plant height, flower number, and fruit number were recorded. In some cases, plants growing in the APCS were allowed to continue undisturbed for an additional 8 weeks with their media replenished in order to promote continued reproductive activities. Correlation coefficients were calculated to compare fresh weight, leaf length and plant height to culture chamber capacity, medium volume employed and culture chamber height when appropriate. Proportional data was analyzed by Fisher's exact test, and other data was tested by standard analysis of variance and Student-Newman-Keuls multiple range test, when appropriate. Cultures were incubated at a constant 26 ± 1 oC under a 16-hr daily exposure to 70 μmol / m^2 / s cool white fluorescent lamps.

RESULTS AND DISCUSSION

Flowering occurred in a number of culture vessels including 25 x 150 mm culture tubes, Magenta containers, 1-pint Mason jars, 1-quart Mason jars, 1-quart Mason jars with APCS, Bio-safes with APCS and mega-vessels with APCS at the end of 8 wks in culture (Fig. 2.2). However, the number of flowers produced in the larger vessels (*e.g.* Bio-Safe or mega- vessels) was consistently greater than in the smaller vessels (*e.g.* tubes and 1-pint Mason jars). Fruit formation only occurred from those vessels employing an APCS. More fruits were produced from plantlets grown in the mega-vessels than using any other culture system tested. Regardless of the culture vessel system employed, plantlets readily exhibited rooting and leaf production. There is a relationship between the number of leaves produced and occurrence of flowering and fruiting. Within the APCS, plantlets grew larger and developed more leaves and flowers than plantlets grown in the non-APCS systems. Flowers were usually initiated after plantlets produced 10-20 leaves. Abscission of flowers was common in all the vessels employed. Within the APCS, about 30 leaves were produced before flowers were retained and developed into fruits. A range from 5 to 15 fruits per plantlet was produced in the mega-vessel (Fig. 2.3A). Fruits were found to achieve sizes of 40 to 50 mm in length after 8 weeks in culture. Fruit as large as 75 to 80 mm in length may be produced eventually. Continued culture of plantlets in the APCS, to 16 weeks, resulted in enlargement of fruits as well as continued initiation of more flowers and fruits (Fig. 2.3A-C). Pea plantlets could be maintained for several months in the APCS by replacing the medium every 8 weeks. After 16 weeks in culture, hundreds of leaves and flowers were produced from each cultured pea plantlet.

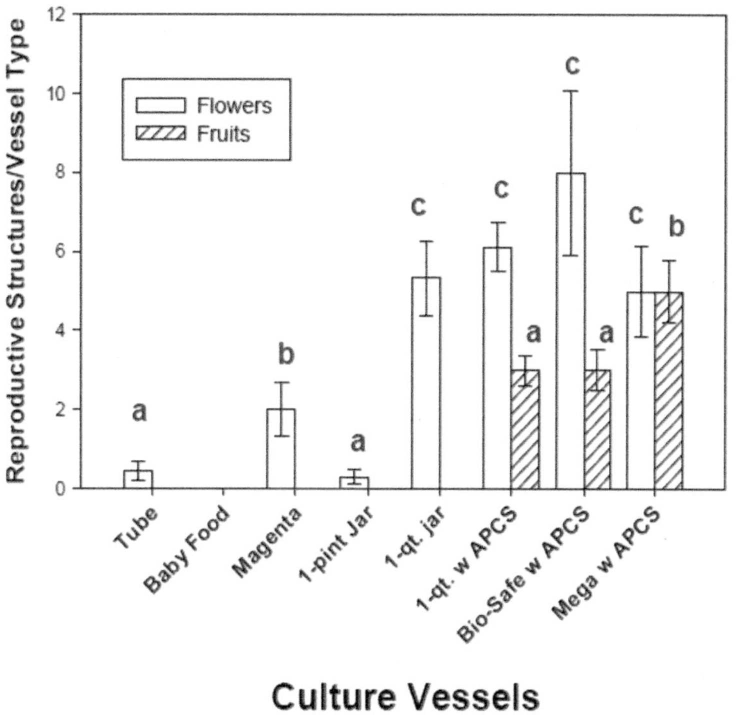

Fig. 2.2. Reproductive responses from pea plantlet tips in various culture vessels after 8 weeks in culture. Note that fruits were only produced from the larger culture vessel systems with APCS. Data were averaged for 10 replications/treatment. Experiments were repeated 3 times and a single representation is presented. Mean separation by Student-Newman-Keuls multiple range test ($P \leq 0.5$). Columns with the same letter on top were not significantly different.

Leafs were short-lived usually surviving 4 to 10 weeks resulting in a continuous production of leaves from the cultured plantlets. As many as 30 to 40 fruits were obtained from a single plantlet cultured in the mega-vessel after 16 weeks. Fruiting was found to be continuous and non-synchronous with older fruits exhibiting senescence adjacent to neighbouring newly initiated fruits on the same plantlet. One to six seeds developed within the fruit pods; however several seeds within a pod may be aborted (Fig. 2.3B). Nevertheless, some seeds were viable and capable of germination as evidenced by their *in vitro* germination during this study. These same plantlets eventually flowered and fruited themselves. Fruits sometimes dehisced in culture and a single seed may germinate from a fruit while still attached to the parent plantlet (Fig. 2.3C). These observations confirm the fertility of pea seeds obtained from sterile fruits by other investigators (Franklin et al., 2000; Ochatt et al., 2008).

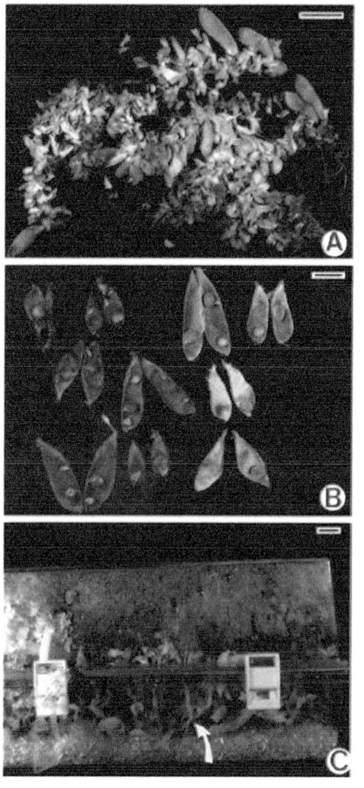

Fig. 2.3. Reproductive responses from 16-week old pea plantlets *in vitro* grown in an APCS.

(A) Pea plantlet exhibiting numerous fruits obtained from the mega-APCS. (B) Examples of bisected fruits revealing aborted and non-aborted seeds. (C) Growth responses of pea plantlets in Bio-Safe vessel, note the arrow indicating the precocious germination of seed from a dehisced fruit while still attached to the parent plant. Bar = 10 mm.

The influence of varying the medium reservoir volume with 150, 250, 350, 500, or 1000 ml reservoir volume was tested using a ½-gallon Mason jar with APCS. The highest vegetative and reproductive responses were obtained from an APCS employing a 1000 ml volume reservoir volume (Fig. 2.4). High positive correlations occurred between media volume employed and culture weight, flower number or fruit number (Fig. 2.4). However, in subsequent experiments (data not shown), when employing medium volumes larger than 1000 ml such as 1500, 2000 or 2500 ml, no improvement of the vegetative or reproductive responses occurred verses using the 1000 ml medium reservoir volume (data not shown). Apparently, the 1000 ml volume is the limit that the medium volume beneficially provides. These results conform to those made by Dickens & Van Staden (1988) who found that vessel size and medium volume influenced flowering in *Kalanchöe blossfeldiana*. Larger vessels,

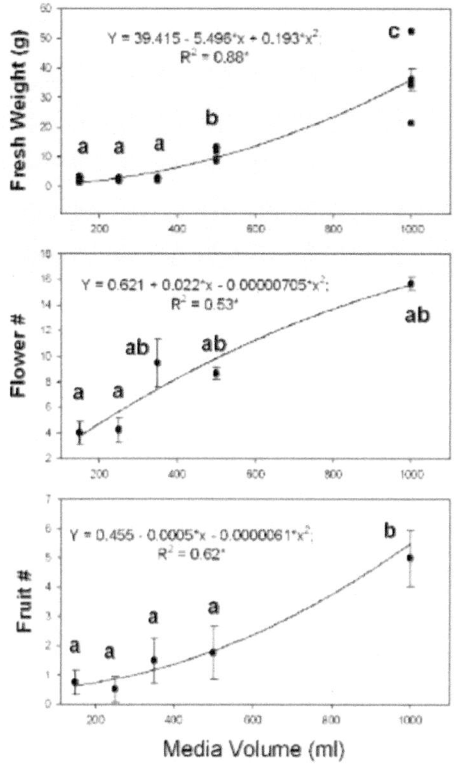

Fig. 2.4. Growth of Pea seedling plantlets in ½-gallon Mason Jars with APCS employing various media volumes. Regression coefficients (R^2) and regression equations between media volume and fresh weight, flower number or fruit number and media volume are given. All correlations are significant at $P \leq 0.05$ if denoted by asterisk. Letters represent statistical comparisons of mean predicted media volumes. Different letters represent non- overlap of the 95% confidence limits.

100 mm^3 (ml) flasks containing 40 ml medium, gave rise to more flowers than those *Kalanchöe* cultures grown in 50 mm^3 flasks or 40 mm^3 test-tubes containing 20 ml medium. In this study, considerably larger vessels and medium volumes were employed and they enhanced flowering and culture growth. In prior studies, Tisserat & Galletta (1995 ;1993) found that an APCS employing large culture chambers (e.g. ± ½-gallon Mason jars) allowed for flowering and fruiting from cucumbers and pepper plantlets. This work confirms these previous findings and quantifies the effect of culture chamber size and medium volume on the production of flowers and fruiting.

The concentration of inorganic salts in the medium influenced vegetative and reproductive responses. Nutrient medium containing ½-strength MS inorganic salts produced cultures of significantly less fresh weight and fewer flowers and fruits than cultures produced on medium containing full-strength MS inorganic salts (data not shown). For example, in terms of culture weight on medium containing ½-strength MS salts, 16.2 ±0.8 g fresh weights was produced while on full-strength MS salts, 36.0 ± 9.0 g fresh weight was produced. Similarly, differences in reproductive activity was influenced by inorganic salt concentration; on ½-strength MS salts, 9.3 ± 1.3 flowers and 2.8 ± 0.5 fruits were produced while on full-strength MS salts, 15.7 ± 4.7 flowers and 6.0 ± 0.6 fruits were produced. Based on these observations, it is apparent that vegetative growth and flowering is dependent on inorganic salt concentration. Dickens and Van Staden (1988) suggest that lack of necessary inorganic salts, especially nitrogen, reduces reproductive activity *in vitro*. In contrast, lowering the salt levels stimulated *Pharbitis* plantlets to flower (Ishioka et al., 1991).

Sucrose was found to be a requirement for both vegetative and reproductive activities in pea *in vitro*. The concentration of externally supplied carbohydrates is an important factor for reproductive activity from cultured pea. No reproductive responses were obtained from cultured pea plantlets grown on medium containing 0, 0.5, 1, or 1.5 % sucrose; reproductive responses only occurred on medium containing ± 3% sucrose. Highest fresh weight and flowering responses occurred using 3 % sucrose; while the highest fruit production occurred in the 3 and 5 % sucrose levels (Fig. 2.5). Less vegetative and reproductive responses (with the exception of the fruiting rate at the 5 % sucrose level) occurred as the sucrose levels increased beyond 3 % (Fig. 2.5). The best concentration for both reproductive and vegetative growth was 3 % sucrose. Similarly, 3 % sucrose has been found to be stimulatory in aiding flowering in other plants (Dickens & Van Staden 1988; Rastogi & Sawhney 1986). However, Ishioka et al., (1991) found that 7.5 % sucrose gave optimum flowering in *Pharbitis* plantlets. These diverse findings suggest that sucrose concentration greatly influences flowering and should be determined empirically for each species cultured.

The effect of pH on either the vegetative or reproductive activities of pea plantlets was not critical. No difference in fresh weights or flowering responses were obtained from cultures grown on any of the pH treatments after 8 weeks in culture (data not shown). These results differ from responses obtained from tobacco floral bud and tomato bud culture where pH greatly influenced flowering (Pasqua et al., 1991; Rastogi & Sawhney, 1986). Presumably, pea plantlets employed in this study were much larger than the tomato bud cultures, and were therefore less affected by media pH.

Testing various concentrations of growth regulators, BA, NAA and GA3 at concentrations varying from 0.1 ,0.01 ,0.0 and 1.0 mg / l revealed that best vegetative and reproductive responses occurred from cultures grown on medium devoid of any growth regulators (Table 2.2). High concentrations of all the growth regulators (*i.e.* 0.1 and 1.0 mg / l) gave decreased flowering responses compared to low concentrations (i.e. 0.01 mg /l). More leafs and flowers were produced from cultures grown on BM without any growth regulators than on medium containing growth regulators. Again this suggests that leaf number is related to flowering. Growth regulators have been found to be beneficial for flowering from immature tissue, organ and shoot cultures grown *in vitro* (Al-Juboory et al., 1991; Rastogi & Sawhney 1986; Ishioka et al., 1991; Lee et al., 1991). In contrast, in this study, larger pea plantlets do not benefit either vegetatively or reproductively from inclusion of exogenous growth regulators. These results were similar to that found for cucumber or pepper where inclusion of growth regulators also repressed flowering *in vitro* (Tisserat & Galletta 1993; 1995).

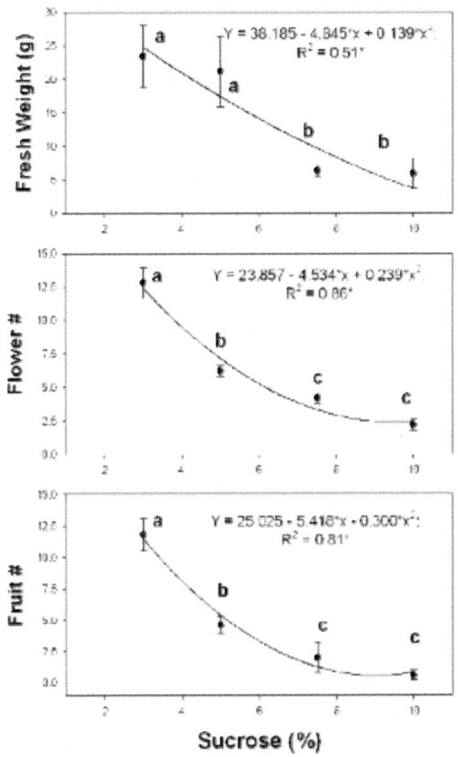

Fig. 2.5. Growth of Pea seedling plantlets in ½-gallon Mason Jars with APCS employing various sucrose concentrations. Regression coefficients (R2) and regression equations between media volume and fresh weight, flower number or fruit number and media volume are given. All correlations are significant at P < 0.05 if denoted by asterisk. Letters represent statistical comparisons of mean predicted media volumes. Different letters represent non- overlap of the 95% confidence limits.

Table 2.2. Growth responses of pea seedlings on agar medium after 8 weeks in culture[a].

Treatment (mg / l)	Leaf #	Flower #	Fruits #
0.1 GA_3	62.8±3.9 a 52.0 ± 1.9 a 49.3 ± 6.4 a	1.3 ±0.4 a 0.7 ± 0.1 b 0.1 ± .1 c	0.7 ± 0.1 a 0.2 ± 0.1 b 0
1.0 GA3	49.95.6± a	0.1 ± 0.1 c	0
0.01 NAA	50.0 ± 2.7 a	0.4 ± 0.4 c	0
0.1 NAA	47.2 ± 5.6 a	0.4 ± 0.2 c	0
1.0 NAA	43.5 ± 2.6 a	0.3 ± 0.1 c	0
0.01 BA	50.5 ± 5.5 a	0.6 ± 0.4 c	0.1 ± 0.1 b
0.1 BA	40.2 ± 4.4 a	0	0
1.0 BA	42.9 ± 7.1 a	0	0

[a]Treatment averages and SE presented. Mean separation within sucrose treatment columns by Student- Newman-Keuls multiple range test ($P < 0.05$).

The influence of co-culturing of several plantlets in a single culture vessel on the vegetative and reproductive responses has not been addressed in the literature. The number of co- cultured plantlets per vessel had a profound influence on the vegetative and reproductive responses exhibited by cultured plantlets. High positive correlations occurred for flowers/culture and vessel capacity, fresh weight/culture or leafs/culture when a single shoot was cultured per vessel (Fig. 2.6). However, when three or five seedlings were employed per vessel less or no correlations occurred. Usually, fewer flowers and fruits were produced from co-cultured plantlets compared to when only a single plantlet was cultured. Clearly, the benefits of employing larger culture vessels are reduced when multiple plantlets are employed.

Pea cv. is important to the number of flowers and fruits produced per plantlet *in vitro* (Table 2.3). 'Bush Snappy' and 'Oregon Sugar Pod II' gave the highest rates of flowering in 8 weeks,

60 and 60 % respectively, while 'Super Snappy' and 'Wando' exhibited 30 and 20 % flowering, respectively (Table 2.2). Similarly, these same two cultivars also gave the highest rates of fruit production, 60 and 20 % respectively. 'Super Snappy' and 'Wando' exhibited less flowering and fruiting but manifested larger shoot lengths than 'Bush Snappy' and 'Oregon Sugar Pod II' (Table 2.2). No difference in terms of the number of leafs produced occurred among the 4 cultivars tested with 23 to 25 leaves/culture usually being produced from all cvs. at the end of 8 weeks. Pea was not influenced by any photoperiod tested in vitro in terms of flowering and fruiting

responses in pea (Table 2.4). Pea is a day-neutral plant and characteristically was unaffected by photoperiod treatments given *in vitro*. Shoot apices of *Pharbitis*, a short-day plant, were unaffected by the photoperiod employed in culture (Ishioka et al., 1991). However, short-day photoperiod was prerequisite for flowering of *Kalanchöe blossfeldiana* and soybean nodal sections (Dickens & Van Staden 1985; 1988). Both *Kalanchöe* and Soybean are short-day plants.

Development of sterile fruits in tissue culture may have some futuristic prospects. Surgery, cancer, AIDs and extreme-allergy patients often require sterile diets to prevent infection complications (Butterweck, 1995; Watson, 2001; Pryke & Taylor, 2008). To supply suitable diets foods must be sterilized through a variety of heat or radiation treatments (Pryke &

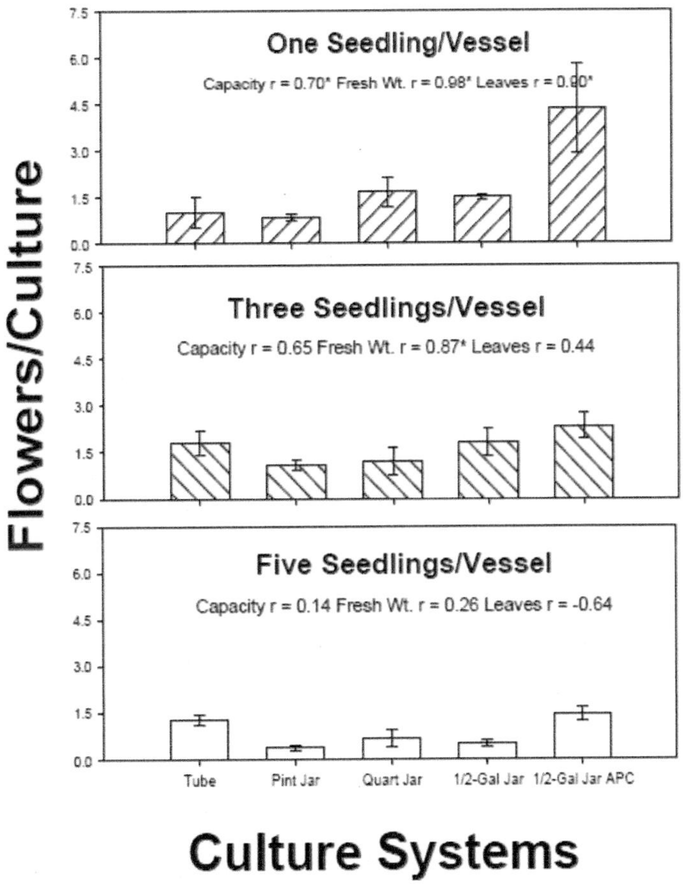

Fig. 2.6. Relationship between numbers of seedlings/vessel, leaves/culture and culture system cap. (mm3) to flowers/culture. Correlation coefficients (r) between flowers/culture and culture chamber cap., fresh weight, or leaves produced are given for number of seedlings cultured/vessel. Correlations significant at P < 0.05 are designated by an asterisk.

Table 2.3. Response of pea cultivars on the vegetative and reproductive responses in vitro after 8 weeks in culture after 8 weeks in culture.

Cultivar	Shoot Length (mm)a	Leaf #a	Flowering %b	Fruiting %b
Bush Snappy	148.3 ± 8.3 a	24.2 ± 0.8 a	60 a	60 a
Oregon Sugar Pod II	149.0 ± 9.7 a	23.0 ± 0.8 a	60 a	20 b
Super Snappy	167.9 ± 7.3 b	25.5 ± 0.9 a	20 b	0
Wando	169.0 ± 5.3 b	25.5 ± 3.2 a	30 b	10 b

[a]Treatment averages and SE presented. Mean separation within sucrose treatment columns by Student- Newman-Keuls multiple range test ($P < 0.05$).

[b]Treatment percentages within columns separated by Fisher's exact test ($P < 0.5$). There were 10 replications per treatment.

Table 2.4. Influence of photoperiod on the morphogenetic responses in vitro in Oregon Sugar Pod II pea after 8 weeks in culture.

Photoperiod (h)	Shoot Length (mm)a	Leaf #a	Flowering %b	Fruiting %b
8	160.0 ± 7.3 a	32.5 ± 3.1 a	90 a	50 a
12	163.0 ± 1.5 a	39.5 ± 2.9 a	70 a	20 a
16	208.8 ± 13.2 b	40.6 ±1.8 a	80 a	20 a
24	185.5 ± 10.6 b	39.0 ± 3.8 a	90 a	30 a

[a]Treatment averages and SE presented. Mean separation within sucrose treatment columns by Student- Newman-Keuls multiple range test ($P < 0.05$).

[b]Treatment percentages within columns separated by Fisher's exact test ($P \leq 0.5$). There were 10 replications per treatment.

Taylor, 2008). The nutritional value of these processed foods is altered somewhat compared to that of the original freshly harvested produce. Sterile fruits produced in vitro may provide high nutritionally viable produce which physically and chemically resemble that from in vivo sources. This study suggests that viable fruiting structures can be procured employing sterile tissue culture technology. Nevertheless much refining work remains to be preformed to develop edible products. In addition, production of sterile fruits with viable seeds can be a means to shorten generational cycles for faster breeding (Ochatt et al., 2002). It has been previously recognized that plant tissue culture offers an avenue to produce food products (Fu et al., 1999; Stafford, 1991). However, little interest has been expended to develop suitable systems that could accommodate fruits and vegetable production in vitro. Presumably, the increased spatial and nutritional demands for flowering and fruiting required by plants to be achieved contributed toward this. A sterile hydroponics system could be employed to achieve sterile food production.

CONCLUSIONS: AT A GALNCE

In vitro reproductive activity has certain advantages, such as year round flowering and fruiting in a controlled environment, and disadvantages, such as the inherent difficulty associated with a sterile environment coupled with poor flowering in certain species and absence of fruiting in most species when cultured *in vitro*, when compared to growing plants in soil. Previous *in vitro* studies often utilized small explants such as nodal sections or shoot tips that were entirely medium dependent for their survival and growth. In this study, large plantlets possessing several leaves and a substantial root system achieved both flowering and fruiting within 8 weeks *in vitro*. Pea is an excellent plant to critically study all reproductive events *in vitro*. The only other plant reported in the literature that flowered in as short a time frame as pea is soybeans which produced flowers in 6 weeks and fruits in 10 weeks in culture (Dickens & Van Staden 1985). The APCS differs from the traditional tissue culture vessels by employing an enlarged culture chamber that accommodates for larger plantlet growth and employs a larger medium reservoir that supplies more nutrients for plantlet growth. With the APCS, no external growth regulators were required for reproductive activities (Tisserat & Galletta 1993; 1995).

Prolonged culture growth is obtained (*e.g.* 16 wks) by replacing the medium reservoir without any culture manipulations. The primary factors controlling the pea reproductive activity were carbohydrate levels and culture vessel size. Absence of or low sucrose levels did not allow cultures to flower. Fruiting could not occur in small vessels (*e.g.* 25 x 150 mm culture tubes or baby food jars). Secondary factors influencing reproductive events were plantlet density, medium volume, inorganic salt concentration, and CV. Factors having no influence on the flowering of pea were pH, growth regulators and photoperiod. Recognition of the merits of these factors will aid in a more thorough understanding of the lowering and fruiting process *in vitro*. The long term goal of manufacturing sterile in vitro hydroponically grown foods can be obtained.

REFERENCES

Anonymous (2004) Contaminated produce top food poisoning culprit, *Journal of Environmental Health,* Vol. 67 No. 2, (September, 2004), pp. 6, ISSN 0022-0892

Al-Juboory, K.; Skirvin; R.M. & Williams, D.J. (1991) Improved flowering of cotyledon- derived shoots of 'Burpless Hybrid' cucumber in vitro, *HortScience*, Vol.26, No.8, (August 1991), pp.1085, ISSN 0018-5345

Bodhipadma, K. & Leung, D.W.M. (2003) In vitro fruiting and seed set of *Capsicum annuum* L. cv. Sweet Banana, *In Vitro Cellular and Developmental Biology—Plant*, Vol.39, No.5, (November 2003), pp. 536-539, ISSN 1054-5476

Butterweck, J.S. (1995) Sterile diets for the immuno-comprised: is there a need? *Radiation Physics and Chemistry*, Vol.46, No.4-6, (November 1995), pp. 601-604, ISSN 0020- 7616

Centers for Disease Control and Prevention (2010) Preliminary Food Net Data on the Incidence of Infection with Pathogens Transmitted Commonly Through Food --- 10 States, 2009, *Morbidity and Mortality Weekly Report*, Vol.59, No.14, (April, 2010), pp. 418-422, ISSN 0149-2195

Centers for Disease Control and Prevention (2011) Making Food Safer to Eat. In: *Vital Signs*. June, 2011, US Department of Health and Human Services, Centers for Disease Control and Prevention, Atlanta, GA, USA, accessed July 07, 2011, Available from:<http://www.cdc.gov/vitalsigns/FoodSafety/index.html>

Dickens, C.W.S. & Van Staden, J. (1985) In vitro flowering and fruiting of soybean explants. *Journal of Plant Physiology*, Vol.120, No.4-6, (November 1995), pp. 83-86, ISSN 0176- 1617

Dickens, C.W.C. & Van Staden, J. (1988) The in vitro flowering of *Kalanchöe blossfeldiana* Poellniz. I. Role of culture conditions and nutrients. *Journal of Experimental Botany*, Vol.39, No.201, (April 1988), pp. 461-471, ISSN 0022-0957

Franklin, G.; Pius, P.K. & Ignacimuthu, S. (2000) Factors affecting in vitro flowering and fruiting of green pea (*Pisum sativum* L.), *Euphytica*, Vol.115, No.1, (September 2000), pp. 65-73, ISSN 1573-5060

Fu, T. J.; Singh, G. & Curtis, W. R. (1999) Plant cell and tissue culture for the production of food ingredients. Springer, ISBN 0-306-46100-5, New York, USA

Hayes, C., Elliot, E., Krales, E. & Downer, G. (2003) Food and water safety for persons infected with human immunodeficiency virus. *Journal of Clinical Infectious Diseases*. Vol.36, Suppl. No.2, pp. S106-S109, ISSN 1058-4838

Hong,Y.C.; Read, P.E.; Harlander, S.K. & Labuza, T.P. (1989) Development of a tissue culture system from immature strawberry fruits, *Journal of Food Scence*, Vol.54, No.2, (March 1989), pp. 388-392, ISSN 1750-3841

Ishioka N.; Tanimoto S. & Harada H. (1991) Roles of nitrogen and carbohydrate in floral-bud formation in *Pharbitis* apex cultures, *Journal of Plant Physiology*, Vol.138, No.5, (March 1991), pp. 573-576, ISSN 0176-1617

Kamps, L. (2004) Is Your Produce Poisoned? *Prevention Magazine*, Vol.56, No.10, (October, 2004), pp. 1-4, ISSN 0032-8006.

Lee, H.S.; Lee, K.W.; Yang, S.G. & Liu, J.R. (1991) In vitro flowering of ginseng (*Panax ginseng* C. A. Meyer) zygotic embryos induced by growth regulators, *Plant &Cell Physiology*, Vol.32, No.7, (October, 1991), pp. 1111-1113, ISSN 0032-0781.

Ochatt, S.J.; Sangwan, R.S.; Marget, P.; Assoumou Ndong, Y.; Rancillac, M.; Perney, P. & Röbbelen, G. (2002) New approaches toward the shorting of generation cycles for faster breeding of protein legumeshh, *Plant Breeding*, Vol.121, No.5, (October, 2002), pp. 436-440, ISSN 0179-9541

Pasqua, G.; Monacelli, B. & Altamura, M.M. (1991) Influence of pH on flower and vegetative bud initiation and development *in vitro*, *Cytobios*, Vol.68, No.2, (March 1991), pp. 111-121, ISSN 0011-4529

Pryke, D.C. & Taylor, R.R. (1995) The use of irradiated food for immune suppressed hospital patients in the United Kingdom, *Journal of Human Nutrition and Dietetics*, Vol.8, No.6, (December 1995), pp. 411-416, ISSN 0952-3871

Rastogi, R. & Sawhney, V.K. (1986) The role of plant growth regulators, sucrose and pH in the development of floral buds of tomato (*Lycopersicon esculentum* Mill.) cultured in vitro, *Journal of Plant Physiology*, Vol.128, No.2, (March 1986), pp. 285-295, ISSN 0176-1617

Stafford, A. 1991. The manufacture of food ingredients using plant cell and tissue cultures. Trends in Food Science & Technology. Vol. 2, No. 5, (May 1991), pp. 116-122, ISSN 0924-2244.

Tisserat, B. (1996) Growth responses and construction costs of various tissue culture systems, *HortTechnology*, Vol.6, No.1, (January-March 1996), pp. 62-68, ISSN 1063- 0198

Tisserat, B. & Galletta, P.D. (1993) Production of cucumber fruits from the culture of 'Marketmore-76' plantlets, *Plant Cell Reports* Vol.13, No.1, (November 1993), pp. 37-40, ISSN 0721-7714.

Tisserat, B. & Galletta, P.D. (1995) In vitro flowering and fruiting of *Capsicum fruitescens* L.,*HortScience*, Vol.30, No.1, (January 1995), pp. 130-132, ISSN 0018-5345

Tisserat, B.; Jones, D. & Galletta,.P.D. (1989a) Growth responses from whole fruit and fruit halves of lemon cultured in vitro, *American Journal of Botany*, Vol.76, No.2, (February 1989), pp. 238-246, ISSN 0002-9122.

Tisserat, B.; Vandercook, C.E. & Berhow, M. (1989b) Citrus juice vesicle culture: a potential research tool for improving juice yield and quality. *Food Technology Magazine*. Vol.43, No.2, (March 1989), pp. 95-100, ISSN 0015-6639.

U.S. Department of Agriculture-Food Safety and Inspection Service (2006) Food safety for People with HIV/AIDS. In: *U.S. Department of Agriculture-Food Safety and Inspection Service*. Washington, D.C., USA, accessed July 11, 2011, Available from:<http://www.fsis.usda.gov/PDF/Food_Safety_for_People_with_HIV.pdf>

Wagner, A.B., Jr. (2011) Bacterial food poisoning. In: *Texas AgriLife Extension Service and Texas A&M University, Texas A&M System*, accessed July 10, 2011, Available from:<http://aggie-horticulture.tamu.edu/extension/poison.html>

Watson, R.R. (2001) *Nutrition and AIDS.* 2nd Ed. CRC Press, IBSN 2-72-9302-084, Boca Raton, Florida. U.S.A.

CHAPTER-3

THE APPLICATION OF HYDROPONIC SYSTEMS IN INVESTIGATING ABIOTIC STRESS TOLERANCE

INTRODUCTION: AN OVERVIEW

Hydroponic systems have become invaluable tools in the realm of scientific research, particularly in the investigation of abiotic stress tolerance in various plant species. Abiotic stressors, such as extreme temperatures, salinity, drought, and heavy metals, pose significant challenges to plant growth and crop productivity. Understanding how plants respond to and mitigate these stressors is essential for developing resilient and stress-tolerant crops. Hydroponics, which involves the cultivation of plants in nutrient-rich water solutions rather than traditional soil, offers researchers precise control over environmental conditions. This controlled environment enables the systematic study of how plants react to and cope with abiotic stress factors. Researchers can manipulate variables such as nutrient availability, water quality, and temperature, allowing them to simulate and analyze specific stress conditions. Consequently, the application of hydroponic systems has led to valuable insights into the physiological, molecular, and genetic mechanisms that underlie a plant's ability to withstand abiotic stress. This knowledge is not only crucial for advancing our understanding of plant biology but also holds great promise for the development of stress-resistant crop varieties, contributing to global food security in the face of changing environmental conditions.

Hydroponics, the 'water culture' of plants, has been used in both research and commercial contexts since the 18th century. Although now used successfully on a large scale by commercial growers of fast-growing horticultural crops such as lettuce, strawberries, tomatoes, and carnations, hydroponics was initially developed as a part of early research into plant nutrition. The idea of hydroponics, its development and improvement, stimulated the interest of plant biologists, and their research provided useful outcomes and scientific knowledge about mechanisms of nutritional toxicity/deficiency and plant development in general. Scientists discovered that plants required only a small number of inorganic elements, in addition to water, oxygen and sunlight, to grow. It was later realised that plants grew better hydroponically if the solutions were aerated. The use of hydroponics enabled plant scientists to identify which elements were essential to plants, in what ionic forms, and what the optimal concentrations of these elements were. It allowed them to easily observe the effects of elemental deficiencies and toxicities and to study other aspects of plant development under more controlled conditions. As scientists' understanding of the requirements for growing plants in hydroponics increased, the system was adopted and refined by commercial producers who found it allowed them to control environmental variables and deliver higher yields of product more reliably. Hydroponics systems are now operated in temperature- and light-controlled glasshouses which allow crop production through all seasons.

Hydroponics remains a fundamental tool for plant research. In this chapter we describe the use of hydroponics in the context of scientific research into plant responses and tolerance to abiotic stresses, drawing on our experiences at the Australian Centre for Plant Functional Genomics (ACPFG) and the University of Adelaide, as well as summarising reports in the published literature. We have also described some of the problems and uncertainties encountered in our hydroponics-based experiments. We hope that the chapter will be of benefit to scientists and other individuals with an interest in this topic.

HYDROPONIC SYSTEMS FOR SCIENTIFIC RESEARCH PURPOSES

Plants growing in hydroponics require oxygen to be delivered to the roots, in addition to the water and nutrients supplied in growth solutions. Without constant aeration, a hydroponics system will become anaerobic and inhibit the growth of most plants. Only a small number of species (including rice) are adapted to grow in submerged environments with minimal oxygen supply. Such plants contain root structures known as aerenchyma, which are large air-filled spaces within the root that accumulate and store oxygen. The majority of plant species, however, require a continuous supply of oxygen into the growth solution for absorption by root cells. Hydroponics can be divided into two basic types depending on the method of aeration employed: (1) Flood-drain systems and (2) Continuous aeration. Both systems are routinely used in our research and in the following sections we will describe each system in more detail. While the nutrient film technique has revolutionised commercial hydroponics, it is not considered to be useful for abiotic stress research and will not be discussed in this chapter.

Flood-Drain System

This type of hydroponics maintains regular mixing of growth solution by repeatedly pumping the solution into the growing vessel and allowing it to drain. Continuous movement of solution during the pumping/drainage cycles facilitates delivery of oxygen to the roots. We have not observed symptoms of depressed plant growth using this system of aeration. A large empty container (usually on the bottom) to store drained growth solution and a pump are required for the system to operate. The volume of growth solution required and the length of the drainage/pumping cycle will depend upon the size and design of the setup. The system illustrated is constructed with a lower storage/pumping tank containing 80 L of growth solution and two containers (40 L each) on the top for growing plants, and employs a 20 minute pump / 20 minute drain cycle (Fig. 3.1). This hydroponics system was initially designed by Mr. R. Hosking and further refined by Mr. A. Kovalchuk from ACPFG.

Fig. 3.1. A flood-drain 80 L supported hydroponics system in use at the Australian Centre for Plant Functional Genomics (ACPFG).

Flood-drain systems also require a supporting material in the growing vessel to hold the plant and to maintain a moist environment around the roots. We use polycarbonate plastic fragments, which we obtain from a local plastics manufacturer (Plastics Granulated Services, Adelaide, Australia). A selection of suitable types of plastic fragments used is illustrated in Fig. 3.2. The most important characteristics of the plastic fragments are: (1) chemical inertness and (2) surface wetness. The first characteristic ensures that there are no changes in the composition of available nutrients provided by the growth solution, and the second characteristic facilitates the continued supply of moisture and nutrients during draining. The plastic fragments should be washed in tap- water several times and finally rinsed with reverse osmosis water before first use to remove traces of manufacturing chemicals. With further regular washing, the fragments can be re-used many times in hydroponics systems.

We employ two methods for growing plants in the plastic fragments: either in separate tubes for individual plants (Fig. 3.3A), or tubs for larger numbers of plants (Fig. 3.3B). Growing a single plant in each tube enables the experimenter to remove individual plants for ongoing analyses (eg. imaging), and also keeps the roots separated for eventual harvest and analysis.

Fig. 3.2. A selection of plastic fragments suitable for use as a supporting substrate in flood- drain hydroponics.

Both the tubes and tubs have a fine mesh fitted to the bottom to hold the plastic fragments in place and allow for the entry/exit of growth solution, which is pumped from below into the outer container (Fig. 3.3). The outer container has an overflow tube for the return of excess growth solution back to the lower storage container during the pumping period.

The optimum age and size of plantlets for transfer to hydroponics is be species-dependent. We pre-germinate wheat and barley seeds for 4 to 5 days in petri dishes or trays lined with damp paper and covered with plastic. Seedlings are transplanted when the roots are between 1 and 2 cm in length. Although the young plants recover from transplanting after 2 to 3 days, we usually allow 7 to 10 days before applying an experimental treatment. It is also possible to germinate seeds directly in the substrate, and this may work well for certain species.

Fig. 3.3. Close view of barley plants in individual tubes (A) and in tubs/ buckets (B) in flood- drain supported hydroponics.

Alternative Substrates for Supported Hydroponics

While we prefer to use plastic fragments as a solid supporting substrate for growing plants in flood-drain hydroponics, there are a range of other suitable supporting materials. The simplest to use are quartz gravel or river sand, and both are widely used in such systems (Boyer et al., 2008; Dreccer et al., 2004; Greenway, 1962; Munns & James, 2003; Rawson et al., 1988a, 1988b). However, small quartz particles can easily block and damage the pumping system. Other suitable substrates may include artificially manufactured expanded clay balls, vermiculite (Forster et al., 1994; Gorham et al., 1990, 1991) or perlite, and rockwool (Gorham, 1990; Gorham et al., 1991). Some substrates are also suitable for small, 'passive' hydroponics systems, where plants are grown in small pots containing the substrate and sitting in a tray, with growth solution supplied via capillary action from below. Expanded clay balls are manufactured in many countries with different brand, for example, Hydroton in Australia, Keramzit in Russia and Blaehton in Germany. An example of this type of hydroponics using expanded clay balls is illustrated in Fig. 3.4. Caution is given that such a system can only be used for relatively short growth periods (less than 4 weeks) and that there is less control over the supply of nutrients to the plants, due to the potential for light-stimulated algal growth and high rates of evaporation from the tray. Unlike plastic fragments, these alternative substrates are also not chemically inert and may leach mineral ions into the growth solution or alter nutrient availability. Such substrates are less suitable for studies of mineral element deficiencies and toxicities.

Continuous Aeration

The engineering requirements of hydroponics with continuous aeration are much less complex than for flood-drain systems. However, this system depends on constant aeration. A selection of aerated hydroponics setups used in our Centre is presented in Fig. 3.5. Good aeration in small volumes (12 L or less) is achieved using commercially available aquarium pumps, plastic tubing and aeration stones. In continuous aeration systems, the roots of germinated seedlings are directly submerged in aerated growth solution and the shoots supported to grow above the solution. One way to do this is with soft pieces of foam wrapped around the seedlings and held in holes drilled into a lid covering the hydroponics container, Fig. 3.5A and 3.5B (Drihem & Pilbeam, 2002; Gorham et al, 1987; Shah et al., 1987). Other alternatives include the use 'rafts' of polystyrene on the surface of the growth solution (Bağci et al., 2007; Ma et al., 2007), also with holes for holding the growing seedlings, the use of rockwool or agar plugs in open-ended Eppendorf tubes held in suitably sized holes drilled into the container lid, and placement of growing seedlings directly in open-ended plastic tubes held in the container lid. The size of hydroponics containers and the type of support can be varied to suit the needs of the researcher. For example, a miniature hydroponics system was designed using a 200 ml pipette tip box, where seedlings are held directly in open-ended 500 µl Eppendorf tubes (Fig. 3.5C). A further alternative uses 4 L lunch-box style containers purchased with plastic rack inserts (Décor, Australia). A layer of plastic mesh is sewn onto the insert, and the level of growth solution is adjusted so as to just reach the surface of the mesh (Fig. 3.5D). Wheat or barley seedlings can be grown directly on plastic mesh, Fig. 3.5D (Watson et al., 2001), acrylic grids (Kingsbury et al., 1984) or anodised aluminium mesh (Shah et al., 1987) without further support.

Fig. 3.4. The use of expanded clay balls (A) as a substrate for a small, 'passive' hydroponics system, showing broccoli (B) and saltbush, *Atriplex* ssp. (C) after a light salt stress (50 mM NaCl). (Figure provided by Ms J. Bovill, ACPFG and students at the University of Adelaide).

A
B
C

Fig. 3.5. Examples of continuous aeration hydroponics setups used by researchers at ACPFG.(A) 12 L boxes with foam supporting growing seedlings, or (B) with 10 ml open-ended plastic tubes. (C) Miniature hydroponics in 200 mL pipette tip boxes (image provided by MrJ. Harris, ACPFG). (D) 4 L lunch boxes with plastic inserts and mesh.

NUTRIENT SOLUTIONS AND ITS COMPOSITION

All nutrient solutions used for hydroponics culture of plants are essentially derived from the original protocol developed by Hoagland and Arnon (1938). A typical growth solution consists of the following essential macro-elements: nitrogen (N), potassium (K), phosphorus (P), calcium (Ca), magnesium (Mg) and sulphur (S); and micro-elements: a soluble form of iron (Fe), boron (B), copper (Cu), manganese (Mn), nickel (Ni), zinc (Zn), molybdenum (Mo) and chlorine (Cl), and, for leguminous species requiring N fixation, cobalt (Co). Sometimes, silicon (Si) and selenium (Se), while not essential elements, are considered beneficial to plant growth and are also included (Epstein, 1994, 1999; Lyons et al., 2009). The standard growth solution used by researchers at ACPFG has previously been published (Genc et al., 2007; Shavrukov et al., 2006). However, we have recently further improved the composition of the growth solution for use with wheat and barley based on tissue nutrient analysis (Table 3.1). We also found that other species, such as rice, have a high requirement for ammonium nitrate (5 mM).

Table 3.1. Composition of growth solutions used in plant nutrition studies by researchers at ACPFG and the University of Adelaide, including an earlier published protocol and an improved protocol for culture of wheat and barley

Elements	Salts used	Final Concentration Published protocol (Genc et al., 2007; Shavrukov et. al., 2006) (mM)	Improved protocol (mM)
N	NH_4NO_3	0.2	0.2
K, N	KNO_3	5.0	5.0
Ca, N	$Ca(NO_3)2$	2.0	2.0
Mg, S	$MgSO_4$	2.0	2.0
P, K	KH_2PO_4	0.1	0.1
Si*	Na2SiO3	0.5	0.5
Fe	NaFe(III)EDTA	(μM)	(μM)
		50.0	100.0
B	H3BO3	50.0	12.5
Mn	MnCl2	5.0	2.0
Zn	ZnSO4	10.0	3.0
Cu	CuSO4	0.5	0.5
Mo	Na2MoO3	0.1	0.1
Ni	NiSO4	0.0	0.1
Cl**	KCI	0.0	25.0

* Silicon is not an essential element and may be omitted from the growth solution, depending on the experiment and plant species grown.

** MnCl2 was reduced to 2μM in the improved protocol to optimize Mn nutrition. However, as MnCl2 is the only source of Cl, additional chloride was supplied as KCl to avoid Cl deficiency.

Maintenance of Ph Range

Most plant species will grow optimally in nutrient solutions of acid to neutral pH (range 5.5– 7.5) (Dreccer et al., 2004; Drihem & Pilbeam, 2002; Dubcovsky et al., 1996; Kronzucker et al., 2006; Munn & James, 2003), although there is some species variability in optimum pH required for growth. The nutrient solutions used in our research do not require pH adjustment to stay within the optimum range if replaced regularly and if silicon is omitted. If Si is included, careful attention must be paid to achieving and maintaining an appropriate solution pH (see Section 3.2 below).

Depending on the experiment, strict maintenance of pH may be required. Occasionally, pH values outside of the optimum range are needed for study of certain abiotic stresses. For example, aluminium toxicity studies are conducted at low pH (4.5 or less) to ensure the presence of soluble, phytotoxic Al^{3+} (eg. Collins et al., 2008; Famoso et al., 2010; Pereira et al., 2010). Maintenance of pH can be achieved by several means. Nutrient solutions can be buffered with low concentrations of a suitable zwitter-ionic buffer that is not phytotoxic, eg. 2 mM MES (-2[N-morpholino] ethane-sulphonic acid)-KOH (Genc et al., 2007). Alternatively, solution pH may be monitored and adjusted frequently. This task can be done either manually or automatically (eg. Jarvis & Hatch, 1985). For automated pH adjustment, an automatic pH controller is attached to each hydroponics unit (eg. Cole-Parmer 5997-20 pH controller, SML Resources International, USA). As solution pH moves above or below the set pH value, acid or base is automatically dispensed into the solution to return the pH to the set value (Deane-Drummond, 1982; Wheeler et al., 1990). However, pH automation is costly, particularly for multiple hydroponics units.

Silicon in Nutrient Solutions

Despite the beneficial effects of Si on plant growth (Epstein, 1994, 1999), there is no consensus on whether or not it should be included in nutrient solutions. In his studies with silicon, Epstein (1994) concluded that "omission of Si from solution cultures may lead to distorted results in studies on inorganic plant nutrition, growth and development, and responses to environmental stress". The author advocated the addition of Si in solution culture to represent its abundance in soil solution (0.1 - 0.6 mM), but he did not acknowledge that there are many environmental sources of silicon, including mineral salts and water and even particulate SiO2 in the atmosphere, which may provide enough Si for plant growth and development.

In our experience, the application of Si to hydroponically-grown wheat and barley did not result in significant differences in plant growth in non-stressed conditions. However, silicon may impact on plant responses to abiotic stresses. For example, we have observed that rice shows different responses to boron toxicity when grown with or without added silicon. One of the genes underlying boron toxicity tolerance in barley, *HvNIP2;1*, encodes a transporter belonging to the aquaporin family which facilitates transport of both B and Si (Chiba et al., 2009; Schnurbusch et al., 2010). This gene is also present in rice (Ma et al., 2006) and may explain the observed interactions between Si nutrition and B toxicity. Furthermore, the addition of silicon has been found to increase levels of observed tolerance to salinity, drought, high and low temperature and metal toxicities (reviewed in Ma & Yamaji, 2006).

Silicon is produced commercially as a crystalline powder, sodium silicate pentahydrate (eg., Chem-Supply, Australia), or as a liquid in the form of sodium silicate solution (water glass; eg., Sigma, USA). Both forms of silicate are suitable for hydroponics. However, careful attention must be paid to achieving and maintaining an appropriate pH for plant growth (acid to neutral range) when Si is used (Dubcovsky et al., 1996). The addition of either form of silicon increases the pH of the growth solution to above 7.0, and may cause the precipitation of other nutrients out of solution.

Nitrogen form and pH Maintenance of Nutrient Solutions

It is well known that nitrogen is one of the most important elements necessary for plant growth. A detailed review of this topic is outside the scope of this chapter. However, we would like to mention a practical issue relating to the use of nitrogen in hydroponics and consequences for nutrient solution pH. Two major forms of nitrogen are used in hydroponics: ammonium (NH_{4+}) and nitrate (NO_{3-}), and usually both forms are included in nutrient solutions. When both molecules are present in growth solution, NH_{4+} cations are often preferentially absorbed by plant roots over NO_{3-} anions (eg. Gazzarrini et al., 1999). The different depletion rates of NH_{4+} and NO_{3-} will result in an altered growth solution pH. For wheat and barley, we found that nutrient solutions containing equimolar (eg. 5 mM) concentrations of ammonium nitrate and potassium nitrate rapidly acidify as NH_{4+} is preferentially taken up by the growing plants and protons (H^+) are released to maintain charge balance. Such high concentrations of ammonium can also be deleterious to plant growth. Britto & Kronzucker (2002) reported that seedlings growth of barley was reduced considerably at 10 mM NH_{4+} compared to 1 mM NH_{4+}. Similarly, we observed that the growth of bread wheat seedlings was significantly reduced (37%) at 5 mM NH_{4+} compared to 1 mM NH_{4+} when grown in nutrient solution containing 5 mM KNO_3 / 2 mM $Ca(NO_3)2$. (Genc et al., unpublished; Fig. 3.6). An optimum ammonium concentration of 0.2 mM NH4+ (with 5 mM KNO_3) was identified for hydroponics solutions for wheat and barley (Genc et al., unpublished). Interestingly, similar ratios of ammonium (20 – 200 μM) to nitrate (1 – 5 mM) have been measured in soil solutions of fertilised agricultural soils (Owen & Jones, 2001). Similar experiments would need to be conducted to optimise ammonium to nitrate N ratios for different species and different compositions of nutrient solution. For example, we use higher concentrations of ammonium N for the culture of rice in hydroponics. When the ratio of ammonium to nitrate N is optimised, solution pH remains relatively stable if replaced regularly to avoid depletion of either form of N.

APPLICATION OF ABIOTIC STRESSES IN HYDROPONICS SYSTEMS

Research into plant tolerance of abiotic stresses, including salinity, drought, toxic concentrations of boron, aluminium or other elements and elemental deficiencies, is of fundamental importance for sustainable and secure agriculture into the future. In this context, hydroponics is a major scientific modelling tool, facilitating precise control over the treatment and consistent observations of treatment effects. Importantly, hydroponics enables observations to be made of intra- and inter-specific genetic variation in plant responses, in terms of levels of tolerance shown and the specific tolerance mechanisms that are employed.

Fig. 3.6. Seedlings of bread wheat (cv Krichauff) after growth in hydroponics for four weeks in the presence of 1 mM and 5 mM of NH_4NO_3. Growth solutions were supplied with 5 mM KNO_3 and 2 mM $Ca(NO_3)$2.

Salinity

Salinity is a major abiotic stress across agricultural regions worldwide, with a significant impact on cereal production (Colmer et al., 2005; Flowers & Yeo, 1995; Rozema & Flowers, 2008; Steppuhn et al., 2005a, 2005b). Salinity occurs either as a result of deforestation and rising saline water tables (dryland salinity) or irrigation with saline water (irrigation salinity) (Rengasamy, 2006). While soil salinity is complex, NaCl is considered the primary contributing salt as it is abundant in many soils and has a very high solubility (Rengasamy, 2002).

Hydroponics is highly suitable for the study of salinity tolerance. However, there are several issues to be considered. The addition of NaCl should not be made in a single application because it will cause osmotic shock and may kill the plants.

Unless osmotic stress tolerance is of interest to the researcher, salt should be added in increments until the final desired concentration is reached, to allow plants to adapt to osmotic stress. Depending on the plant species and aim of the experiment, low, moderate or severe salt stress may be applied. Typically, we add salt twice daily (morning and evening), in 25 mM or 50 mM NaClincrements (Shavrukov et al, 2006, 2009, 2010a, 2010b), in agreement with other salinity research groups (Boyer et al., 2008; Dreccer et al., 2004; Forster et al., 1990, 1994; Gorham, 1990; Munns & James, 2003; Rawson et al., 1988a, 1988b; Shah et al., 1987; Watson et al., 2001). We have found that suitable salt stress levels are typically 100-150 mM NaCl for bread wheat (Dreccer et al., 2004; Gorham et al., 1987; Munns & James, 2003; Shah et al., 1987), 150- 200 mM NaCl for barley (Forster et al., 1990, 1994; Gorham et al., 1990; Rawson et al., 1988a, 1988b; Shavrukov et al., 2010a), and 250-300 mM NaCl for tolerant cereals such as wild emmer wheat, *Triticum dicoccoides* (Shavrukov et al., 2010b), and for saltbush, *Atriplex* ssp. and other halophytes (Flowers et al., 1977). While there are reports of salinity experiments in hydroponics using NaCl concentrations of 300 mM NaCl (Huang et al., 2006), this represents a salinity level of half the strength of sea-water and most plants would be severely stressed in such a treatment.

The addition of NaCl requires supplementation with additional Ca^{2+}. Symptoms of calcium- deficiency are observed when plants are grown in hydroponics under salt treatment and not provided with extra Ca^{2+} (Cramer, 2002; Ehret et al., 1990), and leaves from these plants are calcium-deficient (eg. Francois et al., 1991). Increasing NaCl in hydroponics solutions reduces the activity of Ca^{2+} in solution (Cramer & Läuchli, 1986), and supplementary Ca^{2+} should be added to compensate. The amount of supplementary calcium (as a ratio of $Na^+ : Ca^{2+}$) required varies depending on the concentration of added NaCl and the overall composition and pH of the growth solution, and can be determined using speciation prediction programs such as Geochem-EZ: http://www.plantmineralnutrition.net/ Geochem/ geochem%20 home.htm (Shaff et al., 2010); or Visual MINTEQ: http://www2.lwr.kth.se/English/OurSoftware/vminteq/index.html (Gustaffson, 2008). Recent empirical studies in wheat using the growth solution provided in Table 3.1, found that the most appropriate $Na^+ : Ca^{2+}$ ratio in solution, resulting in tissue Ca^{2+} concentrations of salt-affected pants similar to those of control plants, was 1 : 15 at 100 mM NaCl (Genc et al., 2010). This study also showed that excessive use of supplemental Ca^{2+} could induce additional osmotic stress and nutritional deficiencies such as magnesium, and thus should be avoided.

In salinity research, hydroponics can be used to study sodium accumulation during short- term (7 - 10 days) and long-term (up to maturity) studies. We generally grow wheat and barley in hydroponics in the presence or absence of NaCl for four weeks to determine overall salinity tolerance measured as relative growth (growth in NaCl treatment relative to non-saline conditions). Plants can be easily removed from hydroponics and separated into roots, shoots and leaves (if required) for destructive analysis. Non-destructive measurements can also be made: selected individuals can be removed from hydroponics, their fresh weights obtained, then transplanted to pots of non-saline soil for recovery and cultivation to maturity if seeds are required. We achieve high grain yields from wheat and barley and their wild relatives when transplanted from saline hydroponics to soil-filled pots, provided this is done when the plants are less than four weeks old.

We have also successfully grown wheat and barley plants in flood-drain hydroponics systems to maturity. With regular replacement, hydroponics growth solutions provide the growing plants with sufficient nutrients and there is no inter-plant competition for nutrients as may occur in soil. However, following tillering, the growing plants compete for light and space around the above-ground plant parts. Provided these factors are optimized, plants can

be grown to maturity as shown for bread wheat in Fig. 3.7. This hydroponics study demonstrated both symptoms of growth depression in plants and an increased rate of plant development under salt stress (Fig. 3.7).

Fig. 3.7. Flood-drain supported hydroponics in 20 L containers and tubes, showing the appearance of bread wheat plants grown at different concentrations of NaCl to maturity (from left to right: 0, 50, 75, 100, 150 and 200 mM NaCl).

Drought

Various forms of hydroponics have been used to study drought (water stress) responses by plants, with somewhat limited success (reviewed in Munns et al., 2010). Drought is a particularly complex stress phenomenon that is difficult to model in any growth system. Water deficit may be imposed in hydroponics using osmotica such as mixed salts (eg. high concentrations of macronutrients in nutrient solution), NaCl, mannitol, sorbitol or polyethylene glycol. The applied water stress in hydroponics is more controlled and homogeneous than in soil-based systems. However, small molecules, such as mannitol, are easily absorbed by roots and move to the shoots (Hohl & Schopfer, 1991), and will affect plant metabolism and drought tolerance responses. NaCl or mixed salts may be suitable for short-term studies of water deficit (eg. Tavakkoli et al., 2010), but these will be taken up by the plants with time as well. High molecular weight polyethylene glycol (PEG) is less likely to be absorbed by plants, although uptake of PEG has been observed through damaged roots (Miller, 1987). PEG also increases solution viscosity and reduces the supply of oxygen to plant roots (Mexal et al., 1975). This can be overcome with careful supplemental oxygenation (Verslues et al., 1998). Reasonably consistent ranking of drought tolerance of wheat genotypes has been achieved using both a PEG treatment in hydroponics and drying of pots of soil (Molnár et al., 2004). We use an alternative method utilising hydroponics tostudy terminal drought responses in wheat and barley. Growth solution is withdrawn from the flood/drain system described above, with plants growing in plastic fragments. Usually, up to five days of drying is allowed before plant tissue is sampled for analysis of gene expression changes relative to tissue sampled prior to the withdrawal of growth solution. While we observe variability between experiments using this method, reasonable comparisons can be made between genotypes within a single experiment. It should be noted that the withdrawal of growth solution in hydroponics systems is not suitable for long-term drought experiments because the process of roots drying between plastic fragments is relatively quick and cannot simulate processes of natural drought.

Elemental Toxicities: Boron and Aluminium

Excessive levels of soil boron (B) and aluminium (Al) both reduce plant growth and, in regions where they occur, significantly limit cereal production. Boron toxicity typically occurs in alkaline soils of marine origin, often in conjunction with soil salinity. Boron toxicity may also occur as a consequence of excessive fertiliser application. The effects of B toxicity include reduced root growth and shoot dry matter production, leaf necrosis, and reduced grain yield. Significant yield penalties in southern Australia due to B toxicity have been reported for wheat and barley (Cartwright et al., 1986; Moody et al., 1993). Aluminium toxicity occurs in acid soils where the main form of Al present is the soluble cation, Al^{3+}. Al^{3+} ions severely stunt root growth of cereals and other crop species and, consequently, greatly affect yield. Much of the published research into Al toxicity has relied on hydroponics experiments to measure the effects of Al^{3+} on root growth and exudation.

Boron toxicity in hydroponics is relatively easy to achieve, simply by adding boric acid (H_3BO_3) to basal nutrient solutions. For wheat and barley, we find that between 2 and 5 mM added H_3BO_3 is sufficient to see the development of B toxicity symptoms on leaves, effects on shoot growth and significant B accumulation in intolerant genotypes. Stock solutions of H_3BO_3 below 0.5 M are not adjusted for pH, and treatment levels (up to 5 mM B) do not affect the pH of the nutrient solution, or greatly change its osmolarity. It is to be noted that suitable B concentration ranges for assessing B toxicity tolerance in hydroponics would need to be determined empirically for different plant species and in different hydroponics systems and environmental conditions. For example, monocot and dicot species have different requirements for B (Asad et al., 2001), and species may also differ in B toxicity tolerance (eg. Stiles et al., 2010).

For assessment of wheat varieties for tolerance to B toxicity, we have developed a simplified hydroponics system and use relative root length as a proxy measure of tolerance (Schnurbusch et al., 2008). Seedlings are grown in a solution containing 2.5 μM $ZnSO_{4,}$ 15μM H_3BO_3 and 0.5 mM $Ca(NO_3)2$, supplemented with 10 mM $H_3BO_{3,}$ for 10 – 14 days, and root lengths are measured with a ruler. Relative root length (at 10 mM B compared to low B (15 μM)) is simple to score, and is much less expensive than analysis of shoot B by inductively coupled plasma emission spectrometry. The parameter correlates well with field reports of B toxicity tolerance, and also with the presence of B tolerance alleles on chromosome 7BL (Schnurbusch et al., 2008). We found, however, that this system is not suitable for barley.

Assessment of aluminium toxicity tolerance in hydroponics is complicated by a number of factors. The speciation of Al in solutions depends on both solution pH and total Alconcentration. At low pH, Al predominates as the trivalent cation, Al^{3+}, and this is known to be the major form of Al which is toxic to plants. However, the trivalent cation can complex with anions in solution, rendering it non-toxic. Hydroxyl monomers of Al may also form which are thought to be non-toxic (Parker et al., 1988), and Al readily precipitates out of solution at moderate to high pH. When precipitated, Al is not toxic to plant growth. The use of chemical speciation prediction programs such as Geochem-EZ (Shaff et al., 2010) are necessary to estimate the predicted activity of Al^{3+} in a given hydroponics solution. Careful attention to the maintenance of a low, stable solution pH is also important for obtaining reproducible experimental results. Modified hydroponics solutions (Famoso et al., 2010), or simple solutions, eg. CaCl2 (Ma et al., 2002; Xue et al., 2006), are often used when assessing Al toxicity tolerance, to reduce the likelihood of Al forming complexes or precipitates.

The advantage of using hydroponics to study Al toxicity is that effects on root growth and exudation can easily be measured. Organic acid exudation by the roots is the major mechanism by which plants can tolerate Al and, in hydroponics, these can be collected and measured. Hydroponics has been the medium of choice for screening wheat (Delhaize et al., 1993; Sasaki et al., 2004), barley, rice (Famoso et al., 2010; Nguyen et al., 2001), maize (Magnavaca et al., 1987; Piñeros et al., 2005) and rye (Collins et al., 2008) for Al tolerance.

Boron and aluminium are examples of naturally occurring elemental toxicities and have historically been a focus of elemental toxicity research. However, there have been increasing occurrences of contamination of agricultural soils with arsenic (As) and with heavy metals including cadmium, zinc, nickel, selenium, mercury and lead. There is also a growing awareness of the potential consequences for human health of accumulation of these metals in the food chain. Hydroponics is ideal for investigating the basic mechanisms plants may possess for either reducing or avoiding uptake of as and heavy metals (eg. rice, Ma et al., 2008), or for hyper-accumulation of these elements (eg. *Pteris vittata*, Wang et al., 2002; *Thlaspi caerulescens*, reviewed in Milner & Kochian, 2008). Hyper-accumulation by plants and subsequent harvest for safe disposal is suggested as a means for removing heavy metals from contaminated sites (Salt et al., 1998). In recent research, it is emerging that both hyper- accumulation and avoidance in plants are largely due to transport processes, and these are best studied in hydroponics. Hydroponics also allows the experimenter to contain the metal elements in a closed system to ensure human safety both during the experiment and, with proper disposal and clean-up, following conclusion of the work.

Nutrient Deficiencies

Hydroponics has been instrumental in establishing the essentiality of most of the mineral nutrients required by plants (Jones, 1982; Reed, 1942), from the early development of nutrient solution recipes in the 1860's by the German scientists Sachs and Knop (Hershey, 1994), through to as recently as 1987 when nickel was confirmed as an essential micronutrient for higher plants (Brown et al., 1987). Hydroponics is frequently used to study the effects of mineral nutrient deficiencies on plant growth and physiology. It is particularly useful in identifying visual symptoms or critical deficiency concentrations for diagnostic purposes, characterising physiological functions of mineral nutrients, determining their uptake kinetics, studying root exudates and gene expression changes and also changes in root morphological traits in response to nutrient deficiencies. It is also commonly used to identify germplasm with enhanced nutrient use efficiency (i.e. an ability to produce greater biomass at limited nutrient supply) for breeding programs. However, the use of hydroponics is limited to processes involving efficiency of utilisation or mobilisation within the plant rather than those operating at the root-soil interface (Graham, 1984). Like many other research groups, we routinely use hydroponics to study effects of elemental deficiencies, including phosphorus (Huang et al., 2008) and zinc (Fig. 3.8), on the growth and physiology of important crop species such as wheat and barley. In such studies, particular care must be taken to avoid contamination from external sources of the element of interest. For phosphorus deficiency experiments, for example, the hydroponics setup should be thoroughly washed with a mild acid solution to remove all residual phosphorus, especially as phosphorus is a common ingredient in standard detergents used to wash laboratory glassware and hydroponics equipment.

Fig. 3.8. An experiment designed to study Zn deficiency in barley and wheat, using aerated hydroponics in 1 L pots containing plastic fragments to support the growing plants: (A) Barley plants (cv. Pallas) grown with different concentrations of Zn (from left to right: 0.005, 0.05 and 0.5 µM Zn), and (B) Three bread wheat genotypes (from left to right: Stylet, RAC875-2 and VM506 grown with nil Zn supply.

EXPANDING OPERATIONS: TRANSITIONING FROM HYDROPONIC SYSTEMS TO OUTDOOR CULTIVATION

The main advantages of hydroponics over soil-based systems can be summarised as follows;

- There is a greater degree of control over variables and thus observations are reproducible.

- Effects of nutrient deficiency or toxicity on plant growth can be determined more reliably, and studies on root nutrient uptake and certain root morphological traits are much easier to conduct since in soil-based systems it is often difficult to separate roots from soil particles and accurately measure nutrient concentrations or uptake by roots. This makes hydroponics ideal for studying nutrient toxicities, deficiencies and other abiotic stresses. However, it should be remembered that hydroponics is very much an artificial system, and observations may differ greatly from those made in soil-based systems.

Some reports have demonstrated strikingly similar results in hydroponics and in field trials. For example, identical Quantitative Trait Loci (QTLs) were found on the long arm of chromosome 7A in two unrelated mapping populations in bread wheat (Halberd x Cranbrook and Excalibur x Kukri) for Na^+ accumulation in both hydroponics and in field trials (Edwards et al., 2008; Shavrukov et al., 2011). There are, however, many reported discrepancies between research findings using hydroponics and soil-based systems. Recent studies in barley, for example, suggest that responses to salinity stress (Tavakkoli et al., 2010) and drought stress (Szira et al., 2008) can vary between hydroponics and soil culture. Assessments of P efficiency in wheat also differed greatly when cultivars were grown in hydroponics compared to soil (Hayes et al., 2004). Similarly, despite the effects of B toxicity which we have observed in hydroponics largely reflecting those of glasshouse-based soil experiments as well as observations made in the field (eg. Jefferies et al., 2000 and Table 3.2), there are some inconsistencies. These inconsistencies are likely to be explained by differences in either the physical/chemical characteristics of the growing environment, or root morphology differences created by these characteristics.

Many nutrients in soil do not exist at high concentrations in soil solution, but are instead bound to negatively-charged surfaces of clay or organic matter particles, or are precipitated as mineral salts. Nutrients are only released into solution to replace those taken up by plants. The soil solution is thus strongly buffered, maintaining low but stable nutrient concentrations. By contrast, many nutrients in hydroponics solutions are necessarily supplied at much higher concentrations. This makes studies of nutrient deficiencies particularly difficult. Frequent solution replacements, or large volumes of solution, are necessary to maintain low and relatively stable concentrations of an element of interest. Alternatively, it is possible to try to mimic the buffering ability of soil and maintain stable concentrations of a particular nutrient by adding resins (eg. Asad et al., 2001) or chelating agents (eg. Chaney et al., 1989; Norwell & Welch, 1993; Rengel & Graham, 1996) to the nutrient solution. In soils, there is also a gradient of nutrient concentrations established across the rhizosphere as roots take up and deplete nutrients from their surrounds, so that the nutrient concentration at the root surface is much lower than in the bulk soil solution. It is not possible to establish a similar gradient in nutrient concentrations in well-stirred hydroponics solutions.

Table 3.2. Comparison of boron toxicity tolerance traits observed in hydroponics and soil- based experiments, for the wheat cultivars Halberd (B toxicity tolerant) and Cranbrook (intolerant). While shoot boron concentrations are comparable, relative dry weights respond differently in soil compared to hydroponics. differences in either the physical/ chemical characteristics of the growing environment, or root morphology differences created by these characteristics.

		Hydroponics		Soil	
		+ 2 mM B	+ 5 mM B	+ 10 mg B kg-1	+ 30 mg B kg-1
Shoot B (mg B kg-1)	Halberd	405	2883	900	2600
	Cranbrook	653	3667	1240	3900
Relative DW (%)	Halberd	108%	75%	58%	14%
	Cranbrook	115%	74%	50%	6%
3rd leaf necrosis (%) Halberd		N/A	30%	25%	30%
Cranbrook		N/A	45%	60%	90%

N/A = not obtained

Soils are also heterogeneous environments, with spatial variability in water and nutrient availabilities and physical characteristics. This heterogeneity cannot be replicated in hydroponics. In soil, plants are able to respond to heterogeneity by investing greater root growth in either nutrient- or moisture-rich patches and avoiding hostile micro- environments (Jackson et al., 1990). Research into these types of plant responses can only be done in soil. Mycorrhizal fungi and other soil biota form close associations with plant roots in soil, and these are particularly important for phosphorus, and also zinc and copper uptake, by plants in these environments (Smith & Read, 2008). Nodulation of roots by *Rhizobium* spp. is also vital for nitrogen uptake by leguminous plants (Kinkema et al., 2006). Such interactions between plants and rhizosphere microorganisms, and the implications for abiotic stress tolerance can only be studied effectively in soil.

Although hydroponics allows the experimenter unrestricted access to roots and thus easy assessment of root traits under different stress conditions, it is widely acknowledged that root morphological traits of hydroponically-grown plants may be very different to those of plants grown in soil or other solid media. This may directly affect any conclusions drawn about the tolerance of plants to abiotic stresses. For example, nodal roots originate from either the stem or the mesocotyl between the base of the shoot and the base of the primary root. They are believed to be largely responsible for exploring surface soil layers, and nodal root morphology may contribute greatly to determining the drought tolerance and P uptake efficiency of plants (Ho et al., 2005). However, nodal roots will develop differently in hydroponics and soils because of seed placement differences between the two systems. Development of root hairs also differs between soil- and hydroponically-grown plants. Root hair length and density is reduced in solution culture compared to soil for both maize (Mackay & Barber, 1984) and barley (Gahoonia & Nielsen, 1997; Genc et al., 2007). Root hair length/density is directly correlated with plant uptake of phosphorus (Bates & Lynch, 2000; Gahoonia &Nielsen, 1997) and is also related to zinc uptake efficiency (Genc et al., 2007). Research also suggests that the rate of appearance and maturation of a suberised exodermallayer in roots differs for hydroponically grown plants. The exodermis forms a barrier between the root and the external environment, controlling water and solute influx, and thus is potentially critical for tolerance to water stress (Cruz et al., 1992; Hose et al., 2001). It has been found that there is more rapid suberisation of the exodermal layer in maize roots grown in moist air (aeroponics), vermiculite or stagnant conditions compared to aerated hydroponics (Enstone & Peterson, 1998; Zimmerman & Steudle, 1998). Moreover, in barley, nodal roots are more extensively suberised than seminal roots (Lehmann et al., 2000), and thus hydroponically grown barley roots will have limited suberised exodermal layers compared to equivalent soil-grown plants. Researchers remain unclear as to the morphology of soil-grown roots and how they respond to abiotic stresses in field conditions. Hydroponics, aeroponics, pot and/or field sampling have gone some way towards examining root traits, but the future development of DNA profiling and sophisticated imaging technologies (eg. LemnaTec) will improve our understanding of the role of roots in abiotic stress tolerance. It is likely that root adaptations are a very significant component of stress adaptation.

AT A GLANCE

- Hydroponics is a particularly useful research tool used to study plant responses to abiotic stresses, including salinity, boron and aluminium toxicities, nutrient deficiencies and to a lesser degree, drought. Here we have described two hydroponics systems used by researchers at ACPFG and the University of Adelaide, both of which are suitable for abiotic stress tolerance research. The particular advantages of hydroponics are that treatments can be precisely controlled, and plant responses can be accurately and reproducibly determined. Genetic variation in abiotic stress tolerance, both between and within species, can be assessed with confidence.

- Roots of hydroponically-grown plants are easily accessible, allowing, for example, morphological traits to be examined, short-term uptake experiments to be conducted and root exudates to be collected for analysis. However, it should be remembered that hydroponics is a unique, artificial system for growing plants and is not a substitute for soil. There are many differences between soil- and hydroponically-grown plants, as well as fundamental differences in the supply of water and nutrients, which are important to consider when researching abiotic stress tolerance. We have summarised some of the limitations relating to using hydroponics as a research tool in this chapter, and caution that ultimately, validation of abiotic stress responses and characteristics must be made in soils and in field conditions.

REFERENCES

Asad, A.; Bell, R.W. & Dell, B. (2001) A critical comparison of the external and internal boron requirements for contrasting species in boron-buffered solution culture. *Plant and Soil*, Vol.233, No.1, pp. 31-45

Bağci, S.A.; Ekiz, H. & Yilmaz, A. (2007) Salt tolerance of sixteen wheat genotypes during seedling growth. *Turkish Journal of Agriculture and Forestry*, Vol.31, No.6, pp. 363- 372

Bates, T.R. & Lynch, J.P. (2000) Plant growth and phosphorus accumulation of wild type and two root hair mutants of *Arabidopsis thaliana* (Brassicaceae). *American Journal of Botany*, Vol.87, No.7, pp. 958-963

Boyer, J.S.; James, R.A.; Munns, R.; Condon, T. & Passioura, J.B. (2008) Osmotic adjustment leads to anomalously low estimates of relative water content in wheat and barley. *Functional Plant Biology*, Vol.35, No.11, pp. 1172-1182

Britto, D.T. & Kronzucker, H.J. (2002) NH4 toxicity in higher plants: a critical review. *Journal of Plant Physiology*, Vol.159, No.6, pp. 567-584

Brown, P.H.; Welch, R.M. & Cary, E.E. (1987) Nickel: A micronutrient essential for higher plants. *Plant Physiology*, Vol.85, No.3, pp. 801–803

Cartwright, B.; Zarcinas, B.A. & Spouncer, L.R. (1986) Boron toxicity in South Australian barley crops. *Australian Journal of Agricultural Research*, Vol.37, No.4, pp. 321-330

Chiba, Y.; Mitani, N.; Yamaji, N. & Ma, J.F. (2009) HvLsi1 is a silicon influx transporter in barley. *The Plant Journal*, Vol.57, No.5, pp. 810-818

Chaney, R.L.; Bell, P.F. & Coulombe, B.A. (1989) Screening strategies for improved nutrient uptake and use by plants. *HortScience*, Vol.24, No.4, pp. 565-572

Collins, N.C.; Shirley, N.J.; Saeed, M.; Pallotta, M. & Gustafson, J.P. (2008) An *ALMT1* gene cluster controlling aluminium tolerance at *Alt4* locus of rye (*Secale cereale* L.). *Genetics*, Vol.179, No.1, pp. 669-682

Colmer, T.D.; Munns, R. & Flowers, T.J. (2005) Improving salt tolerance of wheat and barley: future prospects. *Australian Journal of Experimental Agriculture*, Vol.45, No.11, pp. 1425-1443

Cramer, G.R. (2002) Sodium-calcium interactions under salinity stress. In: *Salinity: Environment-Plants-Molecules*, A. Lauchli & U. Luttge, (Eds.), 205-227, Kluwer Academic Publishers, Dordrecht, Netherlands

Cramer, G.R. & Läuchli, A. (1986) Ion activities in solution in relation to Na^{+}-Ca^{2+} interactions at the plasmalemma. *Journal of Experimental Botany*, Vol.37, No.3, pp. 321-330

Cruz, R.T.; Jordan, W.R. & Drew, M.C. (1992) Structural changes and associated reduction of hydraulic conductance in roots of *Sorghum bicolor* L. following exposure to water deficit. *Plant Physiology*, Vol.99, No.1, pp. 203-212

Deane-Drummond, C.E. (1982) Mechanisms for nitrate uptake into barley (*Hordeum vulgare* L. cv. Fergus) seedlings at controlled nitrate concentrations in the nutrient medium. *Plant Science Letters*, Vol.24, No.1, pp. 79-89

Delhaize, E.; Craig, S.; Beaton, C.D.; Bennet, R.J.; Jagadish, V.C. & Randall, P.J. (1993) Aluminum tolerance in wheat (*Triticum aestivum* L.) I. Uptake and distribution of aluminum in root apices. *Plant Physiology*, Vol.103, No.3, pp. 685-693

Dreccer, M.F.; Ogbonnaya, F.C. & Borgognone, M.G. (2004) Sodium exclusion in primary synthetic wheats. *Proceedings of 54th Australian Cereal Chemistry Conference and 11th Wheat Breeding Assembly*, C.K. Black, J.F. Panozzo & G.J. Rebetzke, (Eds.), 118-121, Royal Australian Chemical Institute, Melbourne, Australia

Drihem, K. & Pilbeam, D.J. (2002) Effects of salinity on accumulation of mineral nutrients in wheat growth with nitrate-nitrogen or mixed ammonium: nitrate-nitrogen. *Journal of Plant Nutrition*, Vol. 25, No.10, pp. 2091-2113

Dubcovsky, J.; Santa-Maria, G.; Epstein, E.; Luo, M.C. & Dvorak, J. (1996) Mapping of the K^{+}/Na^{+} discrimination locus *Kna1* in wheat. *Theoretical and Applied Genetics*, Vol.92, No.3-4, pp. 448-454

Edwards, J.; Shavrukov, Y.; Ramsey, C.; Tester, M.; Langridge, P. & Schnurbusch, T. (2008) Identification of a QTL on chromosome 7AS for sodium exclusion in bread wheat, *Proceedings of the 11th International Wheat Genetics Symposium, Brisbane, Australia,* R. Appels, R., Eastwood, E. Lagudah, P. Langridge, M. Mackay, L. McIntyre & P. Sharp, (Eds.), Vol. 3, 891-893, Sydney University Press, Sydney, Australia. Available from: http://ses.library.usyd.edu.au/bitstream/2123/3263/1/P178.pdf

Ehret, D.L.; Redmann, R.E.; Harvey, B.L. & Cipywnyk, A. (1990) Salinity-induced calcium deficiencies in wheat and barley. *Plant and Soil*, Vol.128, No.2, pp. 143-151

Enstone, D.E. & Peterson, C.A. (1998) Effects of exposure to humid air on epidermal viability and suberin deposition in maize (*Zea mays* L.) roots. *Plant, Cell and Environment*, Vol.21, No.8, pp. 837–844

Epstein, E. (1994) The anomaly of silicon in plant biology. *Proceeding of the Naional Academy of Sciences of the United States of America*, Vol.91, No.1, pp. 11-17

Epstein, E. (1999) Silicon. *Annual Review of Plant Physiology and Plant Molecular Biology*, Vol.50, pp. 641-664

Famoso, A.N.; Clark, R.T.; Shaff, J.E.; Craft, E.; McCouch, S.R. & Kochian, L.V. (2010) Development of a novel aluminium tolerance phenotyping platform used for comparisons of cereal aluminium tolerance and investigations into rice aluminium tolerance mechanisms. *Plant Physiology*, Vol.153, No.4, pp. 1678-1691

Flowers, T.J. & Yeo, A.R. (1995) Breeding for salinity resistance in crop plants: Where next? *Australian Journal of Plant Physiology*, Vol.22, No.6, pp. 875-884

Flowers, T.J.; Troke, P.F. & Yeo, A.R. (1977) The mechanism of salt tolerance in halophytes. *Annual Review of Plant Physiology*, Vol.28, pp. 89-121

Forster, B.P.; Phillips, M.S.; Miller, T.E.; Baird, E. & Powell, W. (1990) Chromosome location of genes controlling tolerance to salt (NaCl) and vigour in *Hordeum vulgare* and *H. chilense*. *Heredity*, Vol.65, No.1, pp. 99-107

Forster, B.P.; Pakniyat, H.; Macaulay, M.; Matheson, W.; Phillips, M.S.; Thomas, W.T.B. & Powell, W. (1994) Variation in the leaf sodium content of the *Hordeum vulgare* (barley) cultivar Maythorpe and its derived mutant cv. Golden Promise. *Heredity*, Vol.73, No.3, pp. 249-253

Francois, L.E.; Donovan, T.J. & Maas, E.V. (1991) Calcium deficiency of artichoke buds in relation to salinity. *HortScience*, Vol.26, No.5, pp. 549-553

Gahoonia, T.S. & Nielsen, N.E. (1997) Variation in root hairs of barley cultivars doubled soil phosphorus uptake. *Euphytica*, Vol.98, No.3, pp. 177-182

Gazzarrini, S.; Lejay, L.; Gojon, A.; Ninnemann, O.; Frommer, W.B. & von Wirén, N. (1999) Three functional transporters for constitutive, diurnally regulated, and starvation- induced uptake of ammonium into *Arabidopsis* roots. *The Plant Cell*, Vol.11, No.5, pp. 937-948

Genc, Y.; Huang, C.Y. & Langridge, P. (2007) A study of the role of root morphological traits in growth of barley in zinc-deficient soil. *Journal of Experimental Botany*, Vol.58, No.11, pp. 2775-2784

Genc, Y.; McDonald, G.K. & Tester, M. (2007) Reassessment of tissue Na concentration as a criterion for salinity tolerance in bread wheat. *Plant, Cell and Environment*, Vol.30, No.11, pp. 1486-1498

Genc, Y.; Tester, M. & McDonald, G.K. (2010) Calcium requirement of wheat in saline and non-saline conditions. *Plant and Soil*, Vol.327, No.1-2, pp. 331-345

Gorham, J. (1990) Salt tolerance in the Triticeae: K/Na discrimination in synthetic hexaploid wheats. *Journal of Experimental Botany*, Vol.41, No.5, pp. 623-627

Gorham, J.; Hardy, C.; Wyn Jones, R.G.; Joppa, L.R. & Law, C.N. (1987) Chromosomal location of a K/Na discrimination character in the D genome of wheat. *Theoretical and Applied Genetics*, Vol.74, No.5, pp. 584-588

Gorham, J; Bristol, A.; Young, E.M.; Wyn Jones, R.G. & Kashour, G. (1990) Salt tolerance in the Triticeae: K/Na discrimination in barley. *Journal of Experimental Botany*, Vol.41, No.9, pp. 1095-1101

Gorham, J.; Bristol, A.; Young, E.M. & Wyn Jones, R.G. (1991) The presence of the enhanced K/Na discrimination trait in diploid *Triticum* species. *Theoretical and Applied Genetics*, Vol.82, No.6, pp. 729-736

Graham, R.D. (1984) Breeding for nutritional characteristics in cereals. In: *Advances in Plant Nutrition,* P.B. Tinker & A. Lauchli, (Eds.), Vol.1, 57-102, Praeger, N.-Y. et al.

Greenway, H. (1962) Plant response to saline substrates. 1. Growth and ion uptake of several varieties of *Hordeum* during and after sodium chloride treatment. *Australian Journal of Biological Sciences*, Vol.15, No.1, pp. 16-38

Gustaffson, J.P. (2008) Visual MINTEQ. Version 2.60, Stockholm. Available from http://www2.lwr.kth.se/English/OurSoftware/vminteq/index.html

Hayes, J.E.; Zhu, Y.-G.; Mimura, T & Reid, R.J. (2004) An assessment of the usefulness of solution culture in screening for phosphorus efficiency in wheat. *Plant and Soil*, Vol.261, No.1-2, pp. 91-97

Hershey, D. R. (1994) Solution culture hydroponics - history and inexpensive equipment. *American Biology Teacher*, Vol.56, No.2, pp. 111-118

Ho, M.; Rosas, J.; Brown, K. & Lynch, J. (2005) Root architectural tradeoffs for water and phosphorus acquisition. *Functional Plant Biology*, Vol.32, No.8, pp. 737-748

Hoagland, D.R. & Arnon, D.I. (1938) *The water culture method for growing plants without soil*, University of California, College of Agriculture, Agricultural Experiment Station, Circular 347, Berkeley, USA

Hohl, M. & Schopfer, P. (1991) Water relations of growing maize coleoptiles. Comparison between mannitol and polyethylene glycol 6000 as external osmotica for adjusting turgor pressure. *Plant Physiology*, Vol.95, No.3, pp. 716-722

Hose, E.; Clarkson, D.T.; Steudle, E.; Schreiber, L. & Hartung, W. (2001) The exodermis: a variable apoplastic barrier. *Journal of Experimental Botany*, Vol.52, No.365, pp. 2245- 2264

Huang, Y.; Zhang, G.; Wu, F.; Chen, J. & Zhou, M. (2006) Differences in physiological traits among salt-stressed barley genotypes. *Communications in Soil Science and Plant Analysis*, Vol.37, No.3-4, pp. 557-570

Huang, C.Y.; Roessner, U.; Eickmeier, I.; Genc, Y.; Callahan, D.L.; Shirley, N.; Langridge, P. & Bacic, A. (2008) Metabolite profiling reveals distinct changes in carbon and nitrogen metabolism in phosphate-deficient barley plants (*Hordeum vulgare* L.). *Plant and Cell Physiology*, Vol.49, No.5. pp. 691-703

Jackson, R.B.; Manwaring, J.H. & Caldwell, M.M. (1990) Rapid physiological adjustment of roots to localized soil enrichment. *Nature*, Vol.344, No.6261, pp. 58–60

Jarvis, S.C. & Hatch, D.J. (1985) Rates of hydrogen-ion efflux by nodulated legumes grown in flowing solution culture with continuous pH monitoring and adjustment. *Annals of Botany*, Vol.55, No.1, pp. 41-51

Jefferies, S.P.; Pallotta, M.A.; Paull, J.G.; Karakousis, A.; Kretschmer, J.M.; Manning, S.; Islam, A.K.M.R.; Langridge, P. & Chalmers, K.J. (2000) Mapping and validation of chromosome regions conferring boron toxicity tolerance in wheat (*Triticum aestivum*). *Theoretical and Applied Genetics*, Vol.101, No.5-6, pp. 767-777

Jones, J.B. (1982) Hydroponics: Its history and use in plant nutrition studies. *Journal of Plant Nutrition*, Vol.5, No.8, pp. 1003-1030

Kinkema, M.; Scott, P.T. & Gresshoff, P.M. (2006) Legume nodulation: successful symbiosis through short- and long-distance signalling. *Functional Plant Biology*, Vol.33, No.8, pp. 707-721

Kingsbury, R.W.; Epstein, E. & Pearcy, R.W. (1984) Physiological responses to salinity in selected lines of wheat. *Plant Physiology*, Vol.74, No.2, pp. 417-423

Kronzucker, H.J.; Szczerba, M.W.; Moazami-Goudarzi, M. & Britto, D.T. (2006) The cytosolic Na^+ : K^+ ratio does not explain salinity-induced growth impairment in barley: a dual tracer study using $^{42}K^+$ and $^{24}Na^+$. *Plant, Cell and Environment*, Vol.29, No.12, pp. 2228-2237

Lehmann, H.; Stelzer, R.; Holzamer, S.; Kunz, U. & Gierth, M. (2000) Analytical electron microscopical investigations on the apoplastic pathways of lanthanum transport in barley roots. *Planta*, Vol.211, No.6, pp. 816-822

Lyons, G.H.; Genc,Y.; Soole, K; Stangoulis, J.C.R.; Liu, F. & Graham, R.D. (2009) Selenium increases seed production in Brassica. *Plant and Soil*, Vol.318, No.1-2, pp. 73-80

Ma, J.F.; Shen, R.; Zhao, Z.; Wissuwa, M.; Takeuchi, Y.; Ebitani, T. & Yano, M. (2002) Response of rice to Al stress and identification of Quantitative Trait Loci for Al tolerance. *Plant and Cell Physiology*, Vol.43, No.6, pp. 652-659

Ma, J.F. & Yamaji, N. (2006) Silicon uptake and accumulation in higher plants. *Trends in Plant Science*, Vol.11, No.8, pp. 392-397

Ma, J.F.; Tamai, K.; Yamaji, N.; Mitani, N.; Konishi, S.; Katsuhara, M.; Ishiguro, M.; Murata, Y. & Yano, M. (2006) A silicon transporter in rice. *Nature*, Vol.440, No.7084, pp. 688- 691

Ma, L.; Zhou, E.; Huo, N.; Zhou, R.; Wang, G. & Jia, J. (2007) Genetic analysis of salt tolerance in a recombinant inbred population of wheat (*Triticum aestivum* L.). *Euphytica*, Vol.153, No.1-2, pp. 109-117

Ma, J.F.; Yamaji, N.; Mitani, N.; Xu, X.-Y.; Su, Y.-H.; McGrath, S.P. & Zhao, F.-J. (2008)

Transporters of arsenite in rice and their role in arsenic accumulation in rice grain. *Proceedings of the National Academy of Sciences of the United States of America*, Vol.105, No.29, pp. 9931-9935

Mackay, A.D. & Barber, S.A. (1984) Comparison of root and root hair-growth in solution and soil culture. *Journal of Plant Nutrition*, Vol.7, No.12, pp. 1745-1757

Magnavaca, R.; Gardner, C.O. & Clark, R.B. (1987) Evaluation of inbred maize lines for aluminium tolerance in nutrient solution. In: *Genetic Aspects of Plant Mineral Nutrition*, 255-265, Martinus Nijhoff Publishers, Dordrecht, Netherlands

Mexal, J.; Fisher, J.T.; Osteryoung, J. & Reid, C.P.P. (1975) Oxygen availability in polyethylen glycol solutions and its implications in plant-water relations. *Plant Physiology*, Vol.55, No.1, pp. 20-24

Miller, D.M. (1987) Errors in the measurement of root pressure and exudation volume flow rate caused by damage during the transfer of unsupported roots between solutions. *Plant Physiology*, Vol.85, No.1, pp. 164-166

Milner, M.J. & Kochian, L.V. (2008) Investigating heavy-metal hyperaccumulation using

Thlaspi caerulescens as a model system. *Annals of Botany*, Vol.102, No.1, pp. 3-13 Molnár, I.; Gáspár, L.; Sárvári, E.; Dulai, S.; Hoffmann, B.; Molnár-Láng, M. & Galiba, G.

(2004) Physiological and mophological responses to water stress in *Aegilops biuncialis* and *Tridicum aestivum* genotypes with differing tolerance to drought. *Functional Plant Biology*, Vol.31, No.12, pp. 1149-1159

Moody, D.B; Rathjen, A.J. & Cartwright, B. (1993) Yield evaluation of a gene for boron tolerance using backcross-derived lines. *The 4th International Symposium of Genetic Aspects of Plant Mineral Nutrition*, P.J. Randall, E. Delhaize, R.A. Richards & R. Munns, (Eds.), 363-366, Kluwer Academic Publishers, Dordrecht, Netherlands

Munns, R. & James, R.A. (2003) Screening methods for salinity tolerance: a case study of with tetraploid wheat. *Plant and Soil*, Vol.253, No.1, pp. 201-218

Munns, R.; James, R.A.; Sirault, X.R.R.; Furbank, R.T. & Jones, H.G. (2010) New phenotyping methods for screening wheat and barley for beneficial responses to water deficit. *Journal of Experimental Botany*, Vol.61, No.13, pp. 3499-3507

Nguyen, V.T.; Burow, M.D.; Nguyen, H.T.; Le, B.T.; Le, T.D. & Paterson, A.H. (2001) Molecular mapping of genes conferring aluminum tolerance in rice (*Oryza sativa* L.). *Theoretical and Applied Genetics*, Vol.102, No.6-7, pp. 1002-1010

Norwell, W.A. & Welch, R.M. (1993) Growth and nutrient uptake by barley (*Hordeum vulgare* L.cv.Herta):StudiesusingN-(2-hydroxyethyl)ethylenedinitrilotriacetic acid- buffered nutrient solution technique. I. Zinc ion requirements. *Plant Physiology*, Vol.101, No.2, pp. 619-625

Owen, A.G. & Jones, D.L. (2001) Competition for amino acids between wheat roots and rhizosphere microorganisms and the role of amino acids in plant N acquisition. *Soil Biology and Biochemistry*, Vol.33, No.4-5, pp. 651-657

Parker, D.R.; Zelazny, L.W. & Kinraide, T.B. (1988) Aluminum speciation and phytotoxicity in dilute hydroxy-aluminum solutions. *Soil Science Society of America Journal*, Vol.52, No.2, pp. 438-444

Pereira, J.F.; Zhou, G.; Delhaize, E.; Richardson, T.; Zhou, M. & Ryan, P.R. (2010) Engineering grater aluminium resistance in wheat by over-expressing *TaALMT1*. *Annals of Botany*, Vol.106, No.1, pp. 205-214

Piñeros, M.A.; Shaff, J.E.; Manslank, H.S.; Carvalho Alves, V.M. & Kochian, L.V. (2005) Aluminum resistance in maize cannot be solely explained by root organic acid exudation. A comparative physiological study. *Plant Physiology*, Vol.137, No.1, pp. 231-241

Rawson, H.M.; Long, M.J. & Munns, R. (1988a) Growth and development in NaCl-treated plants. I. Leaf Na^+ and Cl^- concentrations do not determine gas exchange of leaf blades in barley. *Australian Journal of Plant Physiology*, Vol.15, No.4, pp. 519-527

Rawson, H.M.; Richards, R.A. & Munns, R. (1988b) An examination of selection criteria for salt tolerance in wheat, barley and Triticale genotypes. *Australian Journal of Agricultural Research*, Vol.39, No.5, pp. 759-772

Rengasamy, P. (2002) Transient salinity and subsoil constraints to dryland farming in Australian sodic soils: an overview. *Australian Journal of Experimental Agriculture*, Vol.42, No.3, pp. 351-361

Rengasamy, P. (2006) World salinization with emphasis on Australia. *Journal of Experimental Botany*, Vol.57, No.5, pp. 1017-1023

Rengel, Z. & Graham, R.D. (1996) Uptake of zinc from chelate-buffered nutrient solutions by wheat genotypes differing in zinc efficiency. *Journal of Experimental Botany*, Vol.47, No.2, pp. 217-226

Reed, H.S. (1942) *A Short History of the Plant Sciences. Chapter 16. Plant Nutrition,* 241–265, Chronica Botanica Co., Waltham, USA

Rozema, J. & Flowers, T. (2008) Crops for a salinized world. *Science*, Vol.322, No.5907, pp. 1478-1480

Salt, D.E.; Smith, R.D. & Raskin, I. (1998) Phytoremediation. *Annual Reviews in Plant Physiology and Plant Molecular Biology*, Vol. 49, pp. 643-668

Sasaki, T.; Yamamoto, Y.; Ezaki, B.; Katsuhara, M.; Ahn, S.J.; Ryan, P.R.; Delhaize, E. & Matsumoto, H. (2004) A wheat gene encoding an aluminium-activated malate transporter. *The Plant Journal*, Vol.37, No.5, pp. 645-653

Schnurbusch, T.; Langridge, P. & Sutton, T. (2008) The *Bo1*-specific PCR marker AWW5L7 is predictive of boron tolerance status in a range of exotic durum and bread wheats. *Genome*, Vol.51, No.12, pp. 963-971

Schnurbusch, T.; Hayes, J.; Hrmova, M.; Baumann, U.; Ramesh, S.A.; Tyerman, S.D.; Langridge, P. & Sutton, T. (2010) Boron toxicity tolerance in barley through reduced expression of the multifunctional aquaporin HvNIP2;1. *Plant Physiology*, Vol.153, No.4, pp. 1706-1715

Shaff, J.E.; Schultz, B.A.; Craft, E.J.; Clark, R.T. & Kochian, L.V. (2010) GEOCHEM-EZ: A chemical speciation program with greater power and flexibility. *Plant and Soil*, Vol.330, No.1-2, pp. 207-214

Shah, S.H.; Gorham, J.; Forster, B.P. & Wyn Jones, R.G. (1987) Salt tolerance in the Triticeae: The contribution of the D genome to cation selectivity in hexaploid wheat. *Journal of Experimental Botany*, Vol.38, No.2, pp. 254-269

Shavrukov, Y.; Bowner, J.; Langridge, P. & Tester, M. (2006) Screening for sodium exclusion in wheat and barley. *Proceedings of the 13th Australian Society of Agronomy, Perth.* The Regional Institute Ltd. Publishing Web. Available from http://www.regional.org.au/au/asa/2006/concurrent/environment/4581_shavr ukoky.htm#TopOfPage Accessed 11 September 2006

Shavrukov, Y.; Langridge, P. & Tester, M. (2009) Salinity tolerance and sodium exclusion in genus *Triticum*. *Breeding Science*, Vol.59, No.5, pp. 671–678

Shavrukov, Y.; Gupta, N.K.; Miyazaki, J.; Baho, M.N.; Chalmers, K.J.; Tester, M.; Langridge, P. & Collins, N.C. (2010a) *HvNax3* – a locus controlling shoot sodium exclusion

derived from wild barley (*Hordeum vulgare* ssp. *spontaneum*). *Functional and Integrative Genomics*, Vol.10, No.2, pp. 277-291

Shavrukov, Y.; Langridge, P.; Tester, M. & Nevo, E. (2010b) Wide genetic diversity of salinity tolerance, sodium exclusion and growth in wild emmer wheat, *Triticum dicoccoides*. *Breeding Science*, Vol.60, No.4, pp. 426-435

Shavrukov, Y.; Shamaya, N.; Baho, M.; Edwards, J.; Ramsey, C.; Nevo, E.; Langridge, P. & Tester, M. (2011) Salinity tolerance and Na^+ exclusion in wheat: Variability, genetics, mapping populations and QTL analysis. *Czech Journal of Genetics and Plant Breeding*, Vol.47, pp. S85-93

Smith, S.E. & Read, D.J. (2008) *Mycorrhizal Symbiosis*. Academic Press, Amsterdam, Netherlands

Steppuhn, H.; van Genuchten, M.T. & Crieve, C.M. (2005a) Root-zone salinity: I. Selecting a product-yield index and response function for crop tiolerance. *Crop Science*, Vol.45, No.1, pp. 209-220

Steppuhn, H.; van Genuchten, M.T. & Crieve, C.M. (2005b) Root-zone salinity: II. Indices for tolerance in agricultural crops. *Crop Science*, Vol.45, No.1, pp. 221-232

Stiles, A.R.; Bautista, D.; Atalay, E.; Babaoglu, M. & Terry, N. (2010) Mechanisms of boron tolerance and accumulation in plants: A physiological comparison of the extremely boron-tolerant plant species, *Puccinellia distans*, with the moderately boron-tolerant *Gypsophila arrostil*. *Environmental Science and Technology*, Vol.44, No.18, pp. 7089- 7095

Szira, F.; Bálint, A.F.; Börner, A. & Galiba, G. (2008) Evaluation of drought-related traits and screening methods at different developmental stages in spring barley. *Journal of Agronomy and Crop Science*, Vol.194, No.5, pp. 334-342

Tavakkoli, E.; Rengasamy, P. & McDonald, G.K. (2010) High concentrations of Na^+ and Cl^- ions in soil solution have simultaneous detrimental effects on growth of faba bean under salinity stress. *Journal of Experimental Botany*, Vol.61, No.15, pp. 4449- 4459

Verslues, P.E.; Ober, E.S. & Sharp, R.E. (1998) Root growth and oxygen relations at low water potentials. Impact of oxygen availability on polyethylene glycol solutions. *Plant Physiology*, Vol.116, No.4, pp. 1403-1412

Wang, J.; Zhao, F.J.; Meharg, A.A.; Raab, A.; Feldmann, J. & McGrath, S.P. (2002) Mechanisms of arsenic hyperaccumulation in *Pteris vittata*. Uptake kinetics, interactions with phosphate, and arsenic speciation. *Plant Physiology*, Vol.130, No.3, pp. 1552-1556

Watson, R.; Pritchard, J. & Malone, M. (2001) Direct measurement of sodium and potassium in the transpiration stream of salt-excluding and non-excluding varieties of wheat. *Journal of Experimental Botany*, Vol.52, No.362, pp. 1873-1881

Wheeler, R.M.; Hinkle, C.R.; Mackowiak, C.L.; Sager. J.C. & Knott, W.M. (1990) Potato growth and yield using nutrient film technique (NFT). *American Journal of Potato Research*, Vol.67, No.3, pp. 177-187

Xue, Y.; Wan, J.; Jiang, L.; Wang, C.; Liu, L.; Zhang, Y. & Zhai, H. (2006) Identification of Quantitative Trait Loci associated with aluminum tolerance in rice (*Oryza sativa* L.). *Euphytica*, Vol.150, No.1-2, pp. 37-45

Zimmermann, H.M. & Steudle, E, (1998) Apoplastic transport across young maize roots: effect of the exodermis. *Planta*, Vol.206, No.1, pp. 7-19

CHAPTER-4

THE MANAGEMENT OF AUTOTOXICITY IN VEGETABLES AND ORNAMENTAL PLANTS

INTRODUCTION: AN OVERVIEW

The effective management of autotoxicity in vegetables and ornamental plants is a critical aspect of horticulture and agriculture. Autotoxicity refers to the phenomenon where certain plants release allelopathic chemicals into their surrounding soil, which can inhibit the growth of their own species or even related plants. This self-imposed hindrance can have detrimental effects on crop yields and ornamental garden aesthetics, making it essential to understand and address this issue. In this article, we will explore the causes and consequences of autotoxicity in vegetables and ornamental plants, as well as strategies for its successful management. By implementing appropriate techniques and precautions, growers can ensure the health and vitality of their crops and gardens.

Allelopathy comes from the Latin words allelon 'of each other' and pathos 'to suffer' refers to the chemical inhibition of one species by another. The 'inhibitory' chemical is released into the environment where it affects the development and growth of neighboring plants. Allelopathic chemicals can be present in any parts of an allelopathic plant. They can be found in leaves, flowers, roots, fruits, or stems and also in the surrounding soil. Around 300 BC, the Greek botanist Theophrastus was possibly the first person to recognize the allelopathic properties of plants when he observed and recorded that chickpea plants exhausted the soil and destroyed weeds. Later, Pliny the Elder, a Roman scholar and naturalist, noted that walnut trees were toxic to other plants, and that both chickpea and barley ruined crop lands for maize. The term allelopathy was first introduced by a German scientist Molisch in 1937 to include both harmful and beneficial biochemical interactions between all types of plants including microorganisms. Rice (1984) reinforced this definition in the first monograph on allelopathy. Research on the recognition and understanding of allelopathy has been well documented over the past few decades (Rice, 1984; Rizvi & Rizvi, 1992). These include the symptoms and severity of adverse effects of living plants or their residues upon growth of higher plants and crop yields, interactions among organisms, ecological significance of allelopathy in plant communities, replanting problems, problems with crop rotations, autotoxicity, and the production, isolation and identification of allelochemicals in agro ecosystem.

Autotoxicity is a phenomenon of intraspecific allelopathy that occur when a plant species releases chemical substances which inhibit or delay germination and growth of the same plant species (Putnam, 1985; Singh et al., 1999). It been reported to occur in a number of crop plants in agro ecosystem causing serious problems such as growth reduction, yield decline and replant failures (Singh et al., 1999; Pramanik et al., 2000; Asao et al., 2003). Plants when experiences autotoxicity it releases chemicals to its rhizosphere (Singh et al., 1999) through various mechanisms such as leachation (Overland, 1966), volatilization (Petrova, 1977), root exudation (Tang & Young, 1982), and crop residue decomposition (Rice, 1984). Autotoxicity was found to be pronounced if the plants were cultivated consecutively for years on the same land or grown by hydroponic culture without renewal of nutrient solution. One of the principal causes of this autotoxic growth inhibition in the successive culture of plants has been attributed to the effect of exuded chemicals from plant roots. Root extracts and exudations are the common sources of allelochemicals with potent biological activity and are produced by numerous plant species, with great variation in chemical components (Inderjit & Weston, 2003). It represents one of the largest direct inputs of plant chemicals into the rhizosphere environment. The synthesis and exudation of allelochemicals, along with increased overall production of root exudates, is typically enhanced by stress conditions that the plant encounters such as extreme temperature, drought and UV exposure (Inderjit & Weston, 2003; Pramanik et al., 2000). Accumulations of these allelochemicals are immense in reused nutrient solution during hydroponic culture.

Vegetable and ornamental plants generally cultured through hydroponics in Japan and other developed countries and recently closed type hydroponic systems gained popularity for the production of these crops on a commercial basis. However, this managed and viable technique has the autotoxicity constraint. Therefore, we have studied autotoxicity phenomenon in several vegetables crops such as cucumber (*Cucumis sativus*), taro (*Colocasia esculenta*), strawberry (*Fragaria × ananassa* Duch.), some leafy vegetables, and many ornamentals at the glasshouse of Experimental Research Center for Biological Resources Science, Shimane University, and Matsue, Japan using hydroponic culture. We have also investigated the isolation, identification, phytotoxicity evaluation of the allelochemicals and means to recover growth inhibition. In this chapter we illustrate autotoxicity of the above crops in hydroponics, its occurrence, autotoxic substances isolation and phytotoxicity evaluation, and control methods.

MATERIALS AND METHODS

Plant Materials

In our laboratory we have investigated the autotoxicity from root exudates of several vegetable crops such as cucumber, taro, strawberry, some leafy vegetables, and many ornamentals following hydroponic systems. Uniform seedlings or plantlets of similar growth stage produced through seeds or tissue culture means were used as the test plant materials.

Plant Cultivation in Hydroponics

Plant materials under investigation were planted into plastic containers (34 cm × 54 cm × 20 cm) and three containers were used for each treatment (plants with or without Activated charcoal, AC). The containers were filled with 'Enshi' nutrient solution (Table 1) for each crop (Hori, 1966). The nutrient solution concentration employed for each crop was 75% for cucumber, taro and strawberry, and 50% for several ornamentals. Nutrient solution in the container was continuously aerated (3.8 liter/min.) using air pumps with two small air filters each packed with 100 g of AC (Type Y-4P, 4-8 mesh, Ajinomoto Fine Techno Co., Kawasaki, Japan). The same aeration system was maintained for the nutrient solution without AC. The AC was used to trap the chemicals exuded from the plants and was replaced by fresh AC at 2-week intervals until the end of the experiment for efficient adsorption of the chemicals. The used AC was either immediately extracted with alkaline methanol or stored at 4 ºC for later extraction. FeSO4.7 H_2O (0.75 g) was added to each solution container at 2-day intervals since the AC that absorbed Fe-EDTA and Fe^{2+} was rapidly oxidized to Fe^{3+} and less available for the plants (Yu et al., 1993). During cultivation, the water level of the solution containers was kept constant by regularly adding tap water.

Table 4.1. Mineral nutrient concentrations in full strength 'Enshi` nutrient solution (Hori, 1966).

Chemicals	Concentration (μM/l)
$Ca(NO_3)2 \cdot 4H_2O$	4.03
KNO_3	8.02
$MgSO_4 \cdot 7H_2O$	2.03
$NH4H_2PO_4$	1.35
H_3BO_3	0.05

$ZnSO4 \cdot 7 H_2O$	7.64×10^{-4}
$MnSO4 \cdot 5 H_2O$	8.30×10^{-3}
$CuSO4 \cdot 5 H_2O$	2.00×10^{-4}
Na2MoO4	9.71×10^{-5}
NaFe-EDTA	0.06

Nutrient concentrations (NO_{3-}, $PO4_{2-}$, K+, Ca_2+, Mg_{2+}, and Fe_{3+}) in the solution were adjusted as close as possible to the initial concentration at 2-week intervals on the basis of chemical analyses with an atomic absorption spectrometer (AA-630, Shimadzu Co., Kyoto, Japan), a spectrophotometer (UVmini-1240, Shimadzu Co., Kyoto, Japan), and an ion meter (D-23, Horiba, Kyoto, Japan). At the end of the experiment growth parameters, yield and yield components were compared with untreated control.

Gas Chromatography-Mass Spectrometry (GC-MS) Analysis of Root Exudates

The AC used to trap the exudates (organics) were desorbed three-times using 200 ml 1:1 (v/v) methanol (100 ml):0.4 M aqueous NaOH (100 ml) (Pramanik et al., 2001). Each batch of AC (200 g) was gently shaken with the mixture for 12 h at room temperature (25 ºC) with an electric shaker (20 rpm). The three extracts (600 ml) were combined and filtered through Whatman (No. 6) filter paper. The filtrates were neutralized with 6 M HCl and concentrated to 25 ml in a rotary vacuum evaporator at 40 ºC. Organic compounds in the concentrate were then extracted according to Yu & Matsui (1993). The concentrated AC-extract was adjusted to pH 2.0 with 4 M HCl, extracted three times with 35 ml of refined diethyl ether (DE), and a further three times with 35 ml of ethyl acetate (EA). DE2 and EA2 were the pooled DE and EA extract fractions (105 ml), respectively at pH 2.0. DE2 and EA2 fractions were dried over anhydrous CaSO4 and concentrated to 5 ml each in a rotary evaporator at 40ºC. Both concentrated fractions (DE2 and EA2) extracted from the AC were analyzed using a gas chromatograph coupled to a mass spectrometer (GC-MS, Hitachi M80-B, Hitachi, Tokyo, Japan) before or after methylation with diazomethane from *N*-methyl-*N*-nitoso-*p*-toluene sulfonamide. An aliquot of each concentrated fraction (1 or 2 ml) was diluted in 50 ml ether, treated with diazomethane and concentrated to 5 µl in a rotary evaporator then in a N2 stream in a water bath at 35 ºC. One microliter of the concentrated sample was injected into a GC-MS with a capillary column (0.25 mm × 60 m) of TC5- (GL Science, Tokyo, Japan). Helium was used as the carrier gas at a pressure of 78.4 kN/m^2. The column was held initially at 100 ºC for 2 minutes and then raised at 5 ºC/min. to a final temperature of 260 ºC for 10 minutes. The injector temperature was held at 270 ºC. The ionization voltage and temperature in the electron impact (EI) mode were 70 eV and 250 ºC, respectively.

Bioassay with the Identified Chemicals

The bioassays with identified chemicals were carried out according to Asao et al., (1998b). Inhibitions of the test solution were assayed by their effects on seedling growth of the source plant species. Aqueous solutions of the identified compounds at several concentrations between 0 (control) and 400 μM were prepared with nutrient solutions for each crops studied. These test solutions were added to glass flasks (approx. 420 ml) wrapped in black polythene to exclude light from the roots (Asao et al., 1999a). The selected plants were transplanted into each flask with urethane foam as a support. The planted flasks were then placed in a growth chamber at 25ºC with a light intensity of 74-81 $\mu M/m^2/s$ and required photoperiod under fluorescence lights. To minimize the effects of aeration and microbial degradation of the organic acids (Sundin & Waechter-Kristensen, 1994) on the bioassay, the test solutions in the flasks were renewed every 3-4 d. The plantlets were grown for 3 weeks then growth parameters were measured in terms of fresh weight (FW) and dry weight (DW) of the shoots, DW of roots, and the longest root per plant.

Statistical Analysis

A randomized complete block design with three replicates was used for growth chamber bioassay and hydroponic culture in the greenhouse as described above. These experiments were not repeated over time due to the consistent result. Multiple-comparison test were performed by SPSS 11.0J for Windows (SPSS Japan Inc., Tokyo, Japan) as Tukey's test at a level of significance of $P=0.05$. Regression analyses were performed by Statcel2 (Add-in soft for Microsoft Excel, OMS Publishing Inc., Saitama, Japan) at a level of significance of $P=0.05$.

AUTOTOXICITY FROM THE ROOT EXUDATES OF VEGETABLES AND ORNAMENTALS IN HYDROPONICS

Successive culture of the same crop on the same land for years cause soil sickness or replanting injuries (Bonner & Galson, 1944; Davis, 1928; Hirano, 1940; Rice, 1984; Tsuchiya, 1990) resulting reduction in both crop yield and quality. This phenomenon is evidenced in agricultural production especially in the production of horticultural crops (Grodzinsky, 1992; Tsuchiya, 1990; Young, 1984). It leads to resurgence of disease pest, exhaustion of soil fertility, and developing chemical interference in the rhizosphere referring allelopathy (Hegde & Miller, 1990; Komada, 1988; Takahashi, 1984; Young, 1984). Allelopathic effects from crop residues and root exudates have extensively studied in vegetable crops such as in alfalfa (Miller, 1983; Nakahisa et al., 1993, 1994; Chon et al., 2002, Chung et al., 2011), asparagus (Young, 1984; Young & Chou, 1985; Hartung et al., 1990), cucumber (Yu & Matsui, 1994, 1997), watermelon (Kushima et al., 1998; Hao et al., 2007), taro

(Asao et al., 2003), strawberry (Kitazawa et al., 2005), tomato (Yu & Matsui, 1993b), lettuce (Lee et al., 2006) and so on. Therefore, growth of these plants found to be inhibited by the released allelochemicals. So far several methods has been found to be effective in removing or degradating the phytotoxic substances released from plant roots during autotoxicity such asadsorption by activated charcoal (Asao et al., 1998a), degradation by microbial strain (Asao et al., 2004a) or auxin (2,4-D and NAA) supplementation (Kitazawa et al., 2007), electro- degradation of root exudates (Asao et al., 2008) and TiO2 photocatalysis (Sunada et al., 2008).

Similar to successive culture, in closed hydroponics accumulation of phytotoxic chemical leads to the occurrence of autotoxicity. In our lab we have investigated the autotoxic potentials of several vegetables and ornamentals and suggested means to overcome it. Our research history started with the selection of cucumber cultivars suitable for a closed hydroponics system using bioassay with cucumber seedlings (Asao at al., 1998b). Experiments were conducted to clarify why fruit yield decrease during the late growing period of cucumber cultured in non-renewed hydroponic nutrient solution and results were suggested that root exudates had induced the decrease in fruit yield, especially by affecting young fruits, the decrease was reversible through removal of root exudates by AC (Asao et al., 1998a). We found extended harvesting period in a closed nutrient flow system by grafting 'Shogoin-aonaga-fushinari' on 'Hokushin' or 'Aodai' (Asao et al., 1999b) and increased number of harvested fruits by adding AC in the nutrient solution (Asao et al., 2000). Growth inhibiting substances of unknown origin found in the growing nutrient solution of cucumber plants were isolated and identified. The growth inhibitors were adsorbed on AC and extracted by organic solvent (Asao et al., 1999a). We developed a bioassay technique to evaluate toxicity of aromatic acids to cucumber seedlings and to select cucumber cultivars that release little or no 2,4-dichlorobenzoic acid (an autotoxic chemical found in cucumber root exudates), thereby avoiding cucumber autotoxicity in the closed hydroponics system (Asao et al., 1999a). Root exudates, which are detrimental to vegetative growth and yield of cucumber plants, were absorbed by the AC irrespective of dissolved oxygen levels (Asao et al., 1999d).

The number of organic acids and their exudation rates were higher in high temperatures and long photoperiods than that in low temperature with short photoperiod condition and caused higher cucumber autotoxicity in the former conditions (Pramanik et al., 2000). Species differences in the susceptibility to autotoxicity among leaf vegetables were also investigated in hydroponics (Asao et al., 2001a). Autotoxicity of root exudates from taro were showed benzoic acid as the potent growth inhibitor (Asao et al., 2003). A number of aromatic organic acids were identified in several leaf vegetables (Asao et al., 2004b). 2,4- dichlorobenzoic acid (DCBA)-degrading microbial strains, may degrade DCBA including other growth inhibitors exuded from cucumber roots and avoid autotoxicity in cucumber resulting increased fruit yield (Asao et al., 2004a). In strawberry we found vegetative and reproductive growth inhibition due to autotoxicity developed in non-renewed nutrient solution through accumulation of autotoxic root exudates, and the most potent inhibitor was benzoic acid (Kitazawa et al., 2005). Foliar application of auxin such as 1-naphthaleneacetic acid (NAA) avoided the growth reduction of strawberry caused by autotoxicity. NAA at 5.4μM found to be the most effective for alleviating autotoxicity of strawberry and increasing the yield (Kitazawa, et al., 2007). Autotoxicity in some ornamentals were investigated in hydroponics with or without the addition of AC to the nutrient solution and several organic compounds were detected (Asao et al., 2007). Benzoic acid being the strongest growth inhibitor, its removal from the nutrient solution is imperative for sustainable production of taro and strawberry or other crops exudates containing it. Therefore, electro-degradation method have been tried to degrade benzoic acid and it was found to be recovered strawberry yield up to 71% from non-renewed nutrient solution (Asao et al., 2008).

Autotoxicity in Cucumber (*Cucumis Sativus*)

In closed hydroponic culture without renew of the nutrient solution, we found that the fruit yield of cucumber plants decreased significantly in the late reproductive stage (2 weeks ahead of final harvest) and the growth was recovered by the biweekly renewal of nutrients or supplementation of AC to the nutrient solution (Asao et al., 1998a). Shrunken fruits were harvested from the plant grown in non-renewed culture solution (Fig. 4.1). This inhibition has been attributed due to the autotoxicity from root exudates (Yu et al., 1994). The autotoxicity of cucumber also differs among cultivars (Asao et al., 1998b). Fruit harvesting of a susceptible cucumber cultivar grown in a closed nutrient flow system was prolonged by grafting onto a non-autotoxic cultivar (Asao et al., 1999b). Thus, cucumber root exudates from a closed hydroponic system were analyzed and among a number of growth inhibitors detected (Asao et al., 1999c; Pramanik et al., 2000), 2,4-dichlorobenzoic acid (DCBA) was the strongest inhibitor. Microorganisms can degrade chemical substances in soil and water (Markus et al., 1984; Nanbu, 1990; Sundin & Waechter-Kristensen, 1994). Van den Tweel et al. (1987) reported that 2,4-dichlorobenzoate was degraded through reductive dechlorination by microorganisms. Recently, we found that the inhibitory effect of DCBA on cucumber seedlings could be reversed using strains of microorganisms (Asao et al., 2001b). However, the effects of such strains on cucumber reproductive growth in the presence or absence of DCBA have yet to be elucidated. Therefore, in this study we investigated the effects of microbial strains on the autotoxicity of cucumber plants grown with or without DCBA in the nutrient solution.

Fig. 4.1. Fruits of cucumber cv. 'Shogin-aonaga-fushinari', white arrow indicating (A) a normal developing fruit at ten days after anthesis; (B) shrunken fruit.

Cultivation of Cucumber Plants with or Without Microbial Strain in Presence of DCBA

The DCBA-degrading microorganism (microbial strain) was isolated and screened from soil in Aichi prefecture (Asao et al., 2001b). Nutrient solutions with DCBA (10 mg/l) and sucrose (1 g/l) were prepared and sterilized by autoclave. A 200 ml volume of sterile nutrient solution was inoculated with the DCBA-degrading microorganism and shaken continuously by machine at 25 ºC for 9 days to have stock microbial suspension. Cucumber (*Cucumis sativus* L. var. Shougoin-aonaga-fushinari) plants were grown in a greenhouse by hydroponics at different concentrations of DCBA with or without addition of DCBA- degrading microorganisms to the nutrient solution. One-week-old cucumber seedlings raised in vermiculite were transplanted into plastic containers containing 50 l of continuously aerated (3.8 l/min) 75% Enshi nutrient solution having an electrical conductivity (EC) of 2.0 dS/m. Three seedlings with four leaves were transplanted to each container with three replications. The solutions were prepared at concentrations of 0 (control), 2 or 10 μM/l of DCBA with or without bacterial suspension in the nutrient solution.

The solutions were renewed biweekly. Three plants were planted in each container with three repetitions. At the 15-leaf stage, the apical buds of cucumber plants were plucked to maintain 15 leaves on the main stem. The terminal buds of all the developing primary and secondary branches were removed keeping only one node in each branch. The mean air and water temperature during the experiment ranged from 23.0 to 25.5 ºC and from 23.3 to34.4 ºC, respectively. At the end of the experiment, data were recorded on plant growth, dates of anthesis in male and female flowers, number of healthy female flowers, and harvested fruit number.

Cucumber Cultivation with Microbial Suspension in Absence of DCBA

Similar cultivation procedure was followed without addition of DCBA to the nutrient solution in another set of experiments. Three cucumber seedlings with four leaves were transplanted to each container containing the nutrient solution without DCBA and three containers were used for each treatment. During culture, the water level of containers was kept constant by regularly adding tap water. Nutrient contents of the solutions were adjusted to the initial concentrations following procedures described earlier. In all the treatments, the EC and pH in the nutrient solution ranged from 1.4 to 2.8 dS/m and 6.4-7.9, respectively. The microbial suspension was supplied to the nutrient solution added (a) at planting, at the plucking of apical buds and at 2 weeks after initial harvest, (b) at the plucking of apical buds and at 2 weeks after initial harvest, and (c) at 2 weeks after initial harvest. No DCBA was added. An additional cultivation with biweekly renewal of the nutrient solution in the absence of the microbial suspension or DCBA was set up to serve as a control. At the 15-leaf stage, the apical buds of cucumber plants were removed to maintain 15 leaves on the main stem. The terminal buds of the branches were removed keeping only one node on each branch. The mean air and water temperature during the experiment ranged from 24.9 to 31.3 ºC and from 25.8 to 32.1 ºC, respectively. Data were recorded as mentioned for the preceding experiment.

Effects of DCBA with or Without Microbial Suspension on the Growth and Yield of Cucumber

Cucumber plants were grown in hydroponics using different concentrations of DCBA with or without the addition of microorganisms to the nutrient solution. Results reveal that the length of the main stem and primary branches decreased with the increase in DCBA concentration (Table 4.2). The dry weight of stem, leaf, root and primary branches of plants grown with DCBA (10 µM/l) was also decreased by about 60, 30, 26 and 32% of that without DCBA, respectively. This growth inhibition was significantly recovered by the addition of soil microorganisms to the nutrient solution. This indicates that the microbes efficiently degraded the added DCBA in the nutrient solution and thus restored the inhibitory effect of DCBA on the plants.

The date of male flower anthesis was unaffected by the addition of DCBA and the microbial suspension. However, the presence of DCBA at a concentration of 10 μM/l shifted the date of female flower anthesis and harvesting time by about 5 and 16days, respectively. Addition of the microbial suspension to the nutrient solution enhanced early flowering and fruit setting in the cucumber plants treated with DCBA. The number of healthy female flowers and the harvested fruit number per plant also decreased as the DCBA concentration increased, and this decrease was significantly compensated by the microbial suspension.

Table 4.2. Effects of DCBA with or without microbial suspension on the growth and yield of cucumber plants grown in hydroponics

DCBA (μmol/liter)	Microbial suspension	Stem length (cm)	Lateral branch (cm)	DW of stem (g)	DW/ leaf (g)	DW/ leaf (g)	DW of lateral branch/ plant[z]	DW of root (g)	Date of anthesis (month/day)		Beginning of harvest (month/day)	No. of female flowers/plant	Harvested fruits/plant
									male flower	female flower			
0	-	177.2	57.0	11.0	5.3	5.3	70.7	85.9	9/19	9/26	10/5	36.2	20.2
	+	158.3	54.3	11.0	5.2	5.2	70.7	77.7	9/19	9/27	10/7	26.3	18.0
		NS[y]	NS	NS	NS	NS	NS	NS	NS	NS	NS	NS	NS
2	-	165.1	49.1	11.3	5.7	5.7	69.2	69.2	9/19	9/25	10/9	31.2	14.9
	+	178.2	56.5	12.0	5.4	5.4	80.0	80.0	9/20	9/26	10/7	24.6	17.0
		*	**	NS	NS	NS	NS	NS	NS	NS	NS	NS	NS
10	-	136.9	18.9	6.6	1.6	1.6	22.5	22.5	9/20	10/1	10/21	12.8	2.3
	+	147.2	37.9	8.8	3.4	3.4	69	69.0	9/20	9/27	10/16	31.1	9.2
		NS	**	**	**	**	**	**	NS	**	**	**	**
Significance													
Non-microorganism													
Linear		**[x]	**	**	**	**	**	**	NS	**	**	**	**
Quadratic		NS	NS	NS	**	**	NS	NS	NS	NS	NS	NS	NS
Microorganism													
Linear		**	**	**	**	**	NS	NS	NS	NS	**	NS	**
Quadratic		NS	NS	NS	NS	NS	NS	NS	**	NS	NS	NS	NS

[z]stem and leaf; ysignificant at the 5 % level (*), 1 % level(**), and not significant (NS) by T-test;

[x]significant at the 1 % level (**), and not significant (NS) by regression analysis of the oncentrations.

Effects of Microbial Suspension on Cucumber Growth in the Absence of DCBA

The suspension of DCBA-degrading microorganisms was added once at 2 weeks after the initial harvest, twice upon plucking the apical buds and at 2 weeks after the initial harvest, and three times at the beginning of the culture, on plucking the apical buds and at 2 weeks after the initial harvest. There was no significant difference in the growth of cucumber except in the dry weight of roots and fruit number (Table 4.3). Root dry weight increased by about 43% with the addition of the suspension of DCBA-degrading microorganisms once at 2 weeks after the initial harvest. The treatments did not affect the dates of anthesis in male and female flower, the beginning of harvest, or the number of flowering female flowers per plant. The harvested fruit number per plant was the lowest (14.2 per plant) in non-renewed nutrient solution. The number of fruits recovered increased from 14.2 to 17.4 on addition of the suspension to the nutrient solution once at 2 weeks after the initial harvest.

DCBA is one of the growth inhibitors found in cucumber root exudates (Pramanik et al., 2000) and we found it as the most effective inhibitor of the growth of cucumber plants (Asao et al., 1999c). In these experiments we also found that DCBA strongly retarded the growth of cucumber plants (Table 4.2). However, this inhibition was significantly recovered by the addition of DCBA-degrading microbes (Asao et al., 2001b) into the nutrient solution. This result reveals that the microbial strain appreciably deactivated the inhibitory action of DCBA including the other inhibitors in cucumber root exudates in the nutrient solution and thus, the cucumber plant growth was enhanced. The recovery of growth, especially the dry weight of roots and branches, in the plants grown with the microbial suspension, was about three times higher than that of cucumber grown with DCBA alone at a concentration of 10μM/l. Consequently, dates of male and female flower anthesis, and initial harvest were several days earlier. The number of healthy female flowers as well as fruits also significantly increased on the addition of the microbial suspension.

Experiments to clarify the influence of root exudates and microbes on cucumber plant growth were conducted with or without biweekly renew of the nutrient solutions (Table 4.3). Results revealed that the root growth and fruit number of the cucumber plants grown with biweekly renewed nutrient solution were significantly increased than those grown without renew of nutrient solution. The addition of the microbes to the nutrient solutions also increased the growth of cucumber plants compared to the non-renew nutrient solution. However, the microbial suspension added to the nutrient solutions in vegetative stage (at the start of the culture or the plucking of apical buds) did not make significant yield difference from non-renewed solution culture. Addition of DCBA-degrading microbial suspension applied once at 2 weeks after the initial harvest was effective enough to recover yield reduction of cucumber from autotoxicity.

DCBA causing autotoxicity in the cucumber was detected in their root exudates only in the reproductive stage (Pramanik et al., 2000). Apparently it indicates that the growth inhibitors including DCBA would have sufficiently accumulated in the nutrient solution through cucumber root exudation at the reproductive stage. The degrader (applied two or three times) probably became a source of nutrients for other microorganisms. In this case, the degrader did not dominated more than other microorganisms. However, when supplied once at 2 weeks after the start of harvest, the degrader did not become a nutrient source for other microorganisms and degraded the DCBA exuded from cucumber. This was why the DCBA exuded from cucumber sustained the microbial activity. In conclusion, DCBA- degrading microorganisms, if added to the nutrient solution, may degrade DCBA including other growth inhibitors exuded from cucumber roots and avoid autotoxicity in cucumber resulting increase the fruit yield. Addition of the microbial suspension in the reproductive stage of cucumber plants appears to degrade the growth inhibitors efficiently. However, the timing of degrader addition to the nutrient solution for efficient mitigation of cucumber autotoxicity needs further study.

Table 4.3. Effects of addition of the microbial suspension at different growth phases on the growth and yield of cucumber plants grown in the absence of DCBA.

Nutrient solution[z]	Addition of microbial suspension			Stem length (cm)	Lateral branch (cm)	DW of stem (g)	DW /leaf (g)	DW of lateral branch/ plant[w]	DW of root (g)	Date of anthesis (month/day)		Beginning of harvest (month/day)	No. of female flowers/ plant	Harvested fruit number/ plant
	at planting	at plucking[y]	at harvest[x]							male flower	female flower			
+	-	-	-	160.0	44.6	16.8	6.0	80.8	24.5b	7/5	7/9	7/17	29.9	16.2b
-	-	-	-	157.6	46.6	14.7	5.8	77.8	20.3b	7/5	7/9	7/18	31.5	14.2c
-	+	+	+	169.1	48.0	16.9	5.7	88.2	23.5b	7/5	7/9	7/19	33.5	15.6c
-	-	+	+	161.2	41.1	13.4	5.1	70.0	19.1b	7/5	7/10	7/19	30.6	14.4c
-	-	-	+	164.6	43.9	16.2	6.5	92.1	29.0a	7/5	7/9	7/20	26.5	17.4a
				NS[v]	NS	NS	NS	NS		NS	NS	NS	NS	

[z](+)= total renewal of the nutrient solution every other week; (-)= only supplement of the nutrient solution which decreased during culture; [y]at the plucking of apical buds; [x]at two weeks after the initial harvest; [w]stem and leaf; [v]different letters within a column indicate significance at the 5 % level and not significant (NS) by the Tukey test.

Autotoxicity in Strawberry (*Fragaria Ananassa* Duch.)

Closed hydroponics is a system used for plant cultivation in environmentally sensitive areas (Van Os, 1995) where the nutrient solution is not released into the surrounding environment, but recycled (Ruijs, 1994). However, in a closed hydroponic system, plants can suffer autotoxicity, due to the accumulation of toxic exudates from the roots themselves in the nutrient solution (Yu et al., 1993). Recently, closed hydroponics has been considered for strawberry cultivation (Takeuchi, 2000; Oka, 2002; Koshikawa & Yasuda, 2003). However, it was reported that a yield reduction, caused by unknown factors, occurred in closed hydroponic system for strawberry (Oka, 2002). Koda et al., (1977, 1980) reported that a growth reduction in mitsuba (*Cryptotaenia japonica* Hassk.) in hydroponic system was caused by root exudates such as organic acids. Some aromatic acids also accumulated in the nutrient solution during hydroponic cultivation of tomato and had inhibitory activity on growth (Yu & Matsui, 1993). Asao et al., (1998a, 1999c) demonstrated that a reduction in fruit yield in the late reproductive stage of cucumber was induced by root exudates and that the most potent inhibitor was 2,4-dichlorobenzoic acid. In a closed hydroponic culture of rose plantlets, root and shoot growth were reduced by root exudates (Sato, 2004). However, little is known about similar effects in strawberry culture. It was therefore felt necessary to examine the effects of root exudates on the growth of strawberry. Root exudates can be removed by adding activated charcoal (+AC) to the nutrient solution (Koda et al., 1977; Asao et al., 1998a, 1999c; Sato, 2004). In this study, we investigated the effects of non-renewal of the nutrient solution, and of adding AC on vegetative and reproductive growth of strawberry.

Cultivation of Strawberry Plants in Hydroponics

Strawberry (*Fragaria ananassa* Duch. cv. 'Toyonoka') was used for these experiments following the cultivation method explain above (Section 2.2 & Fig. 4.2). Pollination was aided by vibrating the anthers over stigma with a soft brush at 2 d intervals. The fruits were collected when ripe. At harvest growth parameters were measured among the treatments. During cultivation, the number of flower clusters, flowers and fruits per plant were recorded. The allelochemicals in the root exudates of strawberry were identified by GC-MS (section 2.3) and their phytotoxicity were evaluated in growth chamber bioassay (section 2.4) using strawberry plantlets.

Fig. 4.2. Hydroponic system used for strawberry cultivation at the greenhouse of Shimane University, Matsue, Japan.

Effects of Non-Renewed Nutrient Solution on the Growth and Yield of Strawberry

In the treatment '-AC', the number of leaves, FW of shoots, DW of shoots and roots per plant, and the root length decreased by 75%, 59%, 50%, 81% and 45% of control values, respectively. In the '+AC' treatment, each value was 103%, 83%, 75%, 102%

and 98% of control values (Table 4.4). Although the number of flower clusters per plant was not significant by different between treatments, values in '-AC' and '+AC' treatment were 85% and 89% of control values, respectively (Table 4.4). In the '-AC' treatment, the number of flowers per plant decreased significantly to about 74% of control value. In the '+AC' treatment, each value was approx. 102% of the control value and there was no significant difference between the '+AC' treatment and control. In the '-AC' treatment, the number of harvested fruit per plant decreased significantly by approx. 49% of control values. In the '+AC' treatment, this value was about 107% of the control value, and there was no significant difference between the '+AC' treatment and control (Table 4.4). In the '+AC' treatment with non-renewed nutrient solution, the growth and yield of strawberry plants were eventually equivalent to control plants.

Table 4.4. Effects of non-renewed nutrient solution and activated charcoal on the growth, yield components and yield of strawberry in hydroponics

Nutrient solution[z]	AC supplement	No. of leaves /plant	FW of shoots/ plant (g)	DW of shoots/ plant (g)	DW of root/ plant (g)	Root length (cm)	No. of flower cluster/ plant	No. of flower/ plant	No. of fruits/ plant
+	−AC	23.2a[y]	71.8a	15.5a	4.7ab	43.8a	4.6a[b]	21.0a	12.3a
−	−AC	17.5b	42.6b	7.8b	3.8b	19.8b	3.9a	15.6b	6.0b
−	+AC	24.0a	59.8a	11.6a	4.8a	42.8a	4.1a	21.4a	13.1a

[z](+)= complete renewal of the nutrient solution every second week, and (-)= only supplement of the nutrient solution which decreased during culture; [y]values in column followed by a different letter differ significantly by Tukey's test (P = 0.05; n = 9).

The Identified Chemicals Found in Strawberry Root Exudates Exhibit Phytotoxic Properties

GC-MS analysis of strawberry root exudates (Fraction 'DE2') showed more than 20 peaks, whereas the 'EA2' fraction had only a few detectable peaks (Fig. 3). Based on a comparison of retention times and the mass spectra of standard samples, five peaks were identified as methyl esters of lactic acid, benzoic acid, succinic acid, adipic acid and p-hydroxybenzoic acid. The autotoxic effects by the five exudates compounds identified were evaluated by using micro-propagated the plantlet of the same species from which they originated. Lactic, succinic, adipic and p-hydroxybenzoic acid did not significantly reduce the FWs of shoot, or the DWs of shoots and roots (Table 5). No correlation was found between these growth parameters and increasing concentrations of these four exuded compounds.

Benzoic acid, however, significantly reduced both FW and DW of shoots, even at 50 μM and, these growth parameters decreased further with increasing concentration. Benzoic acid also significantly reduced the DWs of roots at all concentrations. Root length was reduced by increasing concentrations of all five compounds, although lactic, succinic and adipic acids reduced root length significantly at concentrations of 100 μM or above, while benzoic and p- hydroxybenzoic acids significantly reduced root length even at 50 μM (Table 4.5).

The effects of non-renewed hydroponic nutrient solution and of adding AC ('+AC') on the vegetative and reproductive growth of strawberry were investigated. Non-renewed nutrient solution resulted in a significant decrease in the growth of strawberry plantlets compared to growth when the nutrient solution was renewed. The number of flower clusters, flowers and fruit harvested all decreased in non-renewed nutrient solution (Table 4.4). Growth and/or yield reductions in plants caused by non-renewed nutrient solution have been reported by many researchers (Koda et al., 1980; Yu & Matsui, 1993). Asao et al., (1998a) reported that fruit yields in cucumber plants decreased significantly at the late reproductive stage (2 weeks before final harvest) when nutrient solution was not renewed and that the reduction was caused by root exudates. In a closed hydroponic system for rose plantlets, the root and shoot growth were reduced by the root exudates (Sato, 2004). Thus, it was thought that vegetative and reproductive growth in strawberry was inhibited by root exudates during non-renewal of the nutrient solution. Growth and/or yield inhibition in cucumber (Asao et al., 1998a), mitsuba (Koda et al., 1980) and rose (Sato, 2004), caused their root exudates could be avoided by adding AC to the nutrient solution. AC added to the nutrient solution adsorbs the organic compounds exuded from plant roots and thereby removes the inhibitory effects of exudates.

In this study, there was no significant difference in vegetative and reproductive growth of strawberry between addition of AC to non-renewed nutrient solution, and renewed the nutrient solution (Table 4.4). Thus, our results suggest that the chemicals exuded from strawberry roots inhibit vegetative and reproductive growth, which can be avoided by adsorption of root exudates using AC. The substances adsorbed on the AC were extracted, analyzed and some identified as phenolic and aliphatic acids (Fig. 3). Growth and/or yield reduction in cucumber plant by root exudates had been confirmed to be due to (Pramanik et al., 2001) and alipathic acids (Yu & Matsui, 1997). In cucumber, 2,4-dichlorobenzoic acid was the most potent inhibitor of growth and yield (Asao et al., 1999c).

Benzoic acid was also the main inhibitor of growth and yield in taro plants (Asao et al., 2003). The phenolic and aliphatic acids identified in strawberry root exudates may have the potential to inhibit growth. The inhibitory potential of root exudates has been tested using germination test of lettuce seeds (Tsuchiya & Ohno, 1992). However, Asao et al. (2001a) suggested that lettuce plants suffer autotoxicity from their own root exudates. Vanillic acid was the most inhibitor of the growth of lettuce (Asao et al., 2004b). Thus, the potential of each compound identified in our study was evaluated using plantlets of the species which they originated. Five compounds including phenolic and aliphatic acid were found to affect plant growth but only benzoic acid significantly inhibited increases in the FW of shoots,

Table 4.5. Effects of purified exudate compounds at different concentrations on the growth of strawberry plantlets.

Compound[z]	(Concentration, µM)	FW of shoot/ plant (g)	DW of shoot/ plant (g)	DW of root/ plant (g)	Root length/ plant (cm)
Control (No compound)[y]	0	0.98a[b]	0.17a	0.07a	17.0a
Lactic acid	50	0.95a	0.16a	0.06a	16.2a
	100	0.88a	0.15a	0.06a	13.7b
	200	0.88a	0.16a	0.07a	11.9bc
	400	0.85a	0.15a	0.06a	11.4c
Benzoic acid	50	0.48b	0.09b	0.03b	12.1b
	100	0.42b	0.09b	0.03b	10.5b
	200	0.42b	0.08b	0.03b	10.3b
	400	0.36b	0.08b	0.03b	9.4c
Succinic acid	50	0.80a	0.14a	0.06a	16.2ab
	100	0.74a	0.13a	0.04a	13.6bc
	200	0.69a	0.19a	0.04a	12.5c
	400	0.87a	0.16a	0.05a	11.7c
Adipic acid	50	0.70a	0.12a	0.03a	14.1ab
	100	0.88a	0.15a	0.05a	12.0bc
	200	0.74a	0.13a	0.05a	10.6c
	400	0.76a	0.15a	0.05a	9.4c
p-hydroxybenzoic acid	50	0.77a	0.14a	0.04a	11.9b
	100	0.67a	0.13a	0.03a	10.8bc
	200	0.77a	0.15a	0.05a	9.1c
	400	0.65a	0.12a	0.04a	10.4bc

[z]compounds identified in the root exudates adsorbed onto the activated charcoal; [y]values in column followed by a different letter differ significantly by Tukey's test ($P = 0.05$; n = 9).

DWs of shoots and roots, and root length at all concentrations (Table 4.5). Therefore, it was thought that benzoic acid was the strongest inhibitor of vegetative and reproductive growth in strawberry. In conclusion, inhibition of the vegetative and reproductive growth in strawberry caused by non-renewed nutrient solution may occur through autotoxic root exudates, and the most potent inhibitor was benzoic acid.

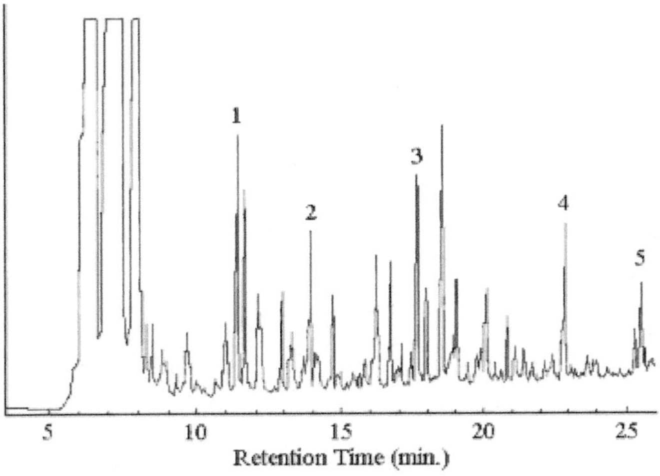

Fig. 4.3. Gas Chromatograms of all components from root exudate of strawberry plants adsorbed and released from AC. Methyl esters of lactic acid (peak 1), benzoic acid (peak 2), succinic acid (peak 3), adipic acid (peak 4) and p-hydroxy benzoic acid (peak 5) are identified based on known standard retention times.

Reduction in the yields of strawberry grown closed hydroponic systems would, therefore, appear to be related to the allelochemicals exuded by the strawberry plant itself. Finally, for strawberry cultivation in a closed hydroponic system, AC should be added to the nutrient solution to relieve the autotoxicity caused by root exudates.

Autotoxicity in Taro (*Colocasia Esculenta* Schott.)

Taro plants do not grow well if cultivated consecutively for years on the same land (Takahashi, 1984). Rotation with other crops for at least three years (Miyoshi et al., 1971a), in combination with organic matter and soil disinfectants (Murota et al. 1984), has been suggested to improve the yield of taro. However, even in a fixed crop rotation system, there was a great difference in the growth and yields of taro plants. This depends upon the kinds of crops in rotation, and the order in which they were rotated. In combination with burdock, the yield of taro was equal to or more than that of taro in the first-year planting, and the extent of corm injury were slight (Murota et al., 1984). Harmful microbes (Atumi, 1956, 1957; Atumi & Nakamura, 1959; Nagae et al., 1971) and nematodes (Miyoshi et al., 1971b; Oashi, 1973; Matsumoto et al., 1973, 1974) in the soil are the main causes of damage in the successive culture of taro. Takahashi (1984) suggested that unknown factors were also involved whereas; Miyaji et al. (1979) found that taro residues in soils after harvest were inhibitory to its growth. Methanol extracts of taro residues alone or of soils with taro residues were found to strongly inhibit the elongation of hypocotyls and radicle growth of turnip. The foregoing results reveal that growth inhibitors from taro were connected with replanting problems. Our laboratory has established used hydroponic culture system to assess autotoxicity in crop plants (Asao et al., 1999c; Pramanik et al., 2000). Thus, an attempt was made to identify the chemicals exuded by taro roots and to evaluate the allelopathic effects of these exudates on the growth and yield of taro through this established hydroponic culture system.

CULTIVATION OF TARO PLANTS IN HYDROPONICS

Taro cv. Aichi-Wase was used for this experiment. Corms were planted in plastic tray (32 cm × 47 cm × 7 cm) containing vermiculite in the green house. At the third leaf stage, taro plantlets were transplanted into plastic containers following the cultivation method described above in section 2.2 (Fig. 4.4). At the end of this experiment, measurements were made on the longest leaf stalk, maximum leaf length and width, leaf number per plant, shoot dry weight and corm yield. Phytotoxic chemicals adsorbed in the ACs were extracted and identified by a Gas-chromatograph coupled with mass-spectrometer as mentioned earlier. Bioassay of the identified acids at concentrations of 0 (control) or 400 µM/l were prepared with a 75 % Enshi nutrient solution (EC 2.0 dS/m) in growth chamber condition. The taro plantlets were grown for 26 days and then the fresh and dry weights of shoots, number of leaves, longest root length and root dry weight were measured. Each treatment was replicated 15 times. We carried out further bioassays following the same procedure with benzoic and adipic acids at concentrations of 0 (control), 25, 50, 100 200 and 400 µM/l. In this case, the taro plantlets were grown for 20 days.

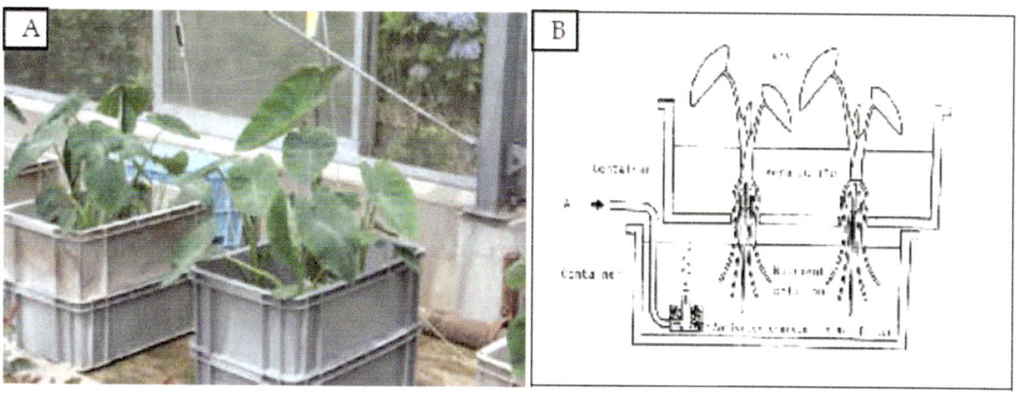

Fig. 4.4. Hydroponic system used for taro cultivation (A) taro plants in plastic containers, (B) sketch of taro hydroponics showing different components.

Effects of Non-Renewed Nutrient Solution on the Growth and Yield of Taro

Results revealed that plants grown without AC had experienced significant shoot growth retardation compared to those grown with AC. The leaf numbers and shoot dry weights of the plants grown without AC decreased to about 90% and 67% of those grown with AC, respectively (Table 4.6). Addition of AC to the nutrient solution also improved yield significantly. The total yield per plant without AC decreased to about 34% compared to that on the addition of AC (Table 4.6). Larger corms were harvested from the nutrient solution with AC.

Table 4.6. Influence of nutrient solutions in the absence and presence of activated charcoal on the vegetative growth, yield and yield components of taro plants grown by hydroponic culture.

harcoal ›plement	No. of leaf/plant	DW of shoot/plant (g)	Total yield/ plant (g)	Yield/plant by corm size[z] (g)					
				2L	L	M	S	2S	3S
–	14.6	10.9	429	72	38	123	74	86	109
+	16.1	16.1	649	154	80	225	97	111	103
	*y	*	**x	**	**	**	NS	NS	NS

[z]2L (>60 g), L (35-59 g), M (20-34 g), S (15-19 g); 2S (10-14 g) and 3S (<10 g); y significant at the 5 % level (*); x significant at the 1 % level (**), and not significant (NS) by T-test.

PHYTOTOXINS IN ROOT EXUDATES OF TARO AND THEIR PHYTOTOXICITY

Analysis of the extracted taro root exudates with GC-MS gave more than thirty peaks (Fig. 4.5). Based on the comparison of retention times and mass spectra with those of authentic samples, seven peaks were assigned as methyl esters of lactic acid, benzoic acid, *m*- hydroxybenzoic acid, *p*-hydroxybenzoic acid, vanillic acid, succinic acid, and adipic acid.

The allelopathic potential of the identified compounds was evaluated using taro plantlets as the test material. Benzoic acid at 400 µM/l induced severe growth inhibition of shoots and roots, while adipic acid at the same concentration reduced only dry weight of roots (Table 4.7). Thus, we further evaluated growth inhibition potential of benzoic acid and adipic acid at concentrations ranged from 0 to 400 µM/l using taro plantlets. Both acids significantly inhibited the growth of plantlets (Table 4.8). Benzoic acid induced growth retardation even at 50 µM/l and growth decreased with increasing concentration of the acid.

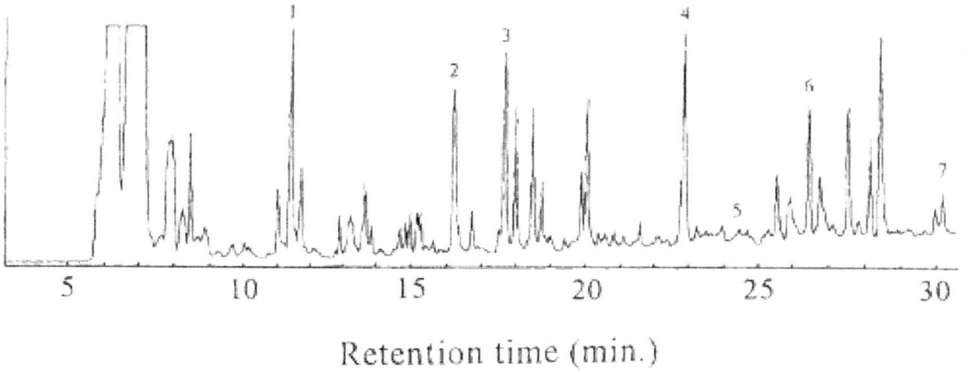

Fig. 4.5. Gas chromatograms of all components from root exudate of strawberry plants adsorbed and released from AC. Methyl esters of lactic acid (peak 1), benzoic acid (peak 2), succinic acid (peak 3), adipic acid (peak 4), *m*-hydroxybenzoic acid (peak 5), *p*- hydroxybenzoic acid (peak 6) and vanillic acid (peak 7) are identified based on known standard retention times.

Benzoic acid at the highest concentration of 400 μM/l reduced fresh weight, shoot dry weight, root length and root dry weight in taro plants to 54 %, 53 %, 54 %, and 75 % of control values, respectively. Adipic acid only at 400 μM/l reduced fresh weight of shoots and root length. Lower concentrations of this acid did not affect shoot or root growth.

Table 4.7. Inhibition potentials of chemicals identified in taro root exudates (400 μM/l) on the growth of taro plantlets.

Allelochemical[z]	FW of shoot/ plant (g)	DW of shoot/plant (g)	No. of leaf/plant	Root length (cm)	DW root/plant (g)
None(control)	26.58a[y]	2.05a	4.3	17.0a	0.52a
Lactic acid	35.40a	2.11a	4.1	18.8a	0.51a
Benzoic acid	20.79b	1.30b	3.6	14.5b	0.29b
m-hydroxybenzoic acid	27.23a	1.80a	4.0	16.5a	0.38a
p-hydroxybenzoic acid	32.87a	2.56a	4.2	16.7a	0.54a
Vanillic acid	27.37a	1.84a	3.5	17.7a	0.45a
Succinic acid	27.17a	1.84a	3.9	19.5a	0.36a
Adipic acid	23.54a	1.61a	4.3	16.7a	0.27b
			NS		

[z]400 μM/l; ydifferent letters within a column indicate significance at the 5 % level and not significant (NS) by the Tukey test.

Table 4.8. Effects of benzoic acid and adipic acid at different concentrations on the growth of taro plantlets.

Allelochemical	Concentrations (μM/l)	FW of shoot/ plant (g)	DW of shoot/plant (g)	No. of leaf/plant	Root length (cm)	DW root/plant (g)
None(control)	0	3.76a[z]	0.235a	4.0	16.0a	24a
Benzoic acid	25	3.46a	0.200a	4.6	13.5a	24a
	50	3.14b	0.167b	4.2	14.2a	19a
	100	2.81b	0.159b	4.0	11.8a	20a
	200	2.16c	0.131c	3.0	10.5b	18b
	400	2.03c	0.125c	4.0	8.6c	18b
Significance						
Linear		*y	*	NS	*	*
Quadratic		*	*	NS	*	*
Adipic acid	25	3.91a	0.251a	4.2	16.6a	31a
	50	3.35a	0.020a	4.0	14.6a	23a
	100	3.43a	0.021a	3.7	12.3a	23a
	200	3.33a	0.195a	4.3	12.0a	27a
	400	3.18b	0.191a	4.3	10.8b	26a
Significance						
Linear		*	NS	NS	*	NS
Quadratic		*	NS	NS	*	NS

[5]Different letters within a column indicate significant at the 5 % level and not significant (NS) by the Tukey test; ysignificant at the 5 % level (*) and non-significant (NS) by regression analysis of the concentrations.

As the nutrient concentrations and growth environment in the hydroponic cultures of taro plants were apparently identical, the significant growth differences between the plants grown with and without AC could be attributed to the variation in the chemical composition of the nutrient solution. These chemicals would have exuded from taro roots. Tsuchiya and Ohno (1992) indicated that water extracts from soils used consecutively for taro cultivation over a period of years inhibited the growth of lettuce. Since the same phenomenon was observed even when the extracts were autoclaved, it was considered that the inhibition was caused by allele chemicals rather than by harmful soil microorganisms. There have been many reports that taro residues exhibited an allelopathic effect on plant growth (Miyaji et al., 1979; Tsuzuki et al., 1995; Pardales & Dingal, 1988). It was made clear here that the vegetative growth and corm yield of taro plants were decreased in the non-renewed culture solution and the loss was recovered by adding AC to the nutrient solution. This result suggests that the chemicals exuded from taro roots had induced the inhibition of growth and reduced yield. This inhibition was prevented by the adsorption of the exuded allele chemicals in AC.

The substances adsorbed on the AC were extracted, analyzed and some of them identified as phenolic and aliphatic acids although many compounds in the root exudates are yet to be detected. The allelopathic potential of each identified compound was evaluated and found that almost all the compounds inhibited the growth of taro plantlets (Table 4.7 & 4.8). Benzoic induced significant growth inhibition in taro plantlets even at concentration of 50 μM/l. Inhibitory effects of phenolic acids (Pramanik et al., 2001) and aliphatic acids (Yu & Matsui, 1997) to plant growth have been well recognized. In a bioassay, Blum (1996) found that 30 % reduction of absolute leaf expansion brought about at 0.23 μM of phenolic acid per gram soil, while it required only 0.05 μM in the presence of 0.06, 0.17, and 0.04 μM of *p*-coumaric, *p*- hydroxybenzoic, and vannilic acids per gram of soil, respectively. Thus, mixture of allele chemicals can below their inhibitory levels. This indicates that taro plants exude a number of compounds (Fig. 4.5) into its surroundings and those inhibit the growth taro plants by synergistic or additive actions. In conclusion, taro roots exude a number of allele chemicals including aromatic acids such as benzoic acid and aliphatic acids such as adipic acid which inhibit the growth of taro plants by additive or synergistic actions. Benzoic acid induced strongest inhibition. Thus, the decline in yield on the successive culture of taro would appear to be related to the allele chemicals exuded from the taro plant itself.

AUTOTOXICITY IN SOME ORNAMENTAL PLANTS

Plants synthesize, store, and exude various kinds of organic compounds in their surroundings as exudates, volatiles, or residues of decomposition (Hale & Orcutt, 1987). Some of the released compounds (allelechemicals) inhibit the growth of the source plants (autotoxicity) or the other species grown in the vicinity of source plants (heterotoxicity). This autotoxicity or heterotoxicity can be treated as allelopathy and the autotoxicity was found to be increased if the plants were cultivated consecutively for years on the same land (Rice, 1984) or grown by hydroponic culture without renew of nutrient solution (Asao et al., 1998a, 2001a). One of the principal causes of this growth inhibition in the successive culture ofplants has been attributed to the effect of exuded chemicals from plants (Pramanik et al., 2000). Growth of some vegetables such as asparagus, taro, cucumber, and tomato was inhibited by allelochemicals found in their root exudates (Asao et al., 1998a, 2003, 2004b; Shafer & Garrison, 1986; Yang, 1982; Yu & Matsui, 1993). Inhibition in growth of apple, peach, rice, strawberry, and sugarcane has been documented for the autotoxicity (Kitazawa et al., 2005; Mizutani et al., 1988; Rice, 1984). This autotoxicity in tomato (Yu et al., 1993) and cucumber (Asao et al., 1998a; Pramanik et al., 2000) has been recovered by addition of AC to the nutrient solution, because the added AC adsorbed the phytotoxic root exudates and thus favored plant growth.

However, research on autotoxicity in ornamentals is limited. Tukey (1969) showed that when chrysanthemum was grown repeatedly in the same place for several years, growth was reduced owing to accumulation of toxic substances in the soil. Kaul (2000) reported on autotoxicity in African marigold, but did not identify the allelochemicals involved. Therefore, we attempted to investigate autotoxicity, if any, in selected ornamentals along with a possible remedial measure to overcome the growth inhibition from autotoxicity.

Cultivation of Ornamental Plants in Hydroponics

Thirty-seven different ornamentals belonging to 16 different families were chosen for this experiment (Table 4.9). Plant cultivation was carried out according to Pramanik et al. (2000) as described above. Seedlings, scions, germinated bulbs, and corms of the plants under study were transplanted to plastic containers (34 cm × 54 cm × 20 cm) in the greenhouse. At the end of the experiment, plant length, number of leaves per plant, maximum root length, flesh and dry weight of shoot and dry weight of root, and number of flowers per plant were recorded.

Bioassay of The Identified Autotoxic Chemicals in Nutrient Solution

Gas chromatography-mass spectroscopy analysis of root exudates adsorbed in activated charcoal identified the responsible autotoxic chemicals. Bioassay of aqueous solutions of the identified compounds was carried out according to Asao et al. (1998b) at concentrations of 0 (control), 50, 100, 200, and 400 μM with 50% Enshi nutrient solution (EC 1.3 dS/m) having ten replications. The plants were grown for 2 weeks and then the fresh and dry weights of shoots were measured.

Bioassay in Soils Amended with Activated Charcoal

Bioassay in soils amended with activated charcoal has also been carried out. Soils were collected from a field successively cultivated with prairie gentian [*Eustoma grandiflorum* (Raf.) Shinn.] for over 10 years in Nagano prefecture, Japan, and was used as medium of growth for the bioassay. Three kilograms of the soil was pulverized and placed in each plastic container (17 cm × 29 cm × 9.5 cm) after amending with AC corresponding to the rate of 0 (control), 30, 60, 120, 240, and 480 kg/10a. Soil collected outside the prairie gentian field was also used as a reference to compare the growth performance of the test plants growth with or without AC (control). The physical and chemical properties of the reference soil were essentially similar to the soil in the prairie gentian field (data not shown). Ten prairie gentian seedlings were planted into the treated containers. Irrigation (500 ml water) was applied to each container at 2-week intervals and 500 ml Enshi nutrient solution (50%) withEC of 1.3 dS/m was applied to each container at 2-week intervals. The cultivation was continued for 8 weeks. At the end of the experiment plant length, number of leaves per plant, maximum root length, shoot dry weight and root dry weight, and number of flowers per plant were recorded.

Table 4.9. Planting materials used for investigating autotoxicity from their root exudates in hydroponics.

Family	Ornamental	Scientific name	Cultivar
Compositae	Pot marigold	*Callendula officinalis* L.	'Gold-star'
	Cornflower	*Centaurea cyanus* L.	'Echo-sultan'
	Chrysanthemum	*Chrysanthemum morifolium* Ramat.	'Shuhou-no-chikara'
	Cosmos	*Cosmos bipinnatus* Cav.	'Dearboro'
	Zinnia	*Zinnia elegans* Jacq.	'Sunbow-orange'
	Thistle	*Cirsium japanicum* DC.	'Rakuonzi-Azami'
	Sunflower	*Helianthus annuns* L.	'Big-smile'
	Safflower	*Carthamus tinctorius* L.	–[z]
	African marigold	*Tagetes erecta* L.	'Orange-isis'
	China aster	*Callistephus chinensis* Nees	'Kurenai'
	Coneflower	*Rudbeckia hirta* L.	'Gloriosa-daisy'
Liliaceae	Tulip	*Tulipa gesneriana* L.	'Blue-champion'
	Thunberg lily	*Lilium* × *elegans* Thunb.	'Iberu-flora'
	Toritelia	*Tritelelia laxa* Benth	'Bridgesii'
	Lily	*Lilium* × *formolongi* Hort.	'Hananomai'
Labiatae	Rocket larkspur	*Delphinium ajacis* L.	'Lilac'
	Love-in-a-mist	*Nigella damascena* L.	'Transformer'
	Scarlet sage	*Salvia splendens* Ker.	'Lavender'
	Fan columbine	*Aquilegia flabellate* Sieb. et Zucc.	'Macana-giant'
Caryophyllaceae	Corn cockl	*Agrostemma githago* L.	'Purple queen'
	Gypsophilla	*Gypsophila elegans* M.B	'Covent-garden'
	Carnation	*Dianthus caryophyllus* L.	'Feminist'
Leguminosae	Sweet pea	*Lathyrus odoratus* L.	'Rolay-lavender'
	Lupine	*Lupine luteus* L.	'Lassell'
Cruciferae	Rape blossoms	*Brassica rapa* L.	'Wase-fushimi-kanzak
	Stock	*Matthiola incana* R.Br.	'Love-me rose'
Onagraceae	Farewell-to-spring	*Godetia amoena* G.Don	'Kyokuhai'
Umbelliferae	Bishop's weed	*Ammi majus* L.	–[z]
Scrophulariaceae	Snapdragon	*Antirrhinum majus* L.	'F1-butterfly-bronze'
Papaveraceae	Corn poppy	*Papaver rhoeas* L.	'Red-sales'
Amaryllidaceae	Narcissus	*Narcissus tazetta* L.	'Fernandesii'
Amaranthaceae	Feather cockscomb	*Celosia argentea* L.	'Red-cupid'
	Globe amaranth	*Gomphrena globosa* L.	'Strawberryfields'
Gentianaceae	Prairie gentian	*Eustoma grandiflorum* (Raf.) Shinn.	'Blue-line I'
Campanulaceae	Balloon flower	*Platycodon grandiflorum* A. DC.	'Samidare-murasaki'
Plumbaginaceae	Statice	*Limonium sinuatum* Mill.	'Marine-blue'
Solanaceae	Chinese-lantern plant	*Physalis alkekengi* L. var. franchetii	'Tanba houzuki' .

[z]Unknown

The Growth Characteristics of Ornamental Plants Cultivated Using Hydroponic Techniques

Thirty-seven ornamentals were grown through hydroponic culture with or without addition of AC in the nutrient solution. Plant growth was significantly affected by the added AC. Performances of the plants were evaluated as percent comparing the growth of the plants grown without AC (control) with those grown with AC. Different plants responded differently to the addition of AC (Table 4.10). Growth in lily was the most severely retarded. Plant length, number of leaves and flowers per plant, root length, and plant dry weight almost all declined significantly in most of the plants grown without AC compared with those grown with AC. However, root growth was found to be more responsive to AC than the other studied parameters possibly for being the roots in direct contract with the exuded chemicals (Pramanik et al., 2000). Root dry weight of lily and rocket larkspur was reduced to approx. 85% and 74%, respectively, followed by prairie gentian with growth reduced to 55%. Root length of lily was reduced to approx. 58%, whereas that in prairie-gentian was reduced to approx. 49%. It appears that lily, prairie gentian, corn poppy, pot marigold, toritelia, and farewell-to-spring were the most sensitive to autotoxicity. Autotoxicity in plants from their own exuded chemicals is also observed in natural ecosystems (Rice, 1984) and was well documented in many crops (Asao et al., 1998a; Kitazawa et al., 2005; Mizutani et al., 1988; Pramanik et al., 2000; Yu et al., 1993). Asao et al. (2001a) detected autotoxicity in some species of Umbellifeae, Compositiae, and Cruciferae. So, autotoxicity in the ornamentals might be incited by the exuded chemicals from their roots. Stimulated growth was observed in the plants such as African marigold, love-in-a-mist, and rape blossoms grown in non-renewed nutrient solution. The exact reasons for this growth stimulation in the latter plants were not discovered. It is well known that a chemical at low concentration acts as a growth stimulant to a plant and the same chemical at high concentration becomes toxic or growth-retardant to the same plant (Rizvi & Rizvi, 1992). Functional activity of an allelochemical depends on its concentration and time exposure to the test plants. So, it is possible that the quality and quantity of root exudates in the nutrient solution in absence of AC might not be sufficient to inhibit growth in the latter ornamental plants, but rather their growth was stimulated.

The Development and Productivity of Prairie Gentian Plants when Cultivated in their Original Planting Soil

Performances of prairie gentian were very poor when successively grown for years in the same land. Significant growth inhibition was noticed in the plants grown in soils from a prairie gentian field without AC compared with those grown in reference soil (Table 4.11). It suggests that soil from a prairie gentian field has some growth inhibitors. In hydroponic culture, we also detected some growth inhibitors in the root exudates of the test plant (Tables 4.12 and 4.13).

Table 4.10. Growth performances of some ornamental plants grown in hydroponics in the presence or absence of activated charcoal (AC) in the nutrient solution (%)[z].

Family	Ornamental	Plant length	No. of leaves	Root length	FW of shoot	DW of shoot	DW of root	No. of Flowers/ plant
Compositae	Pot marigold	89.9*[y]	95.8NS	101.9NS	55.9**	79.9*	70.4**	–
	Cornflower	102.9NS	115.5**	102.1NS	–	111.3NS	86.8NS	–
	Chrysanthemum	103.8NS	–	–	99.9NS	98.9NS	126.6**	–
	Cosmos	–[x]	–	–	119.9NS	120.1NS	111.2NS	–
	Zinnia	93.7NS	–	–	88.6NS	91.7NS	96.8NS	–
	Thistle	114.8NS	–	114.6*	99.9NS	118.1NS	120.8NS	142.9NS
	Sunflower	106.1NS	96.8NS	84.4NS	113.3NS	–	95.8NS	100.0NS
	Safflower	104.8NS	89.7**	79.4**	91.6NS	100.2NS	84.6NS	100.0NS
	African marigold	146.1**	95.5NS	–	146.7**	176.2**	–	100.0NS
	China aster	103.2NS	97.3NS	79.1**	80.7*	82.4*	70.6**	68.4*
	Coneflower	93.7NS	87.2NS	102.8NS	79.2*	84.2*	119.4NS	80.3NS
Liliaceae	Tulip	110.6NS	102.6NS	86.2NS	104.4NS	110.5NS	69.7NS	100.0NS
	Thunberg lily	88.2*	96.0NS	118.2NS	107.3NS	97.1NS	155.3NS	–
	Toritelia	93.1*	100.0NS	55.9**	77.2**	80.2**	74.8**	71.5**
	Lily	37.2**	64.6**	42.1**	13.5**	13.2**	15.6**	–
Labiatae	Rocket larkspur	71.5**	93.8NS	51.4**	25.5**	38.1**	26.3**	88.3NS
	Love-in-a-mist	181.4**	110.3NS	122.7NS	151.6**	127.1*	162.5**	100.0NS
	Scarlet sage	99.5NS	101.0NS	91.6NS	103.6NS	106.1NS	112.5NS	–
	Fan columbine	104.4NS	–	68.1**	74.6*	74.2*	80.3NS	–
Caryophyllaceae	Corn cockl	74.1**	85.4**	62.1**	27.9	33.1**	83.7NS	–
	Gypsophilla	105.3NS	102.6NS	83.9**	99.9NS	118.1NS	121.8NS	100.0NS
	Carnation	42.4**	75.0**	61.2**	34.6**	46.5**	58.5**	–
Leguminosae	Sweet pea	85.1*	105.8NS	–	78.5*	82.2*	79.8NS	–
	Lupine	98.1NS	106.5NS	–	120.3NS	107.2NS	96.3NS	71.9NS
Cruciferae	Rape blossoms	106.1*	100.0NS	95.6NS	121.2**	113.3*	50.2*	–
	Stock	60.3*	89.9NS	101.5NS	62.9**	78.3**	100.0NS	95.3NS
Onagraceae	Farewell-to-spring	78.4**	92.1*	75.1**	44.7**	51.4**	28.3**	56.3**
Umbelliferae	Bishop's weed	91.3*	97.5NS	–	66.3**	69.4*	–	91.1NS
Scrophulariaceae	Snapdragon	72.8**	96.7NS	100.7NS	46.1**	56.3**	79.5NS	73.1*
Papaveraceae	Corn poppy	50.4*	75.3NS	98.1NS	32.1**	52.5*	52.6*	–
Amaryllidaceae	Narcissus	97.1NS	102.0NS	78.8**	96.3NS	89.2NS	97.7NS	100.0NS
Amaranthaceae	Feather cockscomb	92.9NS	80.7*	85.7NS	100.5NS	–	–	100.0NS
	Globe amaranth	102.8NS	100.0NS	102.8NS	84.5**	83.2**	100.0NS	82.7**
Gentianaceae	Prairie gentian	83.8**	107.9*	51.1**	50.8**	60.2**	45.4**	62.2**
Campanulaceae	Balloon flower	117.5*	102.5NS	78.8**	95.7NS	89.4NS	112.5NS	113.2NS
Plumbaginaceae	Statice	109.2NS	94.2NS	98.1NS	94.7NS	97.8NS	68.5*	114.2NS
Solanaceae	Chinese-lantern plant	105.3NS	104.7NS	–	67.6**	64.8**	74.8**	114.7NS

[z]Growth performance (%) = Growth in absence of AC/Growth in presence of AC × 100; [y]significant at 5% level (*), 1% level (**) and not significant (NS) by t-test (n=36); [x]no data.

Those inhibitors should have been adsorbed when the soil was amended with AC. Thus, the growth of the test plants was increased with an increase in amount of AC from 30 to 60 kg/10a followed by a gradual decline at the highest dose of AC (480 kg/10a). This high dose of AC might have affected other chemical properties in soil. Results revealed that the test plant length was increased by 96% over control as a result of the addition of AC (60 kg/10a). Shoot dry weight and root length were increased by 90% and 94%, respectively, over control for the same concentration (60 kg/10a). Flower setting was also increased at 60 kg AC/10a. This indicated that the reduced growth of prairie gentian after prolonged cultivation in a field could be corrected by amending the soil with AC at the rate of 60 kg/10a. In conclusion, of the ornamentals experiencing autotoxicity owing to the chemicals exuded from their roots being more specific, this autotoxicity could be reduced, at least to some extent, using AC in the root media.

Table 4.11. Effects of activated charcoal (AC) on the growth of prairie gentian, an ornamental plant, grown on the soil of prairie gentian field amended with different amount of the AC.

Soil	Addition of AC (kg/10a)	Plant length (cm)	No. of leaves	DW of shoot (g)	Root length (cm)	DW of root (g)	No. of flowers/plant
New (control)	–	50.6a[z]	11.4b	2.06a	19.2a	0.18b	6.7a
Successive	–	39.9c	11.1bc	1.29c	15.6b	0.25a	5.6b
Successive	30	40.8c	11.7b	1.31c	14.6bc	0.18b	5.2c
Successive	60	48.4a	12.2a	1.85a	18.1a	0.19b	6.8a
Successive	120	44.0b	11.4b	1.60b	16.5b	0.19b	6.7a
Successive	240	42.2bc	11.2bc	1.54b	14.5bc	0.20ab	5.8b
Successive	480	40.3c	10.9c	1.35c	10.1c	0.11c	5.4c

[z]Values in a column followed by a different letter differ significant by Tukey›s test (P=0.05; n=10)

Table 4.12. The compounds identified in the exudates of some ornamentals adsorbed on activated charcoal (AC) added in the nutrient solution.

Allelochemicals	Pot marigold	Toritelia	Lily	Rocket larkspur	Sweet pea	Stock	Farewell -to- spring	Bishop's week	Snap-dragon	Prairie gentian
Lactic acid	+[z]	+	–	+	–	+	–	+	–	–
Valeric acid	–	+	–	–	–	–	–	–	–	–
Malonic acid	–	–	–	–	+	+	–	–	–	+
Fumaric acid	–	+	–	–	–	–	–	–	–	–
Maleic acid	–	+	–	–	–	–	–	–	–	+
n-Caproic acid	–	+	+	–	–	–	–	–	+	+

Allelochemicals	Pot marigold	Toritelia	Lily	Rocket larkspur	Sweet pea	Stock	Farewell -to- spring	Bishop's week	Snap- dragon	Prairie gentian
Succinic acid	+	+	–	+	–	+	–	–	–	–
Benzoic acid	+	–	+	–	+	–	–	–	–	+
Malic acid	–	+	–	–	–	–	–	–	–	+
m-Hydroxybenzoic acid	–	–	–	–	–	–	+	–	–	+
p-Hydroxybenzoic acid	–	–	+	–	+	–	–	–	–	+
Adipic acid	–	+	+	–	–	–	–	–	–	–
o-Hydroxyphenylacetic acid	–	–	–	+	–	–	–	–	–	–
p-Hydroxyphenylacetic acid	–	+	–	–	–	–	–	–	–	–
Vanillin	–	–	+	–	–	–	–	–	–	–
3,4-Dihydroxybenzoic acid	–	+	–	–	–	–	–	–	–	–
Vanillic acid	–	–	–	+	+	–	–	–	–	–
n-Capric acid	–	+	–	–	–	–	–	–	–	–

[z]Detected (+) and not detected (-).

Table 4.13. Effects of the identified compounds at different concentrations on the fresh and dry weights (mg) of shoot and root of some ornamental plants.

Allelochemicals	Conc. (μM)	Pot marigold		Lily		Rocket larkspur		Sweet pea		Stock		Prairie gentian	
		FW shoot	DW root	FW shoot	DW root	FW shoot	DW root	FW shoot	DW root	FW shoot	DW root	FW shoot	DW root
None(control)	0	530b[z]	4.7b	1640a	63a	140a	4.4a	1210a	18a	130a	1.1a	560a	23a
Lactic acid	50	420c	3.1c	–	–	130a	3.5a	–	–	150a	1.1a	–	–
	100	420c	3.3c	–	–	140a	3.7a	–	–	120a	0.9a	–	–
	200	430c	3.4c	–	–	160a	4.5a	–	–	120a	1.1a	–	–
	400	420c	3.4c	–	–	150a	4.6a	–	–	110a	0.8a	–	–
Malonic acid	50	–	–	–	–	–	–	1340a	23a	110a	1.1a	530a	24a
	100	–	–	–	–	–	–	1320a	32a	140a	1.2a	510a	24a
	200	–	–	–	–	–	–	1330a	21a	140a	1.1a	510a	25a
	400	–	–	–	–	–	–	1070b	17b	110a	1.5a	490b	29a
Maleic acid	50	–	–	–	–	–	–	–	–	–	–	490b	17b
	100	–	–	–	–	–	–	–	–	–	–	460b	17b
	200	–	–	–	–	–	–	–	–	–	–	420b	18b
	400	–	–	–	–	–	–	–	–	–	–	390c	18b

Allelochemicals	Conc. (µM)	Pot marigold		Lily		Rocket larkspur		Sweet pea		Stock		Prairie gentian	
		FW shoot	DW root	FW shoot	DW root	FW shoot	DW root	FW shoot	DW root	FW shoot	DW root	FW shoot	DW root
n-Caproic acid	50	–	–	1410a	43b	–	–	–	–	–	–	530a	22a
	100	–	–	1380a	41b	–	–	–	–	–	–	580a	25a
	200	–	–	940b	34b	–	–	–	–	–	–	550a	23a
	400	–	–	850b	35b	–	–	–	–	–	–	530a	25a
Succinic acid	50	510b	4.7b	–	–	120a	4.3a	–	–	130a	1.1a	–	–
	100	490b	4.3b	–	–	160a	4.7a	–	–	120a	1.7a	–	–
	200	510b	3.7b	–	–	140a	3.9a	–	–	120a	1.2a	–	–
	400	490b	4.1b	–	–	140a	3.9a	–	–	110a	1.3a	–	–
Benzoic acid	50	470b	4.2b	810b	34b	–	–	1150b	18a	–	–	460b	19b
	100	750a	6.2a	810b	34b	–	–	1110b	18a	–	–	470b	18b
	200	530b	4.3b	800b	34b	–	–	1090b	21a	–	–	480b	17b
	400	440c	3.1c	890b	42b	–	–	1110b	21a	–	–	470b	16b
Malic acid	50	–	–	–	–	–	–	–	–	–	–	510a	22a
	100	–	–	–	–	–	–	–	–	–	–	480b	22a
	200	–	–	–	–	–	–	–	–	–	–	380c	22a
	400	–	–	–	–	–	–	–	–	–	–	390c	22a
m-Hydroxybenzoic acid	50	–	–	–	–	–	–	–	–	–	–	520a	19b
	100	–	–	–	–	–	–	–	–	–	–	510a	18b
	200	–	–	–	–	–	–	–	–	–	–	510a	18b
	400	–	–	–	–	–	–	–	–	–	–	420b	16b
p-Hydroxybenzoic acid	50	–	–	990b	28b	–	–	1330a	21a	–	–	510a	22a
	100	–	–	1170b	35b	–	–	1310a	34a	–	–	550a	23a
	200	–	–	1010b	31b	–	–	980b	15b	–	–	610a	26a
	400	–	–	1010b	35b	–	–	870b	16b	–	–	470b	25a
Adipic acid	50	–	–	1370a	36b	–	–	–	–	–	–	–	–
	100	–	–	1210a	28b	–	–	–	–	–	–	–	–
	200	–	–	910b	29b	–	–	–	–	–	–	–	–
	400	–	–	970b	22c	–	–	–	–	–	–	–	–
o-Hydroxyphenylacetic acid	50	–	–	–	–	110a	3.0b	–	–	–	–	–	–
	100	–	–	–	–	110a	2.8b	–	–	–	–	–	–
	200	–	–	–	–	110a	2.4b	–	–	–	–	–	–
	400	–	–	–	–	60b	2.2b	–	–	–	–	–	–

Allelochemicals	Conc. (μM)	Pot marigold		Lily		Rocket larkspur		Sweet pea		Stock		Prairie gentian	
		FW shoot	DW root	FW shoot	DW root	FW shoot	DW root	FW shoot	DW root	FW shoot	DW root	FW shoot	DW root
Vanillin	50	–	–	1340a	38b	–	–	–	–	–	–	–	–
	100	–	–	1310a	34b	–	–	–	–	–	–	–	–
	200	–	–	1030b	27b	–	–	–	–	–	–	–	–
	400	–	–	1010b	26b	–	–	–	–	–	–	–	–
Vanillic acid	50	–	–	–	–	140a	4.6a	1230a	19a	–	–	–	–
	100	–	–	–	–	140a	4.6a	1190a	19a	–	–	–	–
	200	–	–	–	–	120a	2.8b	1010b	23a	–	–	–	–
	400	–	–	–	–	110a	2.4b	1110b	21a	–	–	–	–

[z]Values in a column followed by a different letter differ significant by Tukey›s test (P=0.05; n=10)

PHYTOTOXICITY OF THE IDENTIFIED ALLELOCHEMICALS

Root exudates from the ornamentals were analyzed and some compounds were detected. The identified chemicals were mainly some small chain aliphatic acids and some simple phenolic acids or phenolic compounds and those varied from extract to extract in the ornamentals that experienced autotoxicity. Eleven organic compounds were detected in the root exudates of toritelia roots and seven in prairie gentian (Table 4.12).

Many compounds in the root exudates of the plants are yet to be identified. However, at least one aliphatic acid or phenolic compound has been detected in the root exudates of the studied plants. A bioassay was carried out to evaluate the inhibition potential of some identified compounds. Different test concentrations were made with the compounds and a bioassay was furnished with some test plants. Almost all the compounds inhibited the growth of tested plants in a concentration dependent manner. Lactic acid significantly reduced fresh shoot weight and root dry weight in pot marigold to 79% and 66% of control, respectively, even at low concentration (50 μM) (Table 4.13). Benzoic and *p*-hydroxybenzoic acid in lily, even at 50 μM, significantly reduced fresh weight to 49% and 60% of over control, *n*-caproic, benzoic, phydroxybenzoic, and adipic acid and vanillin decreased root dry weight to 68%, 54%, 44%, 57%, and 60% of control, respectively. *o*-Hydroxyphenylacetic acid at 50 μM reduced root dry weight in rocket larkspur to 68% of control (Table 4.13). Quantity and quality of exuded allelochemicals varied from plants to plants (Inderjit, 1996) and in cucumber plants, root exudation rate of different chemicals was found to range from 0.20 to 4.17 mg/d per plant (Pramanik et al., 2000). This low concentration is apparently not enough to cause autotoxicity in cucumber plants, but those cucumber plants experienced autotoxicity when grown in absence of AC in the nutrient solution plant (Pramanik et al., 2000). Actually, in natural conditions, occurrence of a chemical at high concentrations (100 μM or more) is rare or absent. However, under field conditions or hydroponic culture, the exuded compounds affect plant growth by additive or synergistic means (Inderjit, 1996) and thus, the compounds even at low concentrations could induce significant growth inhibition in plants, although their threshold inhibition at the individual level is quite high (Rice, 1984). Identical results were found in the experiment (Table 4.13). So, it appears that the identified compounds would be toxic enough to affect growth of the ornamental plants by additive or synergistic effects.

We found a number of ornamental plants with autotoxic potential due to their root exudation. Growth and yield of the ornamentals under investigation were found to be improved in culture solution with AC supplementation where it used to trap the exudates. Therefore, cultivation through hydroponics enables us to isolate the responsible allelochemicals. Most of the autotoxic ornamental plants released mainly aliphatic acids or phenolic compounds to the nutrient solution. Phytotoxicity was evaluated in terms of fresh weight of shoot and dry weight of root and some compounds like lactic, benzoic acid and *p*- hydroxybenzoic acid showed phytotoxicity even at lower concentration (50 μM). The aforesaid results would be useful for sustainable production of ornament plants. We have tried AC to remove the allelochemicals from culture solution; however, more practical measures should be investigated to manage the rhizosphere free of inhibitory exudates.

AT A GLANCE

- Intensive and continuous culture of vegetable crops on the same land for several years causes replanting injuries like outbreak of disease and insect pest, exhaustion of soil fertility, development of chemical interference (allelopathy) leading to growth and yield reduction. Similarly in closed hydroponic system for the commercial cultivation of vegetables, autotoxicity is also evidenced.
- Hydroponic culture solution accumulates root exudates which in terns hamper water and mineral uptake due to root injuries. This managed culture technique has the facility of trapping and isolating the chemicals released through plant roots. Therefore, autotoxicity phenomenon can clearly be investigated through hydroponics. Our lab aims to investigate autotoxicity in a number of vegetables and ornamental crops, to identify potential allelochemicals, their phytotoxicity through bioassay and suggest the control measures. We cultured the plants in hydroponics using plastic containers in growth chamber or in the greenhouse.

- Mineral nutrients were supplied as Enshi nutrient solution and nutrient concentration were adjusted throughout the culture period. Activated charcoals were supplemented in the air filters for trapping allelochemicals. Root exudates were extracted and analyzed through GC-MS. Phytotoxicity of the identified chemicals was assayed at several concentrations and potential growth inhibitors were identified for each crop. We have showed species differences in the susceptibility to autotoxicity among leaf vegetables in hydroponics. Therefore, knowledge regarding autotoxicity in vegetable crops, autotoxic chemicals, and their phytotoxicity with control measures are useful for sustainable crop production.

REFERENCES

Asao, T.; Hasegawa, K.; Sueda, Y.; Tomita, K.; Taniguchi, K.; Hosoki, T.; Pramanik, M.H.R. & Matsui, Y. (2003). Autotoxicity of root exudates from taro. *Scientia Horticulturae*, Vol. 97, No. 3-4, pp. 389–396, ISSN 0304-4238

Asao, T.; Kitazawa, H.; Ban, T. & Pramanik, M.H.R. (2004b). Search of autotoxic substances in some leaf vegetables. *Journal of the Japanese Society for Horticultural Science*, Vol. 73, No. 3, pp. 247–249, ISSN 1347-2658 (in Japanese with English summary)

Asao, T.; Kitazawa, H.; Ban, T. & Pramanik, M.H.R. (2008). Electrodegradation of root exudates to mitigate autotoxicity in hydroponically grown strawberry (*Fragaria × ananassa* Duch.) plants. *HortScience*, Vol. 43, No. 7, pp. 2034-2038, ISSN 0018-5345

Asao, T.; Kitazawa, H.; Tomita, K.; Suyama, K.; Yamamoto, H.; Hosoki, T. & Pramanik, M.H.R. (2004a). Mitigation of cucumber autotoxicity in hydroponic culture using microbial strain. *Scientia Horticulturae*, Vol. 99, No. 3-4, pp. 389-396, ISSN 0304-4238

Asao, T.; Kitazawa, H.; Ushio, K.; Sueda, Y.; Ban, T. & Pramanik, M.H.R. (2007). Autotoxicity in some ornamentals with the means to overcome it. *HortScience*, Vol. 42, No. 6, pp. 1346-1350, ISSN 0018-5345

Asao, T.; Ohba, Y.; Tomita, K.; Ohta, K. & Hosoki, T. (1999d). Effects of activated charcoal and dissolved oxygen levels in the hydroponic solution on the growth and yield of cucumber plants. *Journal of the Japanese Society for Horticultural Science*, Vol. 68, No. 6, pp. 1194-1196, ISSN 1347-2658 (in Japanese with English summary)

Asao, T.; Ohtani, N.; Shimizu, N.; Umeyama, M.; Ohta, K. & Hosoki, T. (1998b). Possible selection of cucumber cultivars suitable for a closed hydroponics system by the bioassay with cucumber seedlings. *Journal of Society of High Technology in Agriculture*, Vol. 10, No. 2, pp. 92-95, ISSN 0918-6638 (in Japanese with English summary)

Asao, T.; Pramanik, M.H.R.; Tomita, K.; Ohba, Y.; Ohta, K.; Hosoki, T. & Matsui, Y. (1999a). Identification and growth effects of compounds adsorbed on activated charcoal from hydroponic nutrient solutions of cucumber. *Allelopathy Journal*, Vol. 6, No. 2, pp. 243-250, ISSN 0971-4693

Asao, T.; Pramanik, M.H.R.; Tomita, K.; Ohba, Y.; Ohta, K.; Hosoki, T. & Matsui, Y. (1999c). Influences of phenolics isolated from the nutrient solution nourishing growing cucumber (*Cucumis sativus* L.) plants on fruit yield. *Journal of the Japanese Society for Horticultural Science*, Vol. 68, No. 4, pp. 847-853, ISSN 1347-2658 (in Japanese with English summary)

Asao, T.; Shimizu, N.; Ohta, K. & Hosoki, T. (1999b). Effect of rootstocks on the extension of harvest period of cucumber (*Cucumis sativus* L.) grown in non-renewal hydroponics. *Journal of the Japanese Society for Horticultural Science*, Vol. 68, No. 3, pp. 598-602, ISSN 1347-2658 (in Japanese with English summary)

Asao, T.; Taniguchi, H.; Suyama, K.; Yamamoto, H.; Itoh, K.; Tomita, K.; Taniguchi, K. & Hosoki, T. (2001b). Reversibility of the action of 2,4-dichlorobenzoic acid on cucumber seedlings by strains of soil microorganisms. *Journal of the Japanese Society for Horticultural Science*, Vol. 70, No. 3, pp. 393-395, ISSN 1347-2658 (in Japanese with English summary)

Asao, T.; Taniguchi, K.; Tomita, K. & Hosoki, T. (2001a). Species differences in the susceptibility to autotoxicity among leaf vegetables grown in hydroponics. *Journal of the Japanese Society for Horticultural Science*, Vol. 70, No. 4, pp. 519–521, ISSN 1347-2658 (in Japanese with English summary)

Asao, T.; Tomita, K.; Taniguchi, K.; Hosoki, T.; Pramanik, M.H.R. & Matsui, Y. (2000). Effects of activated charcoal supplementation in the nutrient solution on the harvested fruit number of cucumbers grafted on the bloomless rootstock in hydroponics. *Journal of Society of High Technology in Agriculture*, Vol. 12, No. 1, pp. 61-63, ISSN 0918-6638 (in Japanese with English summary)

Asao, T.; Umeyama, M.; Ohta, K.; Hosoki, T.; Ito, N. & Ueda, H. (1998a). Decrease of yield of cucumber by non-renewal of the nutrient hydroponic solution and its reversal by supplementation of activated charcoal. *Journal of the Japanese Society for Horticultural Science*, Vol. 67, No. 1, pp. 99-105, ISSN 1347-2658 (in Japanese with English summary)

Atumi, K. & Nakamura, M. (1959). Studies on the soil sickness of taro plants (*Colocasia esculenta* Schott.). III. On the effect of soil sterilization on the recovering of sick soil. *Research Bulletin of Faculty of Agriculture*, Gifu University, Japan, Vol. 11, pp. 32-42 (in Japanese with English abstract)

Atumi, K. (1956). Studies on the soil sickness of taro plants (*Colocasia esculenta* Schott.). *Research Bulletin of Faculty of Agriculture*, Gifu University, Japan, Vol. 7, pp. 34-40 (in Japanese with English abstract)

Atumi, K. (1957). Studies on the soil sickness of taro plants (*Colocasia esculenta* Schott.). II. On the effect of soil sterilization on the recovering of sick soil, *Research Bulletin of Faculty of Agriculture*, Gifu University, Japan, Vol. 8, pp. 39-42 (in Japanese with English abstract)

Blum, H. (1996). Allelopathic interactions involving phenolic acids. *Journal of Nematology*, Vol. 28, No. 3, pp. 259-267, ISSN 0022-300X

Bonner, J. & Galson, A.W. (1944). Toxic substances from the culture media of guayule which may inhibit growth. *Botanical Gazette*, Vol. 106, pp. 185-198, ISSN 0006-8071

Chon, S.U.; Choi, S.K.; Jung, S.; Jang, H.G.; Pyo, B.S. & Kim, S.M. (2002). Effects of alfalfa leaf extracts and phenolic allelochemicals on early seedling growth and root morphology of alfalfa and barnyard grass. *Crop Protection*, Vol. 21, No. 10, pp. 1077- 1082, ISSN 0261-2194

Chung, I.M.; Seigler, D.; Miller, D.A. & Kyung, S.H. (2011). Autotoxic compounds from fresh alfalfa leaf extracts: Identification and biological activity. *Journal of chemical ecology*, Vol. 26, No. 1, pp. 315-327, ISSN 0098-0331

Davis, E.F. (1928). The toxic principle of *Junlans nigra* as identified with synthetic jaglone and its toxic effect on tomato and alfalfa plants. *American Journal of Botany*, Vol. 15, No. 10, pp. 620–629, ISSN 00029122

Grodzinsky, A.M. (1992). Allelopathic effects of cruciferous plants in crop rotation. In: *Allelopathy: Basic and Applied Aspects*, S.J.H. Rizvi & V. Rizvi, (Ed.), 77-86, Chapman & Hall, ISBN 0 412 394006, London

Hale, M.G. & Orcutt, D.M. (1987). Allelochemical stress, In: *The physiology of plants under stress*, M.G. Hale & D.M. Orcutt, (Ed.), 117-127, John Wiley & Sons, ISBN 0471031526, 9780471031529, New York

Hao, Z.P.; Wang, Q.; Christie, P. & Li, X.L. (2007). Allelopathic potential of watermelon tissues and root exudates. *Scientia Horticulturae*, Vol. 112, No. 3, pp. 315-320, ISSN 0304-4238

Hartung, A.C.; Nair, M.G. & Putnum, A.R. (1990). Isolation and characterization of phytotoxic compounds from asparagus (*Asparagus officinalis* L.) roots. *Journal of Chemical Ecology*, Vol. 16, No. 5, pp. 1707-1718, ISSN 0098-0331

Hegde, R. & Miller, D.A. (1990). Allelopathy and autotoxicity in alfalfa: characterization and effects of preceding crops and residue incorporation. *Crop Science*, Vol. 30, No. 6, pp. 1255-1259, ISSN 0011-183X

Hirano, T. (1940). Studies on soil sickness of tomatoes. *Japanese Journal of Soil Science and Plant Nutrition*, Vol. 14, pp. 521-530, ISSN 1747-0765 (in Japanese).

Hori, H. (1966). *Gravel culture of vegetables and ornamental crops*. pp. 60-80, Youkendo Co. Ltd., OCLC No. 673083780, Tokyo, Japan (in Japanese)

Inderjit & Weston, L.A. (2003). Root exudates: an overview. In: *Root ecology*, de Kroon, H. & E.J.W. Visser, (Ed.), 235-255, , Ecological Studies 168, Springer, ISBN 3-540-00185-9, Verlag Berlin, Heidelberg, New York

Inderjit, (1996). Plant phenolics in allelopathy. *Botanical Review*, Vol. 62, No. 2, pp. 186-202, ISSN 0006-8101

Kaul, K. (2000). Autotoxicity in *Tagets erecta* L. on its own germination and seedling growth. *Allelopathy Journal*, Vol. 7, pp. 109-113, ISSN 0971-4693

Kitazawa, H.; Asao, T.; Ban, T.; Hashimoto, Y. & Hosoki, T. (2007). 2,4-D and NAA supplementation mitigates autotoxicity of strawberry in hydroponics. *Journal of Applied Horticulture*, Vol.9, No. 1, pp. 26-30, ISSN 0972-1045

Kitazawa, H.; Asao, T.; Ban, T.; Pramanik, M.H.R. & Hosoki, T. (2005). Autotoxicity of root exudates from strawberry. *The Journal of Horticultural Science & Biotechnology*, Vol. 80, No. 6, pp. 677-680, ISSN 1462-0316

Koda, T.; Ogiwara, S. & Hiroyasu, T. (1977). Effects of addition of activated charcoal into the nutrient solution on the growth of mitsuba (*Cryptotaenia japonica* Hassk.) in hydroponics. *Journal of the Japanese Society for Horticultural Science*, Vol. 46 (Supplement 1), pp. 270-271, ISSN 1347-2658 (in Japanese with English summary)

Koda, T.; Ogiwara, S.; Udagawa, Y. & Hiroyasu, T. (1980). Effects of addition of activated charcoal into the nutrient solution on the growth of mitsuba (*Cryptotaenia japonica* Hassk.) in hydroponics. 2. Effects of organic acids on the growth of mitsuba and its removal. *Journal of the Japanese Society for Horticultural Science*, Vol. 49 (Supplement 2), pp. 224-225, ISSN 1347-2658 (in Japanese with English summary)

Komada, H. (1988). The occurrence, ecology of soil-borne diseases and their control. Takii Seed Co. Ltd., Japan (in Japanese)

Koshikawa, K. & Yasuda, M. (2003). Studies on the bench culture with closed hydroponic system in strawberry (Part 1). *Journal of the Japanese Society for Horticultural Science*, Vol. 72 (Supplement 2), pp. 394, ISSN 1347-2658 (in Japanese with English summary)

Kushima, M.; Kakuta, H.; Kosemura, S.; Yamamura, S.; Yamada, S.; Yokotani-Tomita, K. & Hasegawa, K. (1998). An allelopathic substance exuded from germinating watermelon seeds. *Plant Growth Regulation*, Vol. 25, No. 1, pp. 1-4, ISSN 0167-6903

Lee, J.G.; Lee, B.Y. & Lee, H.J. (2006). Accumulation of phytotoxic organic acids in reused nutrient solution during hydroponic cultivation of lettuce (*Lactuca sativa* L.). *Scientia Horticulturae*, Vol. 110, No. 2, pp. 119-128, ISSN 0304-4238

Markus, A.; Klages, U.; Krauss, S. & Lingens, F. (1984). Oxidation and dehalogenation of 4- chlorophenylacetate by a two-component enzyme system from *Pseudomonas* sp. strain CBS3. *Journal of Bacteriology*, Vol. 160, No. 2, pp. 618-621, ISSN 0021-9193

Matsumoto, M.; Yoshioka, K.; Sumida, S.; Kondo, T.; Takahashi, S.; Nanjo, H. & Manabe, Y. (1973). Harmful effects of consecutive planting of taros on their growth in Ehime prefecture, Japan. *Proceedings of Association for Plant Protection of Shikoku, Japan*, Vol. 8, pp. 57-63, (in Japanese)

Matsumoto, M.; Yoshioka, K.; Sumida, S.; Kondo, T.; Tanbara, K. & Kurihara, H. (1974). Seasonal prevalence of *Pratylenchus coffeae* (Zimmerman.) in taro fields. *Proceedings of Association for Plant Protection of Shikoku, Japan*, Vol. 9, pp. 41-47, (in Japanese)

Miller, D.A. (1983). Allelopathic effects of alfalfa. *Journal of Chemical Ecology*, Vol. 9, No. 8, pp. 1059-1072, ISSN 0098-0331

Miyaji, T. & Shirazawa, N. (1979). Researches on the injuries of continuous cropping and those countermeasures in taro plant. *Bulletin of Kagoshima Agricultural Research Centre*, Vol. 7, pp. 5-15 (in Japanese)

Miyoshi, S.; Yamada, T. & Yoshika, S. (1971a). Studies on soil-borne diseases of taro-plant by continuous cropping. (1) Appearances of diseases by continuous cropping and its control effect of cultivating by alternation of land usage between dry and flooded conditions. *Bulletin of Fukuoka Agricultural Research Centre*, Vol. 9, pp. 45-48 (in Japanese)

Mizutani, F.; Hirota, R. & Kadoya, K. (1988). Growth inhibiting substances from peach roots and their possible involvement in peach replant problems. *Acta Horticulturae*, Book No. 233, pp. 37-43, ISSN 0567-7572, ISBN 978-90-66053-13-7

Murota, M.; Sakamoto, Y.; Tomiyama, K.; Ozaki, M.; Manabe, M.; Iwahashi, T. & Fukukawa, T. (1984). Actual conditions of taro planting in the main producing districts of Miyazaki prefecture and some methods to prevent injury by continuous cropping. *Bulletin of Miyazaki Agricultural Experimental Station*, pp. 39-53 (in Japanese with English abstract)

Nagae, S.; Samesima, T. & Watanabe, B. (1971). Studies on the root rot of taro (I). The causal agents of taro root rot. *Kyusyu Agricultural Research*, pp. 33-90 (in Japanese)

Nakahisa, K.; Tsuzuki, E. & Mitsumizo, T. (1993). Study on the allelopathy of alfalfa (*Medicago sativa* L.). I. Observation of allelopathy and survey for substances

inducing growth inhibition. *Japanese Journal of Crop Science*, Vol. 62, No. 2, pp. 294-299, ISSN 0011-1848

Nakahisa, K.; Tsuzuki, E.; Terao, H. & Kosemura, S. (1994). Study on the allelopathy of alfalfa (*Medicago sativa* L.). II. Isolation and identification of allelopathic substances in alfalfa. *Japanese Journal of Crop Science*, Vol. 63, No. 2, pp. 278-284, ISSN 0011-1848

Nanbu, K. (1990). Fate of pesticides in soil. *Plant Protection*, Vol. 11, pp. 477-482 (in Japanese) Oashi, K. (1973). On the damages of taro caused by *Pratylenchus coffeae* (Zimmerman.) and their chemical control. *Kyusyu Agricultural Research*, Vol. 35, pp. 103-104 (in Japanese)

Oka, S. (2002). Development of the labor-saving cultivation techniques by raising the labor- saving cultivars of vegetables (Part 1). *Bulletin of the National Agricultural Research Center for Western Region*, pp. 26-27 (in Japanese)

Overland, L. (1966). The role of allelopathic substances in the ''smother crops'' barley. *American Journal of Botany*, Vol. 53, No. 5, pp. 423-432, ISSN 00029122

Pardales, J.R.Jr. & Dingal, A.G. (1988). An allelopathic factor in taro residues. *Tropical Agriculture*, Vol. 65, No. 1, pp. 21-24, ISSN 0041-3216

Petrova, A.G. (1977). Effect of phytoncides from soybean, gram, chickpea and bean on uptake of phosphorus by maize. In: *Interaction of Plants and Microorganisms in Phytocenoses,* A.M. Grodzinsky, (Ed.), 91-97, Naukova Dumka, Kiev (in Russian, English summary).

Pramanik, M.H.R.; Asao, T.; Yamamoto, T. & Matsui, Y. (2001). Sensitive bioassay to evaluate toxicity of aromatic acids to cucumber seedlings. *Allelopathy Journal*, Vol. 8, No. 2, pp. 161-170, ISSN

Pramanik, M.H.R.; Nagai, M.; Asao, T. & Matsui, Y. (2000). Effects of temperature and photoperiod on phytotoxic root exudates of cucumber (*Cucumis sativa*) in hydroponic culture. *Journal of Chemical Ecology*, Vol. 26, No. 8, pp. 1953-1967, ISSN 0098-0331

Putnam, A.R. (1985a). Allelopathic research in agriculture: Past highlights and potential. In: *The Chemistry of allelopathy: Biochemical interactions among plants*, A.C. Thompson, (Ed.), pp. 1-8, American Chemical Society, ISBN13: 9780841208865, Washington, D.C.

Rice, E.L. (1984). Allelopathy (Second Edition). Academic Press, ISBN 0125870558, 9780125870559, Orlando, FL

Rizvi, S.J.H. & Rizvi, V. (1992). Allelopathy: Basic and Applied Aspects. pp. 480, Chapman & Hall, ISBN 0 412 39400 6, London

Ruijs, M.N.A. (1994). Economic evaluation of closed production systems in glasshouse horticulture. *Acta Horticulturae*, Book No. 340, pp. 87-94, ISSN 0567-7572, ISBN 978-90-66054-25-7

Sato, N. (2004) Effect of the substances accumulated in the nutrients solution by the rock wool circulated hydro culture to the rose seedlings growth. *Journal of the Japanese Society for Horticultural Science*, Vol. 73 (Supplement 2), pp. 497, ISSN 1347-2658 (in Japanese)

Shafer, W.E. & Garrison, S.A. (1986). Allelopathic effects of soil incorporated asparagus roots on lettuce, tomato, and asparagus seedling emergence. *HortScience*, Vol. 21, No. 1, pp. 82-84, ISSN 0018-5345

Singh, H.P.; Batish, D.R. & Kohli, R.K. (1999). Autotoxicity: concept, organisms and ecological significance. *Critical Review on Plant Science*, Vol. 18, No. 6, pp. 757-772, ISSN 0735-2689

Sunada, K.; Ding, X.G.; Utami, M.S.; Kawashima, Y.; Miyama, Y. & Hashimoto, K. (2008). Detoxification of phytotoxic compounds by TiO2 photocatalysis in a recycling hydroponic cultivation system of asparagus. *Journal of Agricultural and Food Chemistry*, Vol. 56, No. 12, pp. 4819-4824, ISSN 0021-8561

Sundin, P. & Waechter-Kristensen, B. (1994). Degradation of phenolic acids by bacteria from liquid hydroponic culture of tomato. In: *Plant Production on the Threshold of a New Century*, P.C. Struik; W.J. Vredenverg; J.A. Renkama, & J.E. Parlevliet, (Ed.), 473- 475, Kluwer Academic Publishers, Dordrecht, ISBN 0 7923 2903 1, The Netherlands

Takahashi, K., (1984). The replant failures of vegetables. *Research Reports of National Research Institute of Vegetable and Tea Science*, Japan, Vol. 18, pp. 87-99 (in Japanese)

Takeuchi, T. (2000). The nourishment uptake of strawberry cultivar 'Akihime' in rockwool hydroponics with a nutrient solution circulating system. *Bulletin of the Shizuoka Agricultural Experiment Station*, Vol. 45, pp. 13-23 (in Japanese with English summary)

Tang, C.S. & Young, C.C. (1982). Collection and identification of allelopathic compounds from the undisturbed root system of bitalta limpograss (*Helmarthria altissima*). *Plant Physiology*, Vol. 69, No. 1, pp. 155-160, ISSN 0032-0889

Tsuchiya, K. & Ohno, Y. (1992). Analysis of allelopathy in vegetable cultivation. I. Possibility of occurrence of allelopathy in vegetable cultivation. *Bulletin of the National Research Institute of Vegetable, Ornamental Plants and Tea Science*, Vol. 5, pp. 37-44 (in Japanese with English abstract)

Tsuchiya, K. (1990). Problems on allelopathy in vegetable cropping. *Agriculture and Horticulture*, Vol. 65, No. 1, pp. 9-16, ISSN 0369-5247 (in Japanese)

Tsuzuki, E.; Shimazaki, A.; Naivalulevu, L.U. & Tomiyama, K. (1995). Injury by continuous cropping to taro and its related factors. *Japanese Journal of Crop Science*, Vol. 64, No. 2, pp. 195-200, ISSN 0011-1848

Tukey, H.B., Jr. (1969). Implications of allelopathy in agricultural plant science. *Botanical Review*, Vol. 35, No. 1, pp. 1-16, ISSN 0006-8101

van den Tweel, W.J.J.; Kok, J.B. & de Bont, J.A. (1987). Reductive dechlorination of 2,4- dichlorobenzoate to 4-chlorobenzoate and hydrolytic dehalogenation of 4-chloro, 4- bromo-, and 4-iodobenzoate by *Alcaligenes denitrificans* NTB-1. *Applied Environmental Microbiology*, Vol. 53, No. 4, pp. 810-815, ISSN 0099-2240

Van Os, E.A. (1995). Engineering and environmental aspects of soilless growing systems. *Acta Horticulturae*, Book No. 396, pp. 25-32, ISSN 0567-7572, ISBN 978-90-66053-13-7

Yang, H.J. (1982). Autotoxicity of *Asparagus officinalis* L. *Journal of the American Society for Horticultural Science*, Vol. 107, No. 5, pp. 860-862, ISSN 0003-1062

Young, C.C. & Chou, T.C. (1985). Autointoxication in residues of *Asparagus officinalis* L. *Plant and Soil*, Vol. 85, No. 3, pp. 385-393, ISSN 0032-079X

Young, C.C. (1984). Autotoxication in root exudates of *Asparagus officinalis* L. *Plant and Soil*, Vol. 82, No. 2, pp. 247-253, ISSN 0032-079X

Yu, J.Q & Matsui, Y. (1994). Phytotoxic substances in root exudates of *Cucumis sativus* L.*Journal of Chemical Ecology*, Vol. 20, No. 1, pp. 21-31, ISSN 0098-0331

Yu, J.Q. & Matsui, Y. (1993a). Effect of addition of activated charcoal to the nutrient solution on the growth of tomato grown in the hydroponic culture, Soil Science and Plant Nutrition, Vol. 39, No. 1, pp. 13-22, ISSN 1747-0765

Yu, J.Q. & Matsui, Y. (1993b). Extraction and identification of the phytotoxic substances accumulated in the nutrient solution for the hydroponic culture of tomato, *Soil Science and Plant Nutrition*, Vol. 39, No. 4, pp. 691-700, ISSN 1747-0765

Yu, J.Q. & Matsui, Y. (1997). Effects of root exudates of cucumber (*Cucumis sativus*) allelochemicals on ion uptake by cucumber seedlings. *Journal of Chemical Ecology*, Vo. 23, No. 3, pp. 817-827, ISSN 0098-0331

Yu, J.Q.; Lee, K.S. & Matsui, Y. (1993). Effect of addition of activated charcoal to the nutrient solution on the growth of tomato grown in the hydroponic culture. *Soil Science and Plant Nutrition*, Vol. 39, No. 1, pp. 13-22, ISSN 1747-0765

CHAPTER-5

PLANT HYDROPONIC CULTIVATION: SUPPORTING RESEARCH IN THE AREA OF PLANT-MICROBE-ENVIRONMENT INTERACTIONS IN THE FIELD OF BIOLOGY

INTRODUCTION: AN OVERVIEW

Hydroponic cultivation of plants has emerged as a pivotal platform for advancing research in the realm of plant-microbe-environment interactions within the field of biology. This innovative approach offers a controlled and dynamic environment for studying the complex interplay between plants, microorganisms, and their surroundings. As we delve deeper into understanding these intricate relationships, we uncover valuable insights into plant growth, disease resistance, and environmental sustainability.

Plant production is essentially dependant on culture medium and conditions. Vegetable crops are known to grow in soil which, in natural conditions, acts as a mineral nutrient reservoir, but the soil itself is not essential to plant growth. Plant roots are able to absorb the mineral nutrients of the soil dissolved in water. When the required mineral nutrients are introduced into artificially synthesized medium (case of hydroponic cultivation), soil is no longer required for the plant growth. Hydroponic cultivation is a soil-free technology of vegetable growing. First reports about hydroponic culture date from the seventeenth century. Later, it was reported that in the Second World War, American soldiers grew their own food hydroponically while stationed on barren Pacific islands. Hydroponics was a necessity on Wake Island because there was no soil, and it was prohibitively expensive to airlift in fresh vegetables. Hydroponic cultivation is associated to the most revolutionary future technological projects. Indeed, for "future", NASA Scientists are researching optimization of some factor as light, temperature and carbon dioxide, along with plant species for cultivation on planets like Mars.

Technically, hydroponic cultivation is a method of growing plants using mineral nutrient solutions, in water, without soil. The strict definition of "hydroponic" is referred to liquid systems only. Nevertheless, a more large definition could include cultivation in mineral nutrient solution only or in an inert medium, such as perlite, vermiculite, gravel, mineral wool, coconut husk.... Hydroponic systems may similarly be either closed or open with respectively recycling or non-recycling the nutrient solution.

Hydroponic culture is possibly the most intensive method of crop production in today's agricultural industry, mainly for ornamental plants which could be grown even in the absence of specialized spaces (gardens) restricted because of the demographic pressure, and far from their own natural environment. Indeed, Scandinavian citizen could be able to grow in his house tropical vegetables thanks to hydroponic culture systems. It is highly productive, conservative of water, land and space, and protective of the environment.

The principle advantages of hydroponic vegetable cultivation is a more efficient use of water and fertilizers, minimal use of land area, isolation of the crop from the underlying soil which may have problems associated texture and nutrient availability, structure and drainage, disease, salinity... Indeed, in hydroponic medium there are no obstacles for plant roots caused by compact soil structure, which make easier their development and the accessibility of different nutrients.

On the other hand, hydroponic cultivation presents many disadvantages as the high cost of capital and energy inputs required in controlled culture-conditions associated with this mode of plant cultivation. Due to this significantly high cost, successful practical applications (e.g. for food production...) of hydroponic technology are limited to crops of high economic value in specific regions, in country having a sufficient materiel capacities and high degree of competence in plant science and engineering. Unfortunately, these conditions are limited to few countries as the United States, Canada, Europe and Japan that already have no problems with food production. In the developing countries materials and technological conditions did not allow this kind of plant cultivation unless some little exploitation of "exotic" ornamental plants.

Otherwise, hydroponics represents a standard technique in biology research. Indeed this technology of vegetable cultivation enables a more precise control of growth conditions which make easier to study the variables factors or parameters.

The use of Hydroponic cultivation for practical food or ornamental plant production is already detailed in reports and catalogues of societies marketing materials of hydroponic systems. In this chapter, we focus on importance of hydroponic cultivation as a sustainable technology for biological research mainly in studies analyzing plant response to environmental stresses and plant-microbe interactions. Our report will be based essentially on results obtained in the Laboratory of Legumes, Centre of Biotechnology of Borj Cedria, Tunisia, where we analyze the mechanisms, markers and strategy for improvement of legumes production under stressful constraints.

HYDROPONIC CULTIVATION SYSTEMS CAN SERVE AS SIGNIFICANT SOURCES OF BIOMASS PRODUCTION FOR USE IN BIOLOGICAL STUDIES

Studies are conducted on some legume species (*Medicago truncatula, Medicago sativa, Cicer arietinum, Phaseolus vulgaris, Vicia Faba, Lens esculenta...*) with the aim to improve yield and tolerance to biotic and abiotic stresses. To reach these aims, multiples analyses should be carried out to dissect response mechanisms, biochemical and molecular markers for selection of efficient genotypes. Plant cultivation was performed under controlled condition (glasshouse, green house) with minimum variation of culture conditions (light, temperature...). Parameters analyzed require an important "clean" biomass for physiological, biochemical and molecular analyses. Two types of hydroponic cultivation were performed: (i) in liquid medium (aeroponic culture) and (ii) with inert support (sterile gravel, sand, perlite, vermiculite...). These hydroponic media allowed a better biomass production mainly for root part. Indeed chickpea (*Cicer arietinum*), common bean (*Phaseolus vulgaris*), lucern (*Medicago sativa*) and barrel medic (*Medicago truncatula*) plants cultivated in hydroponic media produced root and shoot biomass twice to three time higher than those performed by plants cultivated in soil (Table 5.1). Plants vigor is also better in hydroponic cultivation mode (Figure 5.1) unless some necrosis that could be revealed at the extremities of old leaves.

When harvested for analyses, biological material (roots and shoots) issue from hydroponic culture is in healthy state (absence microorganisms and organic substrate adsorbed to roots) which make easier analysis with minimum interference between variables. The hydroponic systems allow homogeneity of the plant sizes.

Other types of studies are carried out mainly for wild legumes with an aim to analyze genetic diversity and genotype identification. These kinds of analyses required other culture systems, mainly in the first steps of analysis when an important number of samples are expected. Cultivation on agarose media in Petri dishes or in test-tube allows an important sampling (more than 2000 seedlings per experiment) but it is limiting plant development. Indeed, experiments performed in test-tube for *Medicago truncatula* plants allow analysis of an important number of genotypes but with a biomass that not reach 50 mg/plant (Table 5.2) however, these plant produce more than 2 g/plant in hydroponic medium (Table 5.1).

Table 5.1. Examples of biomass production levels g/plant and mg/plant (S; Shoots, R; Roots, N; nodules) of some legumes species (Lucern; *Medicago sativa*, Barrel Medic; *Medicago truncatula*, Chickpea; *Cicer arietinum*, Common bean; *Phaseolus vulgaris* and Faba bean; *Vicia faba*) cultivated under different hydroponic systems compared to soil and in vitro (test-tube) cultivation.

	M. truncatula			*M. sativa*			*Cicer arietinum*			*Phaseolus vulgaris*			*Vicia faba*		
	S	R	N	S	R	N	S	R	N	S	R	N	S	R	N
Liquid	2.5 g	1 g	0.5 g	4.5 g	1.5 g	0.8 g	10g	3.5g	1.5 g	7 g	2 g	1.8 g	12 g	4 g	2.5 g
Sterile Sand	1.2 g	0.35 g	0.15 g	3.5 g	1.2 g	0.28 g				2 g	1.3 g	0.8 g	10 g	4 g	2 g
Sterile Gravel	380 mg	50 mg		450 mg	70 mg		1,8 g	0.46 g	0.25 g	2 g	0.6 g	0.4 g	6 g	3 g	1 g
Perlite	0.7 g	0.4 g	0.2 g	1 g	0.4 g	0.2 g				2 g	0.8 g	0.5 g			
Perlite/Vermiculite	0.8 g	0.3 g	0.2 g	1.2 g	0.3 g	0.2 g									
Perlite/Vermiculite/Sand	1 g	0.3 g		1.5 g	0.5 g										
Soil	1.2 g	0.3 g	0.15 g	2 g	0.8 g	0.2 g	5 g	1.5 g	1 g	6 g	1.5 g	0.8 g	8 g	2 g	1.2 g
Agarose in test tubes	50 mg			70 mg											

Fig. 5.1. Culture systems for hydroponic plant cultivation in liquid medium (A: *Medicago trunctula*, B: *Cicer arietinum*), sterile gravel (C: *Phaseolus vulgaris*) and Perlite/vermiculite (D: *Medicago truncatula*). Plants are grown in glass house with 16/8 photoperiod, 25/18°C day/night temperature, irrigated with nutrient solution (Vadez et al. 1996, Mhadhbi et al. 2005) and aerated with an airflow 400ml/min for liquid medium. Plants in the photos are in main flowering stage (68 days *M. truncatula*, 60 days *Cicer arietinum* and 45 days *Phaseolus vulgaris*).

Table 5.2. Plant biomass production (mg/plant) of different Barrel Medic (*Medicago truncatula*) genotypes (one reference Jemalong 6 and two local Tunisian lines TN8.20 and TN6.18) cultivated "*in vitro*" in test tubes inoculated with different rhizobial strains (*Sinorhizobium meliloti* and *S. medicae*) or depended on mineral nitrogen fertilization. Data denoted with different letters are statistically different according to the Duncun multiple range test at p=0.05

	Jemalong (J6)	TN 8.20	TN 6.18
S. meliloti RCR 2011	12.54[c]	23.44[b]	9.95[c]
S. meliloti KII4	14.79[bc]	23.00[b]	10.20[c]
S. meliloti E4	14.80[bc]	21.83[b]	9.84[c]
S. meliloti TII7	17.54[b]	23.21[b]	12.99[b]
S. meliloti MI4	12.29[c]	11.9[c]	12.00[b]
S. medicae M104	21.62 [b]	19.66 [b]	ND

	Jemalong (J6)	TN 8.20	TN 6.18
S. medicae A321T	17.55 [b]	16.88 [b]	ND
S. medicae SII4	16.36 [b]	17.02 [b]	ND
S. medicae KI 11a	15.42 [bc]	16.56 [b]	ND
S. medicae E5	14.28 [bc]	16.30 [b]	ND
Nitrogen control	38.39 [a]	36.21 [a]	30.60 [a]
Absolute control	8.46 [d]	8.57 [d]	6.61 [d]

Among the different hydroponic media assessed in our experiments, the liquid medium (Table 5.1) associated to artificial aeration (400 ml/min/plant) seems been the adequate medium for a maximum plant biomass production, mainly when plant are cultivated in symbiosis with nitrogen fixing bacteria (Mhadhbi et al., 2004, 2005, 2008, 2009, Jebara et al., 2005, 2010). This system presents more advantages "vigorous and clean biological material" due to the absence of support which make easier the physiological, biochemical and molecular analyses performed on root parts.

REGULATING GROWTH PARAMETERS IN HYDROPONIC SYSTEMS

Analyses performed in biological researches concerning physiological and biochemical metabolisms, proteomic and genomic studies aim the understanding of efficiency and tolerance mechanisms and the selection of plants genotypes. Observations and conclusions performed on these analyses could be precise for the success of selection. Consequently, growth conditions could be the maximum controlled to omit interference with non considered parameters. Indeed, to analyze plant response to biotic agents or environmental constraints, the applied stress could be the only changing factor, which implies the stabilization of the remained factors, as temperature, light intensity, medium acidity, interaction with other microorganisms, equal nutrient provisioning, homogeny plant size.

Plant cultivation in hydroponic media under controlled conditions allow stabilization of these factors since all treatments are under the same condition, in the same support and irrigated with the same nutrient solution. Composition of the nutrient solution could be adjusted according to specific needs of plant species. Indeed, our experience with legumes plants showed that composition of nutrient solution allowing optimal plant growth for some grain legume species (Common bean, Chickpea) (Vadez et al., 1996, Jebara et al., 2001, Mhadhbi et al., 2004, 2008) could be modified for concentration of some elements to be adequate for others legume species (*Medicago truncatula, Medicago sativa*) (Mhadhbi et al., 2005). Control of mineral elements concentrations and forms could be previously adjusted. Indeed, in our experiments, iron is usually added in its sequestered form (commercial "sequesterne"), nevertheless some species grow better when iron is added as Fe-EDTA form, consequently preliminary experiments could be performed to adjust specific concentrations and form of nutrient elements.

Hydroponic cultivation presents some disadvantages related mainly to the fact that, supports as perlite are preferred by saprophytic fungus. Moreover, when studying drought effect hydroponic media, mainly liquid and artificial-support media, are contested for such analyses. Indeed, the first is far from natural conditions where drought means the absence of water note its presence with the inhibition of its absorption using osmoticum as PEG or Mannitol. In the second, it is technically difficult to maintain a stable Field Capacity during long period since vermiculite and mainly perlite are susceptible to evaporation. Among hydroponic media assessed for studying drought stress effect on plant growth, the sterile sand seems been the most adequate. The use of sterile sand as support allow the application of drought constraint and the selection of tolerant genotype or symbioses (Mhadhbi et al., 2011)

MONITORING ROOTS RESPIRATION AND NODULES NITROGEN-FIXATION: EXAMPLES OF SIMPLICITY AND EFFECTIVENESS OF GAS EXCHANGES ANALYSES IN HYDROPONIC SYSTEMS

Plants depend on gas exchanges with atmosphere for surviving. Indeed gas flow is determinant factor for main vital processes (photosynthesis and respiration). Measurement of gas (CO2 and Oxygen) exchange reflects the healthy state of plants. Photosynthesis is performed in leaves, which make its determination independent of culture system. However, roots respiration depends on support texture. Indeed, monitoring of oxygen flow is influenced by soil texture and structure. In hydroponic systems, mainly aeroponic and systems based on non sandy inert supports allowing easy air circulation (gravel, coconut husk...), the measurement of oxygen flow is a "simple" manipulation using specific electrodes and oxygen monitor related to root atmosphere both in closed of open systems. Jebara et al. (2001) reported differential response of symbiotic common bean (*Phaseolus vulgaris*) varieties to salt stress based on ability of nodules to maintain respiration under stressful conditions.

A second example of importance of hydroponic system in gas exchange monitoring is measurement of nitrogen fixation within root nodules of legumes (*Fabacea*) or other symbiotic plants (e.g. Actinorhizale plants). Nitrogen (N2) is the main gas in atmosphere (78%) and it is essential element for protein and pigment biosynthesis. Nevertheless, among living forms, there are only some microorganisms able to fix atmospheric nitrogen. Some nitrogen fixing bacterial genus communally called Rhizobia are able to establish symbiotic association with legume plants (*Fabacea*) which enable plants to benefit of fixed nitrogen (For detailed review, Graham & Vance, 2003, Mhadhbi et al., 2009). Measure of fixed nitrogen is performed by different methods: (i) Khjeldal method which determine the quantity of nitrogen within tissues after plant harvest, there for it is a destructive method, (ii) monitoring fixed nitrogen using N^{15} isotope which require radioactivity manipulation and so special equipments, care and spaces,

An "*In Situ, In Vivo*" method using acetylene (C2H2) as structural analogue of molecular nitrogen (N2) and the result of reaction is ethylene (C2H4) easily determined using gas chromatograph (Hardy et al., 1968). It is an estimative method with limits due to the inhibition of nitrogense enzyme by acetylene (Minchin et al., 1983) but simple and effective mainly for comparative studies (Mhadhbi et al., 2005, 2008).

The principle of this method is the following: Nodule-bearing roots are incubated in 10% C2H2 atmosphere. After "x" min of incubation, the ethylene formation rate was measured using gaseous phase chromatography with Porapak-T column. Statistic replicates of 0.5 ml gas samples were withdrawn from the root atmosphere of each plant, and ethylene production was determined. Pure acetylene and ethylene were used as internal standards.

This method could be performed in soil sample, but error margin is amplified due to the non control of gas diffusion inside soil texture. Aeroponic system is presented as the most adequate system for such kind of measure.

For both examples (respiration rate and nitrogen fixing capacity estimation), hydroponic culture systems allow simple, safety, non destructive (important for kinetic monitoring of some parameters during plant life cycle) measures, which enable more deep and detailed analyses.

ANALYSING PLANT MICROBE INTERACTIONS IN HYDROPONIC SYSTEMS

In the rhizosphere, plants are in continuous interaction with both benefic and pathogenic microorganisms. Thousand years ago, humans discovered some of these interactions and tried to exploit for enhencing food production. Indeed, symbiotic plants as "*Lens* sp." were domesticated 8000 to 9000 year ago (Graham & Vance, 2003). The research analyses targeting the exploitation of benefic microorganisms for improvement of plant growth or the protection against the pathogenic ones require a strict control of culture conditions (mainly sterility to ovoid non specific contamination). These practices could be performed in hydroponic systems which, for example, allow the strict control of sterility conditions and even the use of specific synthetic media limiting the possibility of contamination with non desired microorganisms.

Symbiotic Interactions

Legume plants (*Fabacea*) represent best models for studying plant microbe interaction. Indeed, at least 80% of legume species have the ability to be nodulated by symbiotic atmospheric- nitrogen fixing bacteria belonging to the *Rhizobiaceae* family. This symbiosis lead to formation of a specific organ "nodule" in root parts forming a micro-habitat for nitrogen fixing bacteria. Moreover, legumes are naturally susceptible to biotic agents (pathogenic bacteria and fungus) because of the richness of their roots, leaves and seeds with proteins and other nutritive elements. The most important interaction is the symbiotic interaction that allows the use of abundant atmospheric nitrogen for plant nutrition which decreases the use of mineral fertilizers and consequently reduces the production cost and pollution risks (Graham & Vance, 2003). Valorization of biological nitrogen fixation for sustainable agriculture is the last steps of long procedure having as a final result the proposition (marketing) of a biological fertilizer efficient in enhancing legume and associated culture production. This is a result of studies of the different interactions between plant and bacteria genotypes and with the ambient environment... In such kind of studies, hydroponic cultivation presents an important number of advantages:

- An important root surface allowing more infection points and consequently more important nodule number and biomass, and "clean" healthy roots and nodules for biochemical and molecular analyses susceptible to interference of non specific factors (Figure 5.2).
- Control of culture conditions mainly omitting contamination, nevertheless precaution to discard non specific interaction should be taken in aerated hydroponic medium mainly when more than one bacterial strain is used for the inoculation.
- Possibility of monitoring the different steps of plant microbe interaction and different stages of nodule (the main organ of symbiotic interaction) formation (kinetic of nodule apparition), maturation, functioning steps and senescence of nodules.

As detailed in section 4, the functioning of symbiosis is estimated through the measure of nitrogenase activity (nitrogenase is a nodular enzyme that catalyze the reduction of molecular atmospheric nitrogen to ammoniac). Hydroponic system allows the "*in vivo*" measure of nitrogenase activity via the measure of Acetylene Reducing Activity (ARA) (Hardy et al. 1968, Mhadhbi et al. 2004, 2005, 2009). In this context, aeroponic system allows a maximum of control of external parameters. Other functioning indices could be measured in such system as gas exchange, respiration, nodule occupancy using *gus A* gene transformed bacteria (Mhamdi et al., 2005)

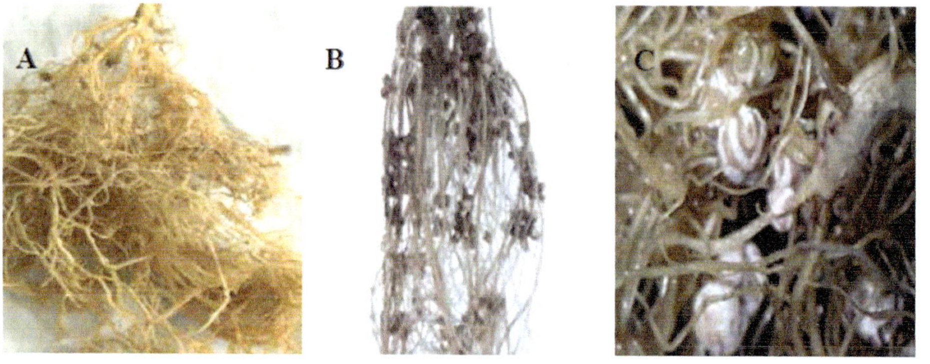

Fig. 5.2. Roots of some legumes species (Chickpea and common bean) issues from aeroponic cultivation showing an important root biomass development (A) and consequently an enormous number of nodules (B) with healthy and vigorous appearance (C).

Interactions Between Plant Pathogens

Plants are susceptible to pathogenic microorganisms mainly bacteria and fungus, which cause dramatic effects on cultivated-plant yields reaching the lost of all production sometimes mainly when environmental conditions are favorable for microorganism development (high temperature and humidity rates). Pathogens attack different parts of and more contested due to its effects on ecosystems' biodiversity and for animal and human health. To use specific pesticide or bio-pesticide we need an exact identification of the pathogen. This identification, mainly at roots level, is not easy to realize in natural conditions since the richness of plant rhizosphere with diverse pathogenic and saprophytic microorganisms that could interfere with the principal agent. Moreover, symptoms of pathogenic attacks as change of root color and root-hair density and quality could be modified by soil quality. As above explained, hydroponic culture allow a maximum control of external parameters (temperature, light, medium sterility...) which enable a precise identification of pathogenic-attack symptoms (Figure 5.3). Pathogen-strain isolation and identification is more easy and rapid allowing precise and quick intervention.

CONSTRAINTS AND DRAWBACKS ASSOCIATED WITH HYDROPONIC CULTIVATION

As mentioned in introduction section, hydroponic cultivation presents multiples limits and disadvantages. The cost of this mode of plant cultivation limits their practice to rich countries (United States, Canada, Europe, Japan...) and discards poor and developing countries which really suffer with environmental constraints, low crop yields and consequent food deficiency. This could exacerbate the lack of equilibrium between the rich and poor countries. Symptoms of this disequilibrium are now (2011) clearly manifested in countries as Somalia, Kenya and Ethiopia where famine causes thousand kids' dies per day. In the other hand, in rich counties, prudence of peoples from risks of "biotechnology products" as GMO (genetically modified organisms) on health lead to an orientation to natural and biological product and avoiding products coming from para-natural practices as hydroponic cultivation.

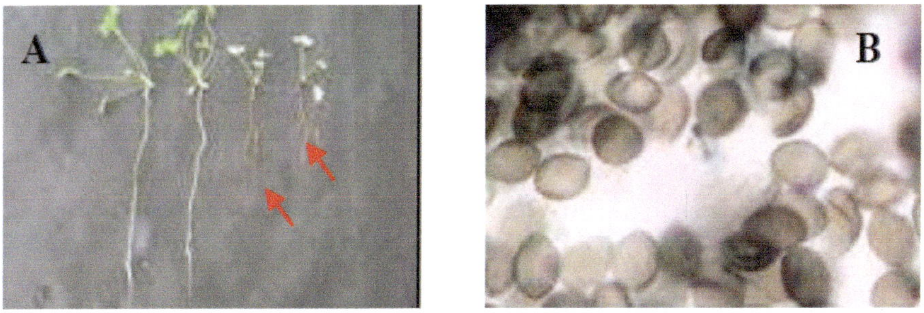

Fig. 5.3. A. Roots of *Medicago truncatula* plants (Jemalong 6) inoculated with *Phoma medicagenis*issues from hydroponic cultivation showing clear symptoms of fungus attack (red arrows).B. fungi spores developed and isolated from infected roots.

At level of hydroponic cultivation dedicated for biological research, the experiment showed an unequivocal best biomass production for analyses. Nevertheless, a special care must be taken for success of such kind of cultivation mode. Indeed, our experiments showed that concentration of mineral elements added in artificial media is easily changeable in hydroponic systems and plants are very sensitive to these variations. A deleterious effect was revealed duc to slight modification of nitrogen or phosphorus concentrations (Mhadhbi et al., 2005). In the hydroponic system the plant growth is totally and strictly dependent on the human control, there is no margin for intrinsic plant adaptation. For aeroponic systems, special care should be taken for an adequate and continuous aeration flow. Indeed, interruption of aeration due to electronic panel during the weekend for example had deleterious effects on experiments, and even if plants recovered from asphyxia, saprophytic contaminations are enhanced by these conditions.

AT A GLANCE

- It could be summarized that hydroponic culture systems represent a suitable support for biological research thanks to some advantages mainly an important clean biomass, a possibility to control external factor influencing experimental conditions even soil structure and texture. Roots and nodules (case of symbiotic plants) biomass production is clearly improved in hydroponic systems mainly aeroponic and those using non-heavy supports (perlite, vermiculite...).

- It is an artificial system adapted to laboratory conditions (exploitation of limited spaces and overcome soil hard-manipulation). In hydroponic systems we are able to analyze interactions between multiples factors influencing plant growth [e.g. analyzing response of legume plant inoculated with its specific rhizobial partner to abiotic (salinity, drought, cold...) and/or biotic (fungi, bacteria...) constraints]. In the other hand, hydroponic cultivation represents a "new" technology promoting intensive crop production for food and ornamental use. Nevertheless, application of this technique remains restricted to technologically developed countries, which deny hope to exploit it in combating poverty in the world. And even in these rich counties an effort is required to convince peoples by differences between hydroponic culture products and those resulting from "non safe" technologies as GMO.

REFERENCES

Barber, D. & Gunn, K. (1974). The effect of mechanical forces on the exudation of organic substrates by the roots of cereal plants grown under sterile conditions. *New Phytol.* 73: 39-45.

Baylis, A.; Gragopoulou, C. & Davidson, K. (1994). Effects of Silicon on the Toxicity of Aluminum to Soybean. *Comm. Soil Sci. Plant Anal.* 25: 537-546.

Bowen, G. & Roveria, A. (1976). Microbial colonization of plant roots. *Annual Reviews of Plant Phytopathology* 14: 121-144.

Bugbee, B. & Salisbury, F. (1985). An evaluation of MES and Amberlite IRC-50 as pH buffers for Nutrient Solution Studies. *J. Plant Nutr.* 8: 567-583.

Bugbee, B. & Salisbury, F. (1989). Controlled Environment Crop Production: Hydroponic vs. Lunar Regolith. In: D. Ming and D. Henninger. (eds) Lunar Base Arriculture. Amer. Soc. Agron. Madison, WI.

Chaney, R., & Coulombe, B. (1982). Effect of phosphate on regulation of Fe-stress in soybean and peanut. *J. Plant Nutr.* 5: 469-487.

Cherif, M.; Menzies, J.; Ehret, D.; Boganoff, C. & Belanger, R. (1994). Yield of Cucumber Infected with *Phythium aphanidermatum* when Grown with Soluble Silicon. *Hort Science* 29: 896-97.

Glenn, E.P. (1984). Seasonal effects of radiation and temperature on growth of greenhouse lettuce in a high isolation desert environment. *Scientific Horticulture* 22: 9-21.

Graham, P.H. & Vance, C.P. (2003). Legumes: importance and constraints to greater use. *Plant Physiology* 131: 872–877

Graves, C.J. (1983). The nutrient film technique. *Horticultural Review* 5: 1-44.

Haller, T. & Stolp, H. (1985). Quantitative estimation of root exudation of the maize plant. *Plant and Soil* 86: 207-216.

Hoagland, D.R. & Arnon, D.I. (1950). The Water-culture Method for Growing Plants Without Soil. Cir. 347, California Agricultural Experiment Station, University of California, Berkeley.

Jebara, M.; Drevon, J.J. & Aouani, M.E. (2001). Effects of hydroponic culture system and NaCl on interactions between common bean lines and native rhizobia from Tunisian soils. *Agronomie* 21: 601–605

Jebara, S.; Drevon, J.J. & Jebara, M. (2010). Modulation of symbiotic efficiency and nodular antioxidant enzyme activities in two *Phaseolus vulgaris* genotypes under salinity. *Acta Physiol. Plant.* 32: 925-932 .

Jebara, S.; Jebara, M.; Limam, F. & Aouani, M.E. (2005). Changes in ascorbate peroxidase, catalase, guaiacol peroxidase and superoxide dismutase activities in common bean (*Phaseolus vulgaris*) nodules under salt stress. *J. Plant Physiol.* 162: 929-936.

Jensen, M.H. (1973). Exciting future for sand culture. *American Vegetable Grower* 21, 11: 33- 34, 72.

Jensen, M.H. (1980). Tomorrow's agriculture today. *American Vegetable Grower*

Jensen, M.H. (1989). 150,000 acres and rising; Greenhouse agriculture in Korea and Japan. Proc. 10th Annual conference on Hydroponics. Hydroponic Society of America, pp. 83-79.

Ma, J.; Nishimura, K. & Takahashi, E. (1989). Effect of Silicon on the growth of the Rice Plant at Different Growth Stages. *Soil Sci. Plant Nutr.* 35: 347-356.

Mhadhbi, H.; Chihaoui, S.; Mhamdi, R.; Mnasri, B.; Jebara, M. & Mhamdi R. (2011). A highly osmotolerant rhizobial strain confers a better tolerance of nitrogen fixation and an enhanced protective activities to nodules of *Phaseolus vulgaris* under drought stress. *African Journal of Biotechnology*, 22: 4555-4563.

Mhadhbi, H.; Djébali, N.; Chihaoui, S.; Jebara, M. & Mhamdi, R. (2011). Nodule senescence in *Medicago truncatula-Sinorhizobium* symbiosis under abiotic constraints: Potential mechanisms involved in maintaining nitrogen fixing capacity. *Journal of Plant Growth Regulation* 30: 480-489

Mhadhbi, H.; Fotopoulos, V.; Djebali, N.; Polidoros, A.N. & Aouani, M.E. (2009). Behaviours of *Medicago truncatula-Sinorhizobium meliloti* symbioses under osmotic stress in relation with symbiotic partner input. Effects on nodule functioning and protection. *J Agron crop. Sci.* 195: 225-231.

Mhadhbi, H.; Fotopoulos, V.; Mylona, P.V.; Jebara, M.; Aouani, M.E. & Polidoros, A.N. (2011). Antioxidant gene–enzyme responses in *Medicago truncatula* genotypes with different degree of sensitivity to salinity. *Physiologia Plant*arum 141: 201-214.

Mhadhbi, H.; Jebara, M.; Limam, F. & Aouani, M.E. (2004). Rhizobial strain involvement in plant growth, nodule protein composition and antioxidant enzyme activities of chickpea-rhizobia symbioses: modulation by salt stress. *Plant Physiol. Biochem.* 42: 717-722.

Mhadhbi, H.; Jebara, M.; Limam, F.; Huguet, T. & Aouani M.E. (2005). Interaction between *Medicago truncatula* lines and *Sinorhizobium meliloti* strains for symbiotic efficiency and nodule antioxidant activities. *Physiologia Plantarum* 124: 4-11.

Mhadhbi, H.; Jebara, M.; Zitoun, A.; Limam, F. & Aouani, M.E. (2008). Symbiotic effectiveness and response to mannitol-mediated osmotic stress of various chickpea-rhizobia associations. *World Journal of Microbiology and Biotechnology* 24: 1027-1035.

Minchin, F.R.; Witty, J.F.; Sheehy, J.E. & Muller, M. (1983). A mayor error in the acetylene reduction assay: decreases in nodular nitrogenase activity under assay conditions. Journal of Experimental Botany 34: 641-649.

Samuels, A.; Glass A.D.M.; Ehret D. & Menzies J. (1991). Mobility and Deposition of Silicon in Cucumber Plants. *Plant Cell and Eenvironnement* 14: 485-492.

Smucker, A. (1984). Carbon utilization and losses by plant root systems. p. 27-46. IN: Roots, nutrient and water influx, and plant growth. Am. Soc. Agron. Special publ. 49, Madison, WI.

Trollenier, G. & Hect-Bucholz, C. (1984). Effect of aeration status of nutrient solution on microorganisms, mucilage and ultrastructure of wheat roots. *Plant and Soil* 80: 381-390.

Valamis, J. & Williams, D. (1967). Manganese and Silicon Interaction in the *Gramineae. Plant and Soil* 28: 131-140.

Winslow, M. (1992). Silicon, Disease Resistance, and Yield of Rice Genotypes under Upland Cultural Conditions. *Crop Sci.* 32: 1208-1213.

CHAPTER-6

UTILIZING HYDROPONICS AS A FUNDAMENTAL APPROACH IN DIVERSE AREAS OF PLANT BIOLOGY RESEARCH

INTRODUCTION: AN OVERVIEW

Hydroponics has emerged as a foundational and versatile approach in various domains of plant biology research. This innovative technique involves cultivating plants without soil, instead, delivering essential nutrients directly to the plant's roots through a nutrient-rich water solution. Researchers across diverse fields of plant biology have increasingly embraced hydroponics due to its ability to provide precise control over the growing environment, facilitating experiments related to nutrient uptake, mineral nutrition, stress responses, and growth patterns. Additionally, hydroponics offers the advantage of minimizing potential confounding variables associated with soil-based research, allowing scientists to isolate and study specific aspects of plant physiology and development with greater precision. As a result, this methodology has become an invaluable tool in addressing critical questions and advancing our understanding of plant biology in various research contexts.

It is simple-mindedness if you think that soilless culture included modern practical knowledge in the world, so that the evidence indicate that from many centuries ago, scientists tried to grow plant in containers aboveground. Whatever it is considerable, the hydroponics system took place in various places and ages in historical context. The paintings in the temple of *Deir el Bahari* show that in 4000 year ago the Egyptians attempted to transfer and grow trees in big pots in media culture (Raviv & Lieth, 2007). The written proofs have shown that *John Woodward* carried out the early hydroponic systems upon plant nutrition researches in 1699, when he attempted to realize whether plants got their nutrients from soil or water.

Water culture used as a method, so that the soil gradually added to water and at the same time, the plant growth and development monitored. The results showed that soil and water provide nutrients for plant growth and development (Jones, 1982). In the nineteenth century, French and German scientists attempted to investigate in terms of plant nutritional requirements and then it has developed by American and English scientists during the first half of the twentieth century (Cooper, 1975; Graves, 1983). Nine elements determined as essential elements for plant growth and developments, and then from 1859 to 1965, *Julius von Sachs* and *Wilhelm Knop* suggested a standard technique of research called solution culture, which nowadays has been applied widespread (Douglas, 1984; Jones, 1982). Except ancient reports of hydroponic activities related to 4000 years ago, the rest of hydroponics activities up to1929 only covered plant biology researches. After 1929 commercial aspects of crop production via nutrition solution began by *William Frederick Gericke* from the University of California at *Berkeley*, and the other hydroponic systems developed subsequently (Hoagland and Arnon, 1950; Hershey 2008).

Obviously due to temperament and various methods of hydroponics or soilless techniques, these methods applied for plant biology researches (Le Bot et al., 1997; Hershey, 1992; Raven et al. 2004; Sonneveld & Voogt, 2009; Hershey 2008). Regarding to hydroponics efficiency, capability of modification and possibility of its development, the use of hydroponic systems has been unavoidable for plant biology researchers.

Soilless is a methodology to use for plant cultivation in nutrient solutions (water containing fertilizers) with or without the use of an organic or inorganic medium (sand, clay-expanded, gravel, vermiculite, rockwool, perlite, peatmoss, coir, coco-peat and sawdust) to provide mechanical support. Water culture system has no organic or inorganic supporting media for plant roots (Jensen, 1999). Additionally soilless methods are classified to open (i.e., once the nutrient solution is delivered to the plant roots, it is not reused) or closed (i.e., surplus solution is recovered, replenished, and recycled) (Raviv and Lieth, 2007). Hydroponic or soilless techniques have been used in many aspects of plant biology researches such as plant nutrition, heavy metals toxicity, identification of elements deficiency, screening for abiotic stresses, screening for aluminum toxicity, root functions, root anatomy and so on(Jones, 1999; Jones, 1982). Choosing a proper soilless method based on research purposes, lead us to achieve credible and informative data.

Based on temperament of soilless culture, many advantages make soilless culture a specified method for plant biology researches. In soilless culture, the researchers are able to manage plant nutrients, control PH and EC in media, monitore the mutual elements interactions, control the micro and macro nutrient concentrations, and investigate the anions and cations absorption (Sonneveld and Voogt, 2009).

Hershey (2008) said, "Growing plants in solution culture is often easier than soil culture because there is no need for dirty soil, there are no soil-borne diseases or pests, irrigation is less frequent in solution culture than in soil culture, solution culture irrigation can be easily automated, roots are visible, and the root zone environment is easily monitored and controlled. Strength of hydroponics is its fascinating history" This chapter will lead researchers and students to identify those aspects of plant biology that competent to run in soilless methods with emphasis on those experiments carried out by soilless methods.

PLANT NUTRITION

Plant nutrition is not only involved with 20 essential elements in a complex process but also some other elements that are either beneficial or harmful to plant metabolism are effective on plant nutrition. The plant nutrition individually influenced by response of the crops due to genome of crop, cultural practices and environmental conditions. In plant nutrition both deficit and excess of elements influence on quantity and quality of yield, so that it seems monitoring of nutrients concentration in crop can be a way of crop management. It is not fulfill that researcher drop all additional factors in rhizosphere and just consider root and nutrient interactions in soil (Harold et al., 2007; Sonneveld and Voogt, 2009). From middle of twenty century, the use of soilless methodology in plant nutrition researches has grown up exponentially. Many researches carried out in hydroponics and the results of those experiments based on the nutrition solution where the plants grown and solutions considered as soil (Sonneveld, and Voogt, 2009). The proper conditions in terms of PH, EC and presence or absence of some elements is monitoring nutrients uptake during the plant nutrition researches (Sonneveld and Voogt, 1990).

Hydroponics provided proper conditions to identify the individual effects of elements on yield quality and quantity, because in hydroponic systems there are fewer interactions relevant to soil properties. In this case, there are many reports in terms of influences of potassium on yield qualitative attributes (Lester et al 2010; da Costa Araujo et al. 2006; Flores et al. 2004; Chapagain and Wiesman 2004; Chapagain et al. 2003).

Results obtained based on hydroponics have shown that existence of sulfate in root zone caused growth and development increase in *Gomphrena globosa L.* due to elevated micro and macro elements absorption. Most of the elements in nutritient solutions formula are in sulfate form, that cause increases the availability of other nutrients and subsequent amelioration in plant growth Figure 6.1 confirms the obtained results (Wang and Zahng 2009).

Assessment of micronutrients and new recommended nutrients in plant tissues and rhizosphere without the use of hydroponics seems to be so difficult. In order to determine of optimal level of silicon (Na2SiO3) for maize growth in nutrient solution under nutrition solution conditions two trials conducted. In the first trial, plants were subjected to two levels of Si; control (0 mM Si) and +Si (3 mM Si). In the second trial, plants were subjected to a wide range of Si concentrations in nutrient solution with seven treatments (0, 0.4, 0.8, 1.2, 1.6, 2.0 and 3.0 mM Si). In the first trial, there was no significant difference between levels of Si accumulation in shoot and root dry matter. In the second trial, plants subjected to 0.8 and1.2 mM Si were superior in terms of fresh and dry matter production, plant height and leaf area. The results showed that 1 mM Si is an optimum concentration for maize growth (Hafiz Faiq et al. 2009).

Tarja et al. (2009) studied influence of silicon on silver birch (*Betula pendula*) seedling grown as a deciduous forest tree in hydroponics. They reported that there was a 6% increase on leaf growth when the plants fed with silicon. The relative advantage in this experiment was that peat soil that used as a hydroponics medium expected has very low silicon levels since the management of silicon and other nutrient were on the hands of researchers.

Absorption of nutrients occurs in a specific range of PH and soil reaction. Determining the nutrients absorption where subjected to limitative materials like lime preferred by hydroponics method. In this case, Sandra and Juan (2009) conducted a trial to study the efficiency of the application of Fe, Mn, Zn and Cu chelates to correct the deficiencies in soybean in nutrition solution in the presence of lime. They reported that the best treatment for all appliances of Mn and Zn was o,pEDDHA. Chemical and commercial forms of fertilizers effect on quality and quantity of crop yields, so proper understanding of suitable forms of nutrients in physiological studies provided clean and adequate yield. In order to examine the influence of various N types on shoot growth, Naoko and Jun, (2010) performed a nutrient solution trial on barley, the results showed when urea was used, shoot biomass production, and particularly tillering declined.

Due to heterogeneity of the materials and organisms in the soils and interaction between them, soilless culture as a simplified experimental method developed and utilized in a broad range of researches (Naoko and Jun 2010). However Naoko and Jun (2010) mentioned that whereas the hydroponic methods contributed to the plant nutrition advancements, but for investigating the interaction between elements and organisms it seems that soil experiments are necessary.

Axton, (2009) through a hydroponic trial has shown that how certain nutrient deficiencies can affect the growth of pumpkin individually. The seedlings of pumpkin grown in four types of water culture (complete nutrition, potassium deficient, nitrogen deficient and iron deficient) solutions compared with controlled nutrient deficiencies. The results showed that there was no significant difference between plants grown in complete nutrition and potassium deficient solutions and the growth rate of seedling in both type of nutrition least, and the nitrogen deficient plant grew slightly more than the iron deficient plant.

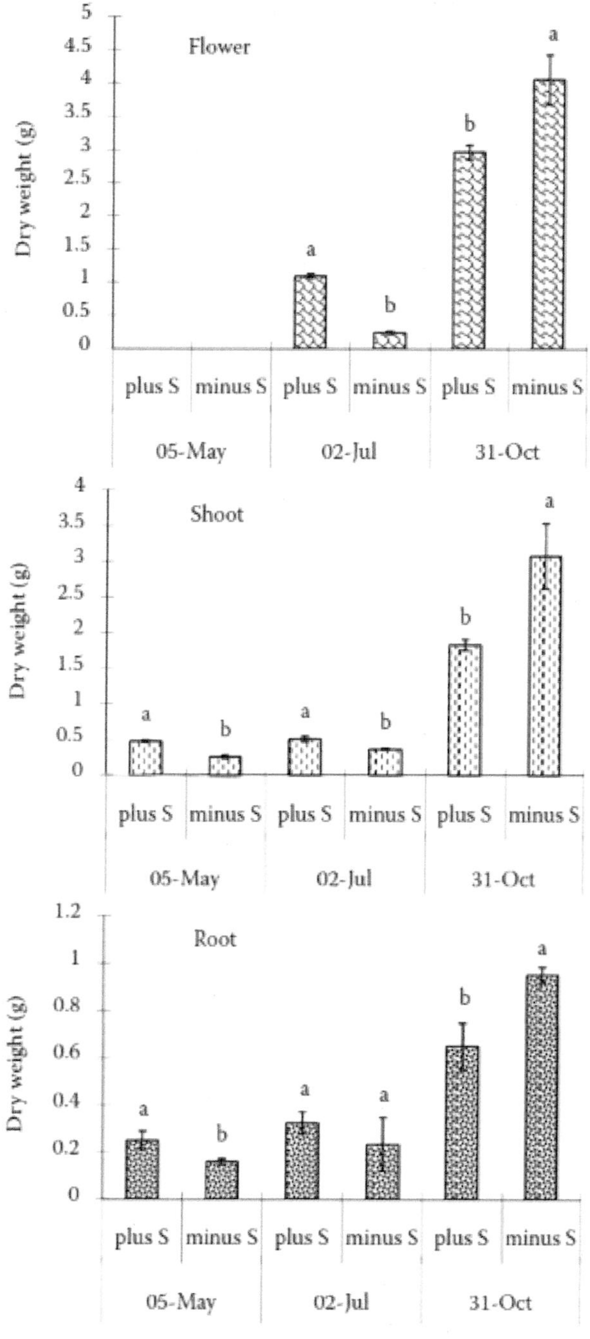

Fig. 6.1. Biomass of globe amaranth with and without root sulfate added (Wang and Zahng 2009)

Beatty et al. (2010) investigated low and high levels of nitrogen uptake and nitrogen utilization at two environments in barley to compare the morphological characteristics and yield of spring barley genotypes grown in the field and in compost or hydroponics growth chambers. They reported that the results obtained from hydroponics system could identify phenotypic and genetic characteristics for improving nitrogen use efficiency in spring barley. Regarding to results observed from hydroponics system they mentioned that there was a significant difference in terms of amino acids among the genotypes of spring barley due to nitrogen utilization.

A hydroponically experiment conducted to find out the effects of various zinc concentrations on sugar beet. The results obtained from controlled hydroponics environment (nutrition solution: 50, 100 and 300 μm) indicated that when the roots of sugar beet were subjected to high concentration of Zn sulphate, the root and shoot fresh and dry masses decreased, and root/shoot ratios increased. Furthermore, plants grown with excess Zn had inward-rolled leaf edges, a damaged and brownish root system, and short lateral roots. The hydroponics system provided the possibility of investigating the interactions between Zn and other elements in terms of absorption. The researchers found that in the high Zn concentration, the concentrations of N, Mg, K and Mn decreased in all plant parts, adversely the concentrations of P and Ca increased. Additionally with increasing of Zn concentration, the chlorophyll and carotenoid concentrations decreased, and carotenoid/chlorophyll and chlorophyll *a*/*b* ratios increased (Sagardoy et al., 2009a).

An experiment conducted in hydroponics system in order to investigate the effects of Zn concentrations on growth and photosynthetic parameters of sugar beet. The obtained results have shown that toxicity of Zn caused a large reduction in photosynthetic rate and subsequently a declination in accumulation of dry matter. The exploration in cause details of net assimilation rate reduction indicated that Zn did not affect leaf photochemistry, but it was due mostly to marked decreases in stomatal conductance (Sagardoy et al., 2009b).

Nowadays has been marked that root discharges affects absorption of nutrients, thus without hydroponics system the identification of these activities is impossible. Hydroponics system provides special condition to monitor root activities, where through an experiment in hydroponics system acid phosphatase produced by roots of white lupin (*Lupinus albus L.*) monitored. Acid phosphatase has a main role in inorganic phosphate (Pi) absorption. The results cleared that APase behavior in the rhizosphere caused release of Pi and provide it to white lupin plants grown under P deficient (Wasaki et al., 2008).

For recognizing the relationship (antagonism and synergism) between nutrients, the hydroponics system provides a desired and suitable condition for more understanding of these issues. In this case Liu et al. (2004) designed a hydroponics trial (nutrient solution) to examine the influence of phosphorus (P) nutrition and iron tablet on root surfaces in arsenate uptake, and its inside translocation in seedlings of three cultivars of rice (*Oryza sativa*). The obtained results have shown that however tablet of arsenate had no significant influence on dry matter of shoots and roots but the rate of arsenate concentrations elevated in tissues, mostly in roots. In absence of P, rice roots appeared reddish after 24 h, showing the creation of iron plaque. The tissues arsenic concentrations in iron tablet were significantly higher in absence of P than when P was included in the nitration solution. This has shown that iron tablet might sequestrate arsenic, and consequently reduce the translocation of arsenic from roots to shoots. Arsenic significantly declined the concentrations of iron (Fe) in roots and shoots and somewhat P concentrations in shoots.

Tracing of nutrients in plants and comparison between plants in terms of nutrient absorption has used since last decade by radiocesium in hydroponics system. Because of interaction between the nutrients in rhizosphere, the use of hydroponics seems necessitous. Tang and Wang (2001) studied the interaction between copper and radiocesium in Indian mustard and sunflower grown in the hydroponic solution. They reported that there was a significant difference between Indian mustard and sunflower in terms of absorption of cooper, so that uptake and translocation of copper by Indian mustard was more than sunflower. In terms of copper immobilization sunflower showed higher copper immobilization significantly.

ALLELOPATHY IN RHIZOSPHERE

Allelopathy is the straight effects of chemicals released from one plant on the growth and development of another plant. The influence has known from many years but newly accepted as a rightful area of research considerably. Researches indicated that allelopathy plays a considerable role in natural ecosystems and has a potential to become an important issue in agro-ecosystems. Interference between plants can be due to allelopathy and competition for nutrients, space and natural resources. It is impossible to distinguish allelopathy in the field, thus determining the interference of allelopathy is so difficult. Explaining the allelopathy needed to identify chemical procedures, biologically active substances and the phytotoxic potential (Olofsdotter 1998).

It seems that identifying all mentioned cases such as chemical materials, activities, interactions and chemical procedures needed to study in some substances except soils, because in the soil researcher cannot distinguish the effects of allelopathic chemical materials and interaction between them. Hydroponics is a proper method to identify quantitative and qualitative determination of allelopathic materials and procedure of interaction between allelopathic materials with other chemical compound.

Tang and Yong (1982) introduced new method of hydroponics to collect the allelopathic chemicals from the undisturbed plant roots. They modified and developed new nutrition solution and sand culture for continuous extraction of quantities of extracellular chemicals from donor plant (*Hemarthria altissima*) as forage grass (Fig. 6.2). The hydroponics system included a sand culture that always nutrition solution circulated continuously through the root system and a column containing XAD4 resin. The resin selectively adsorbed extracellular hydrophobic metabolites, while nutrients circulated to maintain plant growth. Seeds of lettuce used as bioassays for collecting the root exudates. After determining by chromatography, they reported that inhibitors were mainly phenolic compounds. They emphasized that this trapping system should be useful for a wide range of allelopathic studies.

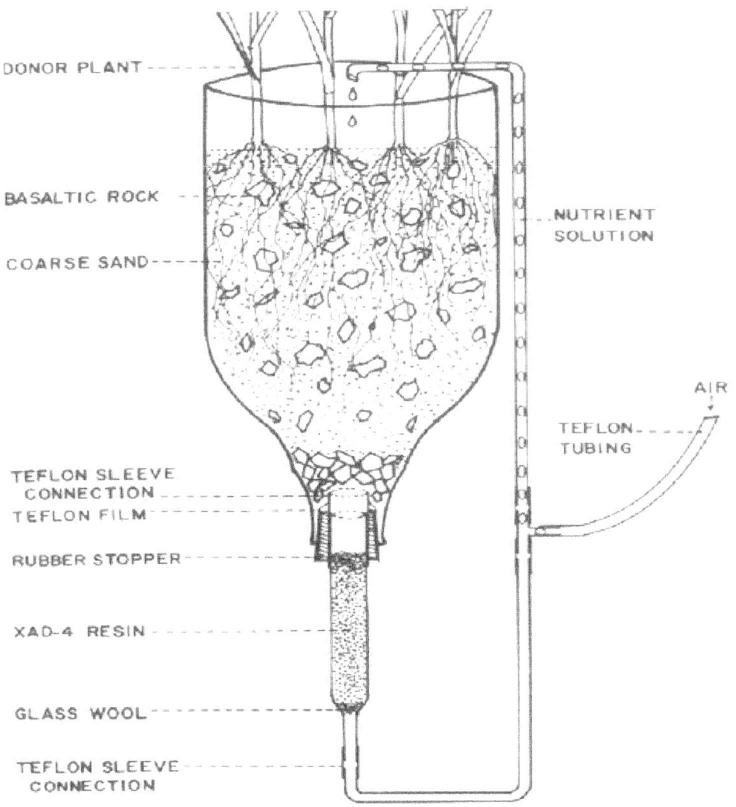

Fig. 6.2. The hydrophobic trapping root exudates in hydroponic system (by Tang and Yong 1982)

Jose and Gillespie (1998) carried out a trial to assess the influence of juglone (5-hydroxy-1, 4- napthoquinone) on the growth and development of corn (*Zea mays L.*) and soybean (*Glycine max L. Merr.*) seedlings in hydroponics system. The treatments were included of three concentrations of juglone (10-6 M, 5-10 M, and 4-10 M). They reported that after the three days the juglone adversely affected on various aspects of growth such as shoot and root relative growth rates, leaf photosynthesis, transpiration, stomatal conductance, and leaf and root respiration. They indicated that soybean was more sensitive than corn relative to use of juglone. Leaf photosynthesis more influenced by toxicity of juglone in compare to other physiological parameters. Overall, the results have shown that both corn and soybean are sensitive to juglone so that growth reductions in corn and soybean in black walnut alley cropping may partly be due to juglone phytotoxicity.

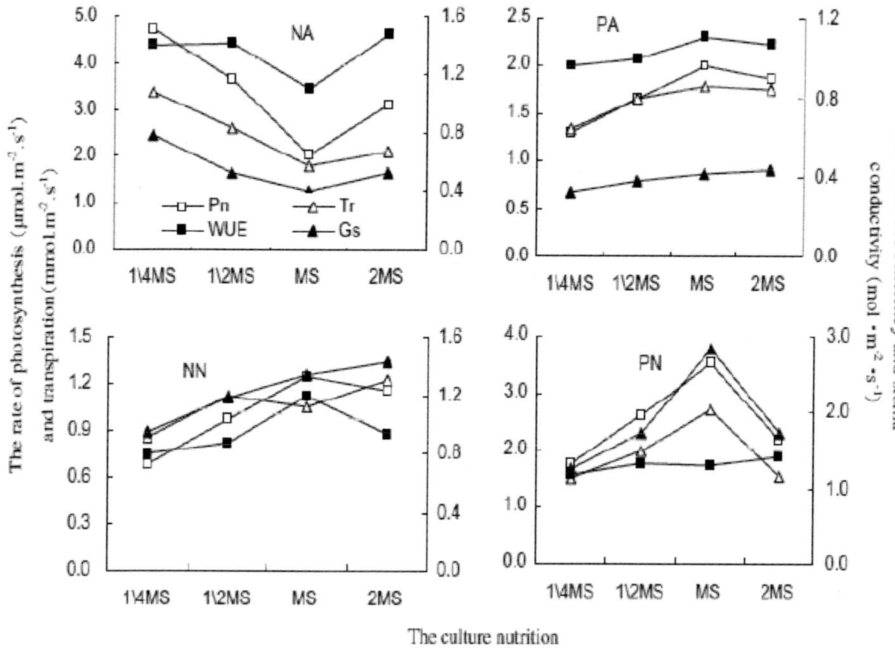

Fig. 6.3. The effect of wheat stubble on photosynthesis of two kinds of transgenic potato. NA: normal seedling under 0.01 mg/ml wheat straw; PA: acclimated seedling under 0.01 mg/ml wheat straw; NN: normal seedling without wheat straw; PN: acclimated seedling without wheat straw (Zeho et al., 2010)

In order to study the allelopathic effects of wheat straw on transgenic potato seedlings an experiment conducted by Zuo et al., (2010). Potato seedlings were cultured on normal medium (normal treatment) and nutrient-deficient medium (acclimated treatment) and then transferred to MS medium, which contained wheat straw powder, where wheat straw powder inhibited potato seedling growth in both treatments. They have shown that the greatest inhibition was for plant fresh weight and least for plant height. The inhibitive effects of wheat straw were greater for seedling roots compared to shoots. Growth inhibition of potato seedlings from the normal treatment increased as the amount of wheat straw powder in the culture medium increased. Calculations indicated that the presence of wheat straw would lead to a 55% reduction in the total biomass of normal potato seedlings compared to a 39% reduction for acclimated seedlings. Parameters such as net photosynthesis rate (Pn) changed as the nutrient content of the culture medium increased or decreased, but the changes in the parameters were smaller for acclimated seedlings compared to normal seedlings. In summary, the results show that previous exposure to pressures such as nutrient deficiency may increase the allelopathic pressure resistance of transgenic potato seedlings (Fig 6.3).

Belz and Hurle (2004) in order to identify the allelopathy materials and screening the grain crops developed a fast and consistent bioassay method. The bioassay carried out in hydroponics culture, and a range of experiments with 2-(3H)-benzoxazolinone, an allelochemical of several grain crops, was carried out to define the basic protocol. The -(3H)- benzoxazolinone influenced on growth, enzyme activities, ascorbate peroxidase, catalase, glutathione S-transferase, peroxidase, phenylalanine ammonia-lyase, and chlorophyll fluorescence, whereby root length was the most reliable response parameter of *Sinapis alba L.* and it was selected as the recipient species. They found that *Secale cereale* L. and *Triticum aestivum* L. as donor species caused optimize the protocol.

Abiotic Stresses

For investigation the abiotic stresses, we need to apply modified conditions that are similar to arid conditions. The hydroponics system can provide modified conditions those capable researchers to elucidate the mechanisms involved in the response of plants to abiotic stresses. Actually, hydroponics can help to understand abiotic stresses tolerant mechanisms; also defines conditions to identify abiotic stress-inducible genes in the tolerant plants (Mohsenzadeh et al. 2009). The abiotic stresses associated with osmotic issues, thus the use of hydroponics system for studying the abiotic stresses such as salinity and drought stresses is unavoidable. In study on abiotic stresses, the adjustment of osmotic pressure in soil due to few factors such as interference of soil chemical and physical properties is difficult or somehow impossible, so that methodology of hydroponics system is unique in these cases (Szira et al. 2008).

For screening the tolerant plants in face to abiotic stresses the hydroponic methods provide proper conditions to achieve right data relevant to physiological and biochemical responses such as proline concentration, K/Na ratio, chlorophyll contents, and photosynthesis rate (Khan and Shirazi, 2009).

Using of nutrient solution to study on abiotic stresses is suitable for marsh plants or for understanding the effects of salt on plant growth, but a hydroponics solution is not representative of the natural habitat of most agricultural plants in seedling and growth stages. Studying on abiotic stresses in most of plants by using inorganic media culture such as sand, perlite and vermiculate culture can be appropriate (peel et al. 2004). In particularly, salinity stress studies involved with various salts that every salt has special effects on plants physiology, so that in order to distinguish individual influences the use of hydroponics system is necessary.

From last decades the researchers were looking to find a proper method for cultivating plants, imposed to a controlled water deficit. Testing genotypes for water deficit in the soil seems so hard or sometimes impossible. Maintainig a uniform pattern, and constant water potential through the whole soil profile is difficult (Munns et al. 2010). Enhancing of drain quickly, perlite or vermiculite as inorganic substrates can play perfect role in water deficit testing in compare with soil (Passioura, 2006).

In a water deficit test in order to reduce water potential, mannitol used as non-ionic osmotic. High-molecular-weight polyethylene glycol (PEG, MW 6000) used to control water potential of imposing plants to water deficit (Munns et al. 2010).

Using of dry soil to simulate a drought or soil water deficit causes reduced access to nutrients by plants, thus plants may suffer from N or P deficiency while in hydroponics system this problem has been solved (Munns et al. 2010).

Hydroponics system is the most appropriate and profitable method for investigation of abiotic stresses on plants. Bouslama (1984) evaluted 20 soybean genotypes for drought tolerance by three screening methods. The methods were included: seed germinating in polyethylene glycol-600 (PEG) at -0.6 MPa osmotic pressure, subjecting seedlings to PEG- 600 at -0.6 MPa osmotic pressure in hydroponic solution for 14 days, and a heat tolerance test based on the cellular membrane thermo-stability. Based on obtained results they reported that there is a high significant correlation between two tested methods including the hydroponic seedlings and heat tolerance tests where the tolerant cultivars in both methods were same. They demonstrated that evaluation of plants for abiotic stresses in germination stage could not be reliable to indentify abiotic stresses. Finally, they indicated that among the three tested methods the hydroponics as a consistent methodology is proper and reliable to assess the genotypes and selecting the tolerant genotypes.

Baranova et al. (2007) investigated utilization of storage starch in the cells of cotyledon mesophyll and root meristem in the course of alfalfa seedling on the nutrient solutions with different concentrations of NaCl, Na2SO4, and mannitol. They showed that salt stress inhibit early development of seedlings because of reduction in the mobilization of reserves. In order to analyze the dynamics of starch mobilization in the seedlings of alfalfa at osmotic stress, they evaluated the effect of two salts and an osmoticum on the cells of cotyledon mesophyll and root meristem. They assumed that injurious effect of salts depends on the osmotic component of their action.

Peng et al. (2008) in their experiment on eco-physiological response of alfalfa seedling to salts (($NaCl$, $Na2SO4$, $NaHCO3$ and $Na2CO3$), and 30 salt-alkaline combinations demonstrated that salt and salt-alkali decreased total biomass and biomass components of seedling definitely and furthermore the interaction between salinity and alkalinity cause changes the root activities due to salinity gradient. They showed that the physiological responses such as leaf electrolyte, leakage rate and proline content varied with different levels of salinity and alkalinity.

Al-Khateeb (2006) studied growth and ions concentration of alfalfa seedlings in response to different concentrations of Na^+ and Ca^{2+}. They indicated that with increasing Na^+, total dry matter including root, stem, and leaves declined, but increasing Ca^{2+} alleviated the detrimental effects of Na^+.

Assessment of barley germplasms in face to drought stress carried out by using 47 Tibet annual wild barley genotypes in hydroponics system. The observations included chlorophyll content, plant height, and biomass of shoot/root were significantly reduced (Zhao 2010).

Lafitte et al. (2007) in a study evaluted genotypes of rice in face of drought stress in different stages by various methods such as soil and hydroponic culture. Effects of drought stress on rice genotypes in hydroponic method was more clearly visable where in hydroponics system the level of abcisic acid as index of drought stress was high in plants in compare with plants cultivated by other methods.

PLANT ROOTS

The root is located underground in most plants which anchor the plant, store nutritious products, absorb water and nutrient elements from soil and conduct to other parts of the plant. Roots make and secrets lots of exudates and help to produce organic matters in the soil. Roots also alter the chemical and physical properties of the soil. Hydroponics can make easier the study of the root anatomy, morphology and root/shoot ratio in plants under controlled conditions. Hydroponics can be used to study root and liquid interactions, symptoms of nutrient deficiencies in different plants without interference of limiting factors in the soil, the enzymatic activities of the roots, the role of root exudates in availability of essential nutrients for plants and microorganisms activities in the rhizosphere and the effects of toxic elements on roots, efficiently. Investigating the architecture of root system is necessary because crop productivity is influenced by the availability of water and nutrients that are distributed in the soil heterogeneously (Cichy et al 2009;Gregory etal.2009).Changes in soil water exploitation by plants due to changes in root system architecture influence the accumulation of crop biomass(hammer et al.;2009). Several non soil-filled techniques are available for measuring and visualizing the root architecture and finding its relationship to plant productivity (Gregory et al. 2009).Hydroponics is an ideal method for observing the hidden parts of the plant and study root growth and development over time in different conditions. This article refers to some studies that reflect the different aspects of hydroponic applications in studying the plant roots.

Chen et al. (2011) evaluated the reliability and efficiency of a semi-hydroponic bin system in screening root traits of 20 wild genotypes of lupin (*Lupinus angustifolius* L.) and found significant differences in both architectural and morphological traits among genotypes. They concluded that this system is a desirable tool for evaluating root architecture of deep root systems and a large number of plants in a small space. Hermind Reinoso et al.(2005) studied the effects of increasing concentrations of Na2So4(0-531mmol L-1) on anatomical changes in the roots of Prosopis strombulifera seedlings were grown hydroponically in Hoagland's solution with addition of 19mmol L-1 Na2So4 every 48h until final salt concentration of 531 mmol L-1 were reached. The anatomical changes were noticeable and different with other treatments at 531 mmol L-1including more cortical layers in primary root and earlier development of endodermis in younger root zones and decreased hypocotyls cambial activity in older root zones. Anderson et al.(1991) compared the effects of NH4 nutrition with NO_3 nutrition on root morphology of corn(B73×LH51) that was cultured in complete nutrient solutions containing NO_3 or NH4-N. They demonstrated thicker roots and better growth of plants in NH4 nutrition than NO_3 nutrition. Swenen et al.(1986) reported large differences in root development of different genotypes of juvenile Musa SPP plants grown under hydroponic conditions. Bacilio-Jimenez et al. (2003) collected the root exudates from rice plantlets cultured under hydroponic conditions. Then, they compared the root exudates effects on the chemotaxis of two strains of endophytic bacteria collected from the rice rhizosphere with chemotactic effects of two strains, one collected from the corn rhizosphere and another one isolated from the rice hizosphere. Their results showed that rice exudates caused a higher hemotactic response for endophytic bacteria than for other strains of bacteria presented in the rice rhizosphere. McQuattie et al.(1990) studied the effects of excess aluminum on changes in root anatomy of one year red spruce(Picearubens Sarg.) cultured in a nutrient solution containing 0,50,100 or 200 mg/L aluminum. They reported well developed aluminum toxicity symptoms in the roots including decreased root length but increased root diameter and the number of cell layers in the root cap. The root tips examination by light and electron microscope showed a great cellular changes in roots, premature vacuolation, phenolic material accumulation, intercellular spaces formation, cellular membrane disruption and cytoplasm injury.Ergul Cetin(2009) studied the effects of Boron (B) excess and deficiency on anatomical structure of alfalfa (*Medicago Sativa* L.) in –B,control,+B hydroponic cultures. The results showed that B deficiency (-B) effected the root cortex parenchyma.In +B treatment there was a significant decrease in the diameter of roots. The degree of xeromorphy calculation showed that *M. Sativa* L. developed xeromophic structure after +B treatment.

HEAVY METALS

The soils naturally contaminated by heavy metals are found in areas where metal ores come to the surface and deteriorate due to weathering. These kinds of soils are found in some regions around the world (Malaisse et al., 1999; Proctor, 2003;) Rajakaruna et al., 2009). In regions with originally low heavy metal, human activities such as ore- mining and ore – refining industry nearby, are the main reasons of soil and water contamination (Barcan, 2002; Mandal et al., 2002). But In many areas the main reason of heavy metal contamination of soil and ground water is applying metals in agriculture. Examples are the application of copper and zinc –based pesticides, copper in vineyard, mineral fertilizers and sewage sludge in crop production (Komarek et al., 2010 ; He et al., 2005).

Molecular mass>5.0 g cm-3 of heavy metals is distinctly higher than the average particle density of soils(2.65 g cm-3).Certain heavy metals such as Manganese(Mn), Zinc(Zn), Iron(Fe), Copper(Cu), Cobalt(Co), or Molybdenum(Mo) are necessary for the growth of organisms. Others have a single role in some organisms such as Vanadium(V) in some peroxidases and in V – nitrogenases and Nickel (Ni) in hydrogenases. The rest of the heavy metals such as Cadmium(Cd), Lead(Pb), Uranium(U), Tallium(Tl), Chromium(Cr), Silver(Ag),and Mercury(Hg)are always toxic to organisms. Arsenic (As) and Selenium (Se) are not heavy metals but they are often called "metalloids". With the exception of Iron, all heavy metals with the concentration above 0.1% in the soil are toxic to plants.

Plants depend on soil for up taking water and nutrients. Thus, the soils contain excess levels of heavy metals decrease crop productivity under certain conditions. The heavy metals pollution of soil, ground water and surface water is a serious threat to the environment and human health. Therefore, heavy metal toxicity, screening and selecting heavy metals tolerant and remediative plants and mechanisms involved in their detoxification have been subjects of many researches in the entire world.

Many researchers reported the advantages and limitations of hydroponics in heavy metal studies. Meharg(2005) reported that uptake the heavy metals by plants and any subsequent phytotoxic reaction should be analyzed under controlled conditions and suggested hydroponics as a suitable way to screen and identify the heavy metal tolerant and remediative plants before field studies. For example, Kupper et al.(2000) Found that Arabidopsis halleri (Brassicaceae) and Sedum alfredii (Crassulaceae) can accumulate Cd under hydroponics condition. Hongli and Wei (2009) investigated 23 Amaranth cultivars from different regions in hydroponics' culture and observed that Amtaranthus cv. Tianxigmi is a Cd hyperaccumulator. Duskenhov et al. (1995) demonstrated that Brassica juncea can accumulate a remarkable amounts of Cd in its tissues under hydroponic conditions. Varad and Freitas(2003) reported that Sun flower can remove cesium, strontium, uranium and lead from hydroponic solution. Yang et al. (2002) reported that Elsholtzia splendens can be used as a phytoremediation for Cu-contaminated soils. Moreno-Jimenez etal.(2009) investigated Hg and As tolerance and bioaccumulation in Pistacia lentiscus and Tamarix gallica plants under hydroponic conditions and showed that T. gallica accumulated more Hg and As and had higher tolerance to both Hg and As than Pistacia lentiscus.

Whilst hydroponics is useful for screening and selecting heavy metals tolerant and remediative plants but there is evidence that plants react in a different way when they are grown under hydroponic conditions or soils (Zabtudowska et al.;2009). Therefore extrapolating results from hydroponics to field conditions are too optimistic and unrealistic without considering these limitations (Dickinson et al.:2009).For example, Moreno-Jimenez et al. (2010) compared phytoavailability of Arsenic(As) in mine soils and hydroponics and reported that As had a low solubility in mine soils and the extrapolation of data obtained in hydroponics to mine soils were unrealistic and proposed that hydroponics data only in As levels in the range 0-10μM that plants are exposed in soils can be extrapolated to soil conditions confidently. Qing Cao et al.(2007) investigated toxicity of Arsenic(As) and Cadmium(Cd) individually and in combination on root elongation of wheat seedlings (Triticum aestivum.L.) in hydroponics and in soils. In their hydroponic and soil experiments As and Cd showed synergetic and antagonistic effects on wheat root elongation, respectively. They concluded that phytoability of Cd is increased by forming uptake- facilating Chloride complexes in a soil with high level of Chloride, also, soil characteristics (PH, organic matter, absorption capacity and moisture) affect the availability and toxicity of Cd in soil matrix. On the contrary, hydroponics can be used to evaluate suitable plants extract contaminants from groundwater, surface water, wastewater, and retain within their roots by the way of rhizofiltration, confidently. For example, Jing et al.(2007) suggested rhizofiltration a suitable way for extracting Pb, Cd, Cu, Ni, Zn, and Cr, from contaminant aqueous sources. Dushenkov et al. (1995) found out that hydroponically grown plants such as Indian mustard (B. juncea L.) and Sunflower (H. annuus L.) can extract the toxic metals like, Pb, Cu, Cr, Cd, Zn, and Ni, from aqueous solutions efficiently. Karkhanis et al.(2005) in an experiment in green house, used pistia, duckweed, and water hyacinth(Eichornia rassipes) to remove heavy metals from aquatic environment polluted by coal ash containing heavy metals.

Coiso et al(2006) and hernandez-Allica etal(2007) suggested hydroponics the only way for eliminating mass transfer limitations and elucidating free metal ion and uptake and translocation metal-chelate within the plant. For example, January et al.(2007) conducted experiments to find out the effectiveness of Sunflower (Helianthus annus) for the phytoremediation of mixed heavy metals including : Cd, Cr, Ni,As and Fe. The experiments were conducted under hydroponic conditions in order to eliminate mass transfer problems that would be present in soil experiments. The results showed that heavy metal uptake preference was Cd = Cr > Ni,Cr> Cd >Ni >As.

Hydroponics can be used to identify and characterize the mechanisms of tolerance to excess heavy metals in plants. For example: Wei et al.(2008) compared a more tolerant plant species (sorghum sudanense) with a less tolerant plant species (Chrysanthemum coronarium) were grown in hydroponic culture in the presence of Cu excess(up to 50 mM) to find out the possible role of their cell walls in relation to tolerance. The authors observed that, total Cu concentrations in C. coronarium shoots and cell walls were much higher than in S. sudanense. However, the percentage of cell wall bound Cu in C. coronarium was much lower, but the percentage of water soluble Cu was higher than S. sudanense. In contrast, the amount of Cu in the roots and the percentage of cell wall bound Cu in S. sudanense were substantially higher than C. coronarium. As a result, the amount of Cu load on protoplast of S. sudanense was much lower than in C. coronarium.

Hydroponics can be used to study interesting properties from a physiological point of view efficiently. For example, in a study conducted in hydroponic cultures Shi (2008) investigatedthe role of root exudates of Elsholtzia splendens on activating Cu in soils and reported that the amount of root exudates in Cu dissolution from contaminated soil was higher than the amount dissolved in deionized water significantly. Thus, concluded that root exudates of Elsholtzia splendens can activate Cu in the soil strongly. Lou et al. (2004) studied the seed germination of Elsholtzia splendens in Hydroponic cultures containing excessive Cu and found out that lower than 250 mmol L_1 of Cu in culture solution stimulate seed germination, but at higher concentrations (750 mmol L_1) of Cu, seed germination decreased.

TOXICITY OF ALUMINUM

Metals are found naturally in soils that may be useful or toxic for plants or environment. Many researchers reported metal toxicity in plants (Brown etal., 1975; Chaney etal.,1977; Foy et al.,1978). Aluminum (Al) is not an essential element for plant growth, but low concentrations can increase plant growth or cause other desirable effects (Foy, 1983; Foy etal.,1978). Aluminum toxicity is the most common limiting factor for plants grown in acid soils. Excess Al interrupts cell division in roots, increases the cell wall rigidity, fixes phosphorous in soils, reduces root respiration, interferes with uptake, transport and use of several essential elements such as Ca, Mg, K, Fe and P (Foy,1992).

Improving aluminum tolerant plants and identification the mechanisms involved in tolerance of plants are the most suitable ways to decrease destructive effects of aluminum toxicity. Plants greatly differ in tolerance to Al, available literature data showed a great differences in the experimental conditions that determine the Al toxicity and plant reactions as follows: The concentration of Al and Al form in the medium, PH of substrate and PH stability, exposure time, ionic strength of the culture medium, soil type, availability of essential nutrients, plant age and plant variety (Poschenrider et al.; 2008). Since tolerance is genetically controlled, screening and selecting of Al tolerant genotypes under hydroponics condition can lead to improve tolerant cultivars to excess aluminum because hydroponics provides easy access to root system, exact control over available nutrients and PH and non - destructive measurements of tolerance(Carver and Ownby,1995). Additionally, studies have shown that the primary response to Al stress occurs in the roots, therefore root growth measurements under Al stress can be used in several Al tolerance screening techniques (Kim et al., 2001). Hydroponics is a very suitable technique for measuring the root growth under Al stress conditions for example Makau et al (2011) ydroponically investigated maze (Zea mays), garden pea (Pisum sativum),bean (Phaseolus vulgaris L.) and cucumber (Cucumis sativus) tolerance to excess Al in the rhizosphere during the early plant growth stages and introduced the root elongation as a single most significant index in ranking the genotypes for Al tolerance. Hydroponics has been used to evaluate a large number of soybean genotypes for aluminum tolerance quickly (Silvia et al., 2001). Foy et al. (1965) compared an aluminum tolerant cultivar with an aluminum sensitive cultivar of Triticum aestivum in nutrient solutions with or without aluminum and found out that aluminum tolerant cultivar was able to increase PH in nutrient solutions.wood etal. (1984) reported that the concentration of salts or ionic strength in nutrient medium related to the critical level for tolerance to aluminum. Speher (1994) with studies under hydroponics condition selected tolerant soybean genotypes to excess aluminum. Ma etal. (1997) studied 600 barley lines with a rapid hydroponic screening way and reported that ninety lines had medium tolerance to excess aluminum.Krizec etal.(1997) conducted an experiment for evaluating twocultivars of Coleus blumei in nutrient media containing 0 to 24 mg/l aluminum and on an acid Al-toxic. Sivaguru and Paliwal(1993) tested twenty two rice cultivars for studying their tolerance in nutrient solution containing aluminum in toxicity level at PH 4.1.They reported six tolerant cultivars on the basis of root and shoot tolerance indices. They also demonstrate the mechanism of tolerance to toxic levels of aluminum on the basis of mineral uptake and mineral utilization. The tolerant plants in the presence of aluminum took up and utilized Ca and P efficiently, but the aluminum sensitive cultivars took up and utilized less Ca and P (Sivaguru etal.,1993).

AT A GLANCE

- Many advantages make hydroponics a specific technique for plant biology researches.In plant nutrition researches hydroponics allows researchers manage nutrients and control PH, EC, macro and microelements concentrations in nutrients media, monitor mutual-element interactions and investigate the anions and cations absorption. Hydroponics provides proper conditions to determine the individual effects of nutrient elements on quality and quantity of crop yields. Absorption of plant nutrients occurs in a specific range of PH and soil reaction, thus determining the nutrients absorption where subjected to limitative materials like lime preferred by hydroponics method.however, for investigating the interaction between nutrient elements and organisms it seems that soil experiments are necessary.
- In allelopathy researches, interference between plants can be due to allelopathy, competition for nutrients, space and natural resources. Therefore, separation the individual effects of allelopathy from the other interferences is easier in hydroponics. Research in terms of influence of water stress on plants in the soil involve by deficit of nutrients while in hydroponics method not only the effect is due to water deficit but also it is possible to adjust water potential.
- Some of researchers believe that study in terms of salt stress except hydroponics system, will be associate with many problems in the results because not only regulation of osmotic potential in rhizosphere is complicate but also the separation the effects of various salt is impossible. In plant roots studies hydroponics can make easier the study of the root anatomy, morphology and root/shoot ratio in plants under controlled conditions.
- Hydroponics can be used to study root and liquid interactions, symptoms of nutrient deficiencies in different plants without interference of limiting factors in the soil, the enzymatic activities of the roots, the role of root exudates in availability of essential nutrients for plants and microorganisms activities in the rhizosphere, the effects of toxic elements on roots and root architecture efficiently. Many researchers reported

the advantages and limitations of hydroponics in heavy metal studies. Meharg(2005) reported that uptake the heavy metals by plants and any subsequent phytotoxic reaction should be analyzed under controlled conditions and suggested hydroponics as a suitable way to screen and identify the heavy metal tolerant and remediative plants before field studies. Whilst hydroponics is useful for screening and selecting heavy metals tolerant and remediative plants but there is evidence that plants react in a different way when they are grown under hydroponic conditions or soils (Zabtudowska et al.;2009). Therefore extrapolating results from hydroponics to field conditions are too optimistic and unrealistic without considering these limitations (Dickinson et al.:2009).

- Improving aluminum tolerant plants and identification the mechanisms involved in tolerance of plants are the most suitable ways to decrease destructive effects of aluminum toxicity. Since tolerance is genetically controlled, screening and selecting of Al tolerant genotypes under hydroponicscondition can lead to improve tolerant cultivars to excess aluminum because hydroponics provides easy access to root system, exact control over available nutrients and PH and non - destructive measurements of tolerance(Carver and Ownby,1995). Additionally, studies have shown that the primary response to Al stress occurs in the roots, therefore root growth measurements under Al stress can be used in several Al tolerance screening techniques(kim et al.,2001). Hydroponics is a very suitable technique for measuring the root growth under Al stress conditions.

REFERENCES

Al-Khateeb, S. (2006). Effect of calcium/sodium ratio on growth and ion relations of Alfalfa (*Medicago sativa L.*) seedling grown under saline condition. Journal of Agronomy, 5(2), 175-181.

Anderson D. S. Teyker R. H. RayBurn A. L. (1991). Nitrogen form effects on early corn root morphological and anatomical development; Journal of Plant Nutrition. Vol.14. PP: 1255-1266

Axton, J. (2009). The Effect of Nutrient Deficiencies on the Growth of Plants. *http://www.k12science.missouristate.edu/Junior_Academy/MJAS%20Docs/State%202009/Papers%202009/MS_CHE/Axton_Joelle_MS.pdf*

Baranova, E., Gulevich, A., & Polyakov, V. Y. (2007). Effect of NaCl, Na 2 SO 4, and mannitol on utilization of storage starch and formation of plastids in the cotyledons and roots of alfalfa seedlings. Russian Journal of Plant Physiology, 54(1), 50-57.

Bacilio-Jimanez, M., Aguilar-Flores, S., Ventura-Zapata, E., Pí©rez-Campos, E., Bouquelet, S., & Zenteno, E. (2003). Chemical characterization of root exudates from rice (Oryza sativa) and their effects on the chemotactic response of endophytic bacteria. Plant and Soil, 249(2), 271-277.

Barcan, V. (2002). Nature and origin of multicomponent aerial emissions of the copper- nickel smelter complex. *Environment international, 28*(6), 451-456.

Bar-Yosef, B., Mattson, N., & Lieth, H. (2009). Effects of NH< sub> 4</sub>: NO< sub> 3</sub>: urea ratio on cut roses yield, leaf nutrients content and proton efflux by roots in closed hydroponic system. Scientia Horticulturae, 122(4), 610-619.

Beatty, P. H., Anbessa, Y., Juskiw, P., Carroll, R. T., Wang, J., & Good, A. G. (2010). Nitrogen use efficiencies of spring barley grown under varying nitrogen conditions in the field and growth chamber. Annals of Botany, 105(7), 1171.

Belz, R. G., & Hurle, K. (2004). A novel laboratory screening bioassay for crop seedling allelopathy. Journal of chemical ecology, 30(1), 175-198.

Bernhard B., Fritz B., Nicolaus V. (2009). Influence of nitrogen form on tillering, cytokinin translocation yield in cereal crop plants. *http://www.escholarship.org/uc/item/49s757g4*

Bouslama, M. S. (1984). Stress tolerance in soybeans. I. Evaluation of three screening techniques for heat and drought tolerance1. Crop Science, 24(5), 933.

Brown, J., & Jones, W. (1975). Heavy-metal toxicity in plants. 1. A crisis in embryo. *Commun. Soil Sci. Plant Anal.;(United States), 6*(4).

Carver, B. F., & Ownby, J. D. (1995). Acid soil tolerance in wheat. advances in Agronomy, 54, 117-173.

Cetin, E. (2010). Effects of Boron Stress on the Anatomical Structure of Medicago sativa L. IUFS Journal of Biology, 68(1), 27-35.

Chaney, R. L. Giordano, P. M. (1977). Macroelements as related to plant deficiencies and toxicities, in Elliot L.F., Stevenson F.J. (Eds.), Soils for Management and Utilisation of Organic Wastes and Waste Waters. Madison, Am. Soc. Agron. pp. 233–280.

Chapagain BP, Wiesman Z, (2004). Effect of potassium magnesium chloride in the fertigation solution as partial source of potassium on growth, yield and quality of greenhouse tomato. Scientia horticulturae, 99(3-4):279-288.

Chapagain B, Wiesman Z, Zaccai M, Imas P, Magen H. (2003). Potassium chloride enhances fruit appearance and improves quality of fertigated greenhouse tomato as compared to potassium nitrate. Journal of plant nutrition, 26(3):643-658.

Chen, Y. L., Dunbabin, V. M., Diggle, A. J., Siddique, K. H. M., & Rengel, Z. (2011). Development of a novel semi-hydroponic phenotyping system for studying root architecture. Functional Plant Biology, 38(5), 355-363.

Cichy, K. A., Caldas, G. V., Snapp, S. S., & Blair, M. W. (2009). QTL Analysis of Seed Iron, Zinc, and Phosphorus Levels in an Andean Bean Population.

Cooper, A. (1975). Crop production in recirculating nutrient solution. Scientia Horticulturae, 3(3), 251-258.

Cosio, C., Vollenweider, P., & Keller, C. (2006). Localization and effects of cadmium in leaves of a cadmium-tolerant willow (Salix viminalis L.):: I. Macrolocalization and phytotoxic effects of cadmium. Environmental and Experimental Botany, 58(1-3), 64-74

da Costa Araujo R, Bruckner CH, Prieto Martinez HE, Chamhum Salom¡£o LC, Alvarez VH, de Souza AP, Pereira WE, Hizumi S. (2006) Quality of yellow passionfruit (Passiflora edulis Sims f. flavicarpa Deg.) as affected by potassium nutrition. Fruits, 61(02):109-115.

Dickinson, N. M., Baker, A. J. M., Doronila, A., Laidlaw, S., & Reeves, R. D. (2009). Phytoremediation of inorganics: realism and synergies. International Journal of Phytoremediation, 11(2), 97-114.

Dushenkov, V., Kumar, P. B. A. N., Motto, H., & Raskin, I. (1995). Rhizofiltration: the use of plants to remove heavy metals from aqueous streams. *Environmental science & technology, 29*(5), 1239-1245.

Ergul C. (2009). Effects of Boron Stress on the Anatomical Structure of Medicago sativa L. IUFS Journal of Biology; 68(1): 27-35.

Fan, H. L., & Zhou, W. (2009). Screening of Amaranth Cultivars (Amaranthus mangostanus L.) for Cadmium Hyperaccumulation. *Agricultural Sciences in China, 8*(3), 342-351.

Flores P, Navarro JM, Garrido C, Rubio JS, Mart¡nez V, (2004). Influence of Ca2+, K+ and NO_3 â^' fertilisation on nutritional quality of pepper. Journal of the Science of Food and Agriculture, 84(6):569-574.

Foy, C. D. (1983). The physiology of plant adaptation to metal stress, Iowa State J. Res. 57: 355–391.

Foy C. D. (1992). Soil chemical factors limiting plant root growth, in: Hatfield J.L., Stewart B.A. (Eds.),Advances in Soil Sciences: Limitations to Plant Root Growth, Vol. 19, Springer Verlag, New York, pp. 97–149.

Foy C. D. Burns G. R. Brown J. C. Fleming A. L. (1965) Differential aluminum tolerance of two wheat varieties associated with plant-induced PH changes around their roots, Soil Sci. Soc. Am. Proc. 29: 64–67.

Foy, C., Chaney, R., & White, M. (1978). The physiology of metal toxicity in plants. *Annual Review of Plant Physiology, 29*(1), 511-566.

Graves, C. J. (1983). The nutrient film technique. Horticultural Reviews, 1-44.

Gregory, P. J., Bengough, A. G., Grinev, D., Schmidt, S., Thomas, W. B. T. B., Wojciechowski, T., et al. (2009). Root phenomics of crops: opportunities and challenges. Functional Plant Biology, 36(11), 922-929.

Hafiz Faiq, B., Stefan, H., Sven, S. (2009). Optimal level of silicon for maize (Zea mays L. c.v. AMADEO) growth in nutrient solution under controlled conditions. *http://escholarship.org/uc/item/7n94x8pv*

Hammer, G. L., Dong, Z., McLean, G., Doherty, A., Messina, C., Schussler, J., et al. (2009). Can changes in canopy and/or root system architecture explain historical maize yield trends in the US Corn Belt. Crop Sci, 49, 299-312.

Harold C. Passam • Ioannis C. Karapanos • Penelope J. Bebeli2 • Dimitrios Savvas (2007). A review of recent research on tomato nutrition, breeding and post-harvest technology with reference to fruit quality. The European Journal of Plant Science and Biotechnology, 1(1): 1-21

He, Z. L., Yang, X. E., & Stoffella, P. J. (2005). Trace elements in agroecosystems and impacts on the environment. *Journal of Trace Elements in Medicine and Biology, 19*(2-3), 125- 140.

Hernandez-Allica, J., Garbisu, C., Barrutia, O., & Becerril, J. M. (2007). EDTA-induced heavy metal accumulation and phytotoxicity in cardoon plants. Environmental and Experimental Botany, 60(1), 26-32.

Hershey D. R. (2008) Solution Culture Hydroponics: History & Inexpensive Equipment. *URL: http://www.jstor.org/stable/4449764*

Hershey D.R. (1992) Plant nutrient solution pH changes. Journal of Biological Education 26:107-111.

Hoagland, D. R., & ARNON, D. I. (1950). The water-culture method for growing plants without soil. Circular. California Agricultural Experiment Station, 347(2nd edit).

Jadia, C. D., & Fulekar, M. (2010). Phytoremediation of heavy metals: Recent techniques. *African journal of biotechnology, 8*(6)

January, M. C., Cutright, T. J., Keulen, H. V., & Wei, R. (2008). Hydroponic phytoremediation of Cd, Cr, Ni, As, and Fe: Can Helianthus annuus hyperaccumulate multiple heavy metals? Chemosphere, 70(3), 531-537.

Jensen, M. H. (1999). Hydroponics worldwide. Acta Horticulturae, 2, 719-730.

Jing, Y., He, Z., & Yang, X. (2007). Role of soil rhizobacteria in phytoremediation of heavy metal contaminated soils. Journal of Zhejiang University-Science B, 8(3), 192-207.

Jones Jr, J. B.. (1999). Advantages gained by controlling root growth in a newly-developed hydroponic growing system. Proc. Int. Sym. Growing Media and Hydroponics, 22, 1.

Jones J.B. (1982) Hydroponics: its history and use in plant nutrition studies. Journal of plant Nutrition 5:1003-1030.

Jose, S., & Gillespie, A. R. (1998). Allelopathy in black walnut (Juglans nigraL.) alley cropping. II. Effects of juglone on hydroponically grown corn (Zea maysL.) and soybean (Glycine maxL. Merr.) growth and physiology. Plant and Soil, 203(2), 199- 206.

Karkhanis, M., Jadia, C., & Fulekar, M. (2005). Rhizofilteration of metals from coal ash leachate. Asian J. Water, Environ. Pollut, 3(1), 91-94.

Khan, M., & Shirazi, M., (2009). Role of proline, K/Na ratio and chlorophyll content in salt tolerance of wheat (Triticum aestivum L.). Pak. J. Bot 41(2): 633-638

Khan, A. G. (2005). Role of soil microbes in the rhizospheres of plants growing on trace metal contaminated soils in phytoremediation. *Journal of Trace Elements in Medicine and Biology, 18*(4), 355-364.

Kim, B., Baier, A., Somers, D., & Gustafson, J. (2001). Aluminum tolerance in triticale, wheat, and rye. Euphytica, 120(3), 329-337.

Komarek, M., Cadkova, E., Chrastny, V., Bordas, F., & Bollinger, J. C. (2010). Contamination of vineyard soils with fungicides: a review of environmental and toxicological aspects. *Environment international, 36*(1), 138-15

Krizek, D. T., Foy, C. D., & Mirecki, R. M. (1997). Influence of aluminum stress on shoot and root growth of contrasting genotypes of Coleus. *Journal of plant nutrition, 20*(9), 1045-1060.

Küpper, H., Lombi, E., Zhao, F. J., & McGrath, S. P. (2000). Cellular compartmentation of cadmium and zinc in relation to other elements in the hyperaccumulator Arabidopsis halleri. *Planta, 212*(1), 75-84.

Lafitte, H., Yongsheng, G., Yan, S., & Li, Z. (2007). Whole plant responses, key processes, and adaptation to drought stress: the case of rice. Journal of experimental botany, 58(2), 169.

Le Bot, J., Adamowicz, S., & Robin, P. (1997). Modelling plant nutrition of horticultural crops: a review. Scientia Horticulturae, 74(1), 47-82.

Lester GE, Jifon J, Makus D (2010). Impact of potassium nutrition on food quality of fruits and vegetables: A condensed and concise review of the literature. Better Crops, 94(1):18-21.

Liu, W. J., Zhu, Y. G., Smith, F. A., & Smith, S. (2004). Do phosphorus nutrition and iron plaque alter arsenate (As) uptake by rice seedlings in hydroponic culture? New Phytologist, 162(2), 481-488.

Lou, L., Shen, Z., & Li, X. (2004). The copper tolerance mechanisms of *Elsholtzia haichowensis*, a plant from copper-enriched soils. *Environmental and Experimental Botany, 51*(2), 111-120.

Bacilio-Jiménez, M., Aguilar-Flores, S., Ventura-Zapata, E., Pérez-Campos, E., Bouquelet, S., & Zenteno, E. (2003). Chemical characterization of root exudates from rice (Oryza sativa) and their effects on the chemotactic response of endophytic bacteria. Plant and Soil, 249(2), 271-277.

Ma, J. F., Zheng, S. J., Li, X. F., Takeda, K., & Matsumoto, H. (1997). A rapid hydroponic screening for aluminium tolerance in barley. *Plant and Soil, 191*(1), 133-137.

Makau, M., Masito, S., & Gweyi-Onyango, J. (2011). A rapid hydroponics screening of field and horticultural crops for aluminum tolerance. African Journal of Horticultural Science, 4(1).

Malaisse, F., Baker, A. J. M., & Ruelle, S. (1999). Diversity of plant communities and leaf heavy metal content at Luiswishi copper/cobalt mineralization, Upper Katanga, Dem. Rep. Congo. *Biotechnol Agron Soc Environ, 3*(2), 104-114.

Mandal, R., Hassan, N. M., Murimboh, J., Chakrabarti, C. L., Back, M. H., Rahayu, U., et al. (2002). Chemical speciation and toxicity of nickel species in natural waters from the Sudbury area (Canada). *Environmental science & technology, 36*(7), 1477-1484

McQuattie, C. J., & Schier, G. A. (1990). Response of red spruce seedlings to aluminum toxicity in nutrient solution: alterations in root anatomy. Canadian journal of forest research, 20(7), 1001-1011.

Meharg, A. A. (2005). Venomous Earth: How arsenic caused the world's worst mass poisoning: Macmillan Houndmills,, England.

Mohsenzadeh, S., Sadeghi, S., Mohabatkar, H., & Niazi, A. (2009). Some responses of dry farming wheat to osmotic stresses in hydroponics culture. Journal of Cell and Molecular Research, 1(2), 84-90.

Moreno-Jimanez, E., Manzano, R., Esteban, E., & Pealosa, J. (2010) The fate of arsenic in soils adjacent to an old mine site (Bustarviejo, Spain): mobility and transfer to native flora. Journal of Soils and Sediments, 10(2), 301-312.

Moreno-Jimenez, E., Pealosa, J. M., Manzano, R., Carpena-Ruiz, R. O., Gamarra, R., & Esteban, E. (2009). Heavy metals distribution in soils surrounding an abandoned mine in NW Madrid (Spain) and their transference to wild flora. Journal of hazardous materials, 162(2-3), 854-859.

Munns R, James RA, Sirault XRR, Furbank RT, Jones HG. (2010). New phenotyping methods for screening wheat and barley for water stress tolerance. Journal of Experimental Botany 61, 3499-3507.

Naoko O., and Wasaki J., (2010) Recent progress in plant nutrition research: cross-talk between nutrients, plant physiology and soil microorganisms. Plant Cell Physiol. 51(8):1255–1264

Olofsdotter, M. (1998). Allelopathy in rice: Int. Rice Res. Inst. (IRRI)

Passioura JB. (2006). The perils of pot experiments. *Functional Plant Biology* 33, 1075-1079.

Peel, M. D., Waldron, B. L., Jensen, K. B., Chatterton, N. J., Horton, H., & Dudley, L. M. (2004). Screening for salinity tolerance in alfalfa: a repeatable method.

Peng, Y. L., Gao, Z. W., Gao, Y., Liu, G. F., Sheng, L. X., & Wang, D. L. (2008). Eco physiological Characteristics of Alfalfa Seedlings in Response to Various Mixed Salt alkaline Stresses. Journal of Integrative Plant Biology, 50(1), 29-39.

Poschenrieder, C., Guns, B., Corrales, I., & Barcel, J. (2008). A glance into aluminum toxicity and resistance in plants. Science of the total environment, 400(1-3), 356-368.

Proctor, J. (2003). Vegetation and soil and plant chemistry on ultramafic rocks in the tropical Far East. *Perspectives in plant ecology, evolution and systematics, 6*(1-2), 105-124.

Rajakaruna, N., Harris, T. B., & Alexander, E. B. (2009). Serpentine geoecology of eastern North America: A review. *Rhodora, 111*(945), 21-108.

Raven J.A., Handley L.L., Wollenweber B. (2004) 7 Plant nutrition and water use efficiency. Water use efficiency in plant biology:171.

Raviv, M., & Lieth, J. H. (2007). Soilless culture: theory and practice: Elsevier Science Ltd. Reinoso, H., Sosa, L., Reginato, M., & Luna, V. (2005). Histological alterations induced by sodium sulphate in the vegetative anatomy of Prosopis strombulifera (Lam.) Benth. World Journal of Agricultural Sciences, 1, 109-119.

Sagardoy, R., Morales, F., Ló pez Millan, A. F., A. Abadia, J. (2009a). Effects of zinc toxicity on sugar beet (Beta vulgaris L.) plants grown in hydroponics. Plant Biology, 11(3), 339 350.

Sagardoy, R., Flexas, J., Ribas-Carbó, M., Morales, F., & Abadia, J. (2009b). Stomatal conductance is the main limitation to photosynthesis in sugar beet plants treated with Zn excess.

Sandra L. R., Juan L. (2009) Zn and Mn o,p-EDDHA chelates for soybean nutrition in hydroponics in high pH conditions. http://escholarship.org/uc/item/96887750

Schier, G. A. (1985). Response of red spruce and balsam fir seedlings to aluminum toxicity in nutrient solutions. *Canadian journal of forest research, 15*(1), 29-33.

Shi, J., Wu, B., Yuan, X., Chen, X., Chen, Y., & Hu, T. (2008). An X-ray absorption spectroscopy investigation of speciation and biotransformation of copper in Elsholtzia splendens. *Plant and Soil, 302*(1), 163-174.

Sholto Douglas, J. (1984). Beginner's guide to hydroponics: Soilless gardening: Pelham Books (London).

Silva, I. R., Smyth, T. J., Israel, D. W., & Rufty, T. W. (2001). Altered aluminum inhibition of soybean root elongation in the presence of magnesium. *Plant and Soil, 230*(2), 223- 230.

Sivaguru, M., & Paliwal, K. (1993). Differential aluminium tolerance in some tropical rice cultivars. I: Growth performance. *Journal of plant nutrition, 16*(9), 1705-1716.

Sonneveld, C., & Voogt, W. (2009). Plant nutrition of greenhouse crops: Springer Verlag.

Sonneveld, C., & Voogt, W. (1990). Response of tomatoes (*Lycopersicon esculentum*) to an unequal distribution of nutrients in the root environment. Plant Soil 124, 251–256.

Spehar, C. (1994). Aluminium tolerance of soya bean genotypes in short term experiments. *Euphytica, 76*(1), 73-80.

Swennen, R., De Langhe, E., Janssen, J., & Decoene, D. (1986). Study of the root development of some Musa cultivars in hydroponics. Fruits (France).

Szira, F., Blint, A., Bo rner, A., & Galiba, G. (2008). Evaluation of Drought Related Traits and Screening Methods at Different Developmental Stages in Spring Barley. Journal of Agronomy and Crop Science, 194(5), 334-342.

Tang, C. S., & Young, C. C. (1982). Collection and identification of allelopathic compounds from the undisturbed root system of bigalta limpograss (Hemarthria altissima). Plant Physiology, 69(1), 155.

Tang, C. S., Wang, X. (2001). Interaction between copper and radiocesium in Indian mustard and sunflower grown in the hydroponic solution. Journal of Radioanalytical and Nuclear Chemistry, Vol. 252, No. (1) 9–14

Tarja, L. Kamrul Md, H. Pedro, J. (2009). Effects of silicon on growth of silver birch (*Betula pendula*). *http://escholarship.org/uc/item/38c9z7cm.*

Vara Prasad, M. N., & de Oliveira Freitas, H. M. (2003). Metal hyperaccumulation in plants: Biodiversity prospecting for phytoremediation technology. *Electronic Journal of Biotechnology, 6*(3), 285-321.

Wang, M., Wu, L., & Zhang, J. (2009). Impacts of root sulfate deprivation on growth and elements concentration of globe amaranth (Gomphrena globosa L.) under hydroponic condition. Plant, Soil and Environment, 55(11), 484-493.

Wasaki, J., Kojima, S., Maruyama, H., Haase, S., Osaki, M., & Kandeler, E. (2008). Localization of acid phosphatase activities in the roots of white lupin plants grown under phosphorus deficient conditions. Soil Science & Plant Nutrition, 54(1), 95- 102.

Wei, L., Luo, C., Li, X., & Shen, Z. (2008). Copper Accumulation and Tolerance in Chrysanthemum coronarium L. and Sorghumsudanense L. *Archives of environmental contamination and toxicology, 55*(2), 238-246.

Wood, M., Cooper, J., & Holding, A. (1984). Soil acidity factors and nodulation ofTrifolium repens. *Plant and Soil, 78*(3), 367-379.

Yang, M. J., Yang, X. E., & Romheld, V. (2002). Growth and nutrient composition of Elsholtzia splendens Nakai under copper toxicity. Journal of plant nutrition, 25(7), 1359-1375.

Zabludowska, E., Kowalska, J., Jedynak, L., Wojas, S., Sklodowska, A., & Antosiewicz, D. (2009). Search for a plant for phytoremediation-What can we learn from field and hydroponic studies? Chemosphere, 77(3), 301-307.

Zhao, J., Sun, H., Dai, H., Zhang, G., & Wu, F. (2010). Difference in response to drought stress among Tibet wild barley genotypes. Euphytica, 172(3), 395-403.

Zuo, S. P., Li, X. W., & Ma, Y. Q. (2010). Response of transgenic potato seedlings to allelopathic pressure and the effect of nutrients in the culture medium. Acta Ecologica Sinica, 30(4), 226-232.

CHAPTER-7

HYDROPONIC GROWTH SYSTEMS FOR INVESTIGATING THE IONOME OF BOTH ROOTS AND SHOOTS IN ARABIDOPSIS THALIANA

INTRODUCTION: AN OVERVIEW

Hydroponic growth systems have become invaluable tools in the exploration of the ionome in both the root and shoot systems of Arabidopsis thaliana. These systems provide researchers with precise control over the nutrient composition of the growth medium, enabling the systematic study of the elemental composition in different plant tissues. By employing hydroponics, scientists can investigate how Arabidopsis plants uptake, transport, and accumulate essential ions and trace elements. This methodology has proven instrumental in unraveling the intricacies of nutrient acquisition and distribution within the plant, shedding light on critical aspects of plant physiology and nutrition. Moreover, it facilitates the identification of genes and regulatory mechanisms that govern ion homeostasis, aiding our understanding of how plants adapt to varying environmental conditions and nutrient availability. Consequently, hydroponic growth systems play a pivotal role in advancing our knowledge of the ionome in Arabidopsis thaliana and beyond, contributing to broader insights in plant biology and nutrient management.

Plants provide an important source of minerals for human and animal consumption. There is a general worldwide deficiency of human intake of several essential minerals, including iron, zinc, copper, calcium, magnesium, selenium, and iodine (White & Broadley, 2009). This deficiency exists not only in developing countries, but also in the developed world. On the other hand, excessive amounts of heavy metals in crop plants can endanger the health of humans and livestock. It is, therefore, important to develop crop plants with balanced levels of minerals. Understanding of the genes and processes that control the mineral content of plants can provide the basis for engineering crops with the optimal concentrations of minerals in various organs and tissues. In recent years, there has been increasing interest in determining the elemental composition (ionome) of various plant species and identifying the genes that dominate this composition [reviewed by Baxter (2009), Salt et al. (2008), and Williams & Salt (2009)]. The agriculturally related plants whose ionomes were investigated include the crops rice (Norton et al., 2010) and *Brassica napus* (Liu et al., 2009), as well as the plant *Lotus japonicus*, which serves as a model species for legume crops (Chen et al., 2009).

Although *Arabidopsis thaliana* is not a crop plant, the availability of a wide range of genetic and genomic tools in this species makes it a useful model organism for identifying the genes that determine the plant ionome. This can be done by correlating ionomic data with genetic information and data about gene expression in genetically divergent *A. thaliana* lines. The commonly utilized genetically divergent *A. thaliana* populations are natural accessions, recombinant inbred lines (RIL), and mutant lines. Such populations were used in several high-throughput studies, which resulted in the identification of many quantitative trait loci (QTLs) controlling the *A. thaliana* ionome (Buescher et al., 2010; Ghandilyan et al., 2009; Lahner et al., 2003; Prinzenberg et al., 2010). Due to the high-throughput nature of these studies, the *A. thaliana* plants were grown in soil in many cases, and the elemental composition was determined in shoots but not in roots. Since minerals are absorbed into plants through roots, it is important to gain knowledge on the root ionome. Hydroponic growth systems make it possible to obtain clean roots suitable for mineral analysis.

Moreover, hydroponic systems enable the application of accurate and relatively stable concentrations of minerals. RIL populations of *A. thaliana* grown in hydroponic systems were utilized to identify QTLs that correlate the mineral composition of roots, leaves, and seeds with phytate content and growth-related traits (Ghandilyan et al., 2009; Prinzenberg et al., 2010). The parental *A. thaliana* accessions utilized in these studies were Ler-3, Kas-2, Kon, An-1, and Eri-1.

In this chapter, we describe in detail the system we use for the growth of *A. thaliana* plants in hydroponics. We then provide data on the mineral composition of roots, rosette leaves, and inflorescence stems (the latter two are referred to hereafter as leaves and stems, respectively) of the Col-0 accession of *A. thaliana*. Col-0 is widely used as a model accession in a variety of genetic, genomic, and gene-expression studies of *A. thaliana*. We discuss the inter-and intra- experimental variations we observed in the hydroponic system. We then compare the results obtained in this system with those reported in a high-throughput analysis of plants grown in soil. In previous studies, the *A. thaliana* stem ionome was not investigated *per se*. We discuss the differences in the composition of the leaf and stem ionome, and the significance of this point for the design of ionomic studies. We also provide data on the differences between the root and leaf ionome of the Col-0 and C24 accessions of *A. thaliana,* as determined using a hydroponic growth system.

A HYDROPONIC SYSTEM FOR THE GROWTH OF *A. THALIANA* PLANTS

Hydroponic systems for the growth of *A. thaliana* plants were described in several publications (for example, David-Assael et al., 2005; Gibeaut et al., 1997; Hermans & Verbruggen, 2005; Laganowsky et al., 2009). Although these systems are basically similar, there are some variations between them. Hydroponic growth media are usually based on different dilutions of the media described by Hoagland & Arnon (1938), Johnson et al. (1957), or others [reviewed by Laganowsky et al. (2009)]. We describe here the system and media we use for the growth of *A. thaliana* plants in hydroponics, but other systems and growth media can be utilized as well.

Fig. 7.1. The hydroponic system

Twenty-four-day-old A. *thaliana* plants grown in a 1:1 mixture of sand and perlite. B. A typical 5-liter container. C. A mature plant that was grown hydroponically. D. A polystyrene surface including about 50 plants floating in a 5-liter container, with aluminum foil covering the gaps between the surface and the container. The arrows indicate two Pasteur pipette connected rubber hoses. E. A polystyrene surface including about 20 plants floating in a 1.3 liter container. F. A container containing 50 A. *thaliana* plants photographed at different days after germination. A to E are Col-0 plants, and F are C24 plants.

The *A. thaliana* plants are germinated in Jiffy pellets or in MS plates. After about 10 days, when the roots can still be easily removed from the Jiffy pellets or agar, the seedlings are transferred into a 1:1 mixture of sand (salt-free sand designed for agricultural use) and perlite (course particles), and watered in the x1 basic hydroponic solution (see below). After about 10-12 days (Fig. 7.1A), the seedlings are carefully removed from the sand and perlite mixture, and their roots are manually washed in the hydroponic solution to eliminate all sand or perlite particles. The plants are transferred into a hydroponic growth system consisting of 5- or 1.3-liter containers containing the same basic mineral solution. The plants are inserted into holes in polystyrene surfaces freely floating in the containers. Each 5- or 1.3-liter container can contain 25-50 or 10-20 plants, respectively. To prevent algal growth, the containers and polystyrene surfaces should be black or brown (Fig. 7.1B), and the gaps between the polystyrene surfaces and the containers should be covered in aluminum foil (Fig. 7.1D, E). Aquarium pumps connected to rubber hoses and Pasteur pipettes are used to maintain aeration (Fig. 7.1D). The solutions are refilled each day according to their consumption, and are changed completely at least once a week. Fig. 7.1C shows a mature, hydroponically grown plant, and Fig. 7.1E shows a 5-liter container containing 50 plants photographed at 5-6–day intervals.

PREPARATION OF THE HYDROPONIC MEDIUM

Preparation of x4000 Micronutrient Stock

Mineral	Gram for separate stocks of 500 ml	Concentration of each separate stock (mM)	Amount from each separate stock for 1 liter of x4000 micronutrient stock
$CoCl_2{\cdot}6H_2O$	0.475	4	1 ml
$CuSO_4{\cdot}5H_2O$	62.500	500	1 ml
H_3BO_3	19.325	625	40 ml
$MnSO_4{\cdot}H_2O$	70.500	844	6 ml
$Na_2MoO_4{\cdot}2H_2O$	40.500	335	1 ml
$ZnSO_4{\cdot}7H_2O$	143.750	1000	2 ml
KCl			3.728 gram

Preparation of 10 Liters of x20 Solution I

Mineral	Stock concentration	ml from stock
K_2HPO_4	1 M	50
K_2SO_4	0.5 M	100
KNO_3	1 M	600
$MgSO_4 \cdot 7H_2O$	1 M	100
Fe-EDDHA (see notes 1, 2)	20 mM	50
x4000 micronutrient stock	x4000	50

Add dH_2O to reach a final volume of 10 liters.

Preparation of 10 Liters of x20 Solution II

Dissolve 17.2 g $CaSO_4$ $2H_2O$ in dH_2O to reach a final volume of 10 liters.

Preparation of x1 Basic Hydroponic Solution

Add 9 liters of dH_2O into a 10-liter bottle, then add 500 ml of x20 Solution I and mix, then add 500 ml of x20 Solution II. Mix and adjust pH to 5.5 with ~40% H_2SO_4.

Final concentrations in the x1 basic hydroponic solution

Mineral	Final concentration
$CaSO_4$	0.5 mM
K_2HPO_4	0.25 mM
K_2SO_4	0.25 mM
KNO_3	3 mM
$MgSO_4$	0.5 mM
Fe-EDDHA	5 µM
$CoCl_2$	0.001 µM
$CuSO_4$	0.125 µM
H_3BO_4	6.25 µM
KCl	12.5 µM
$MnSO_4$	1.27 µM
Na_2MoO_4	0.084 µM
$ZnSO_4$	0.5 µM

Notes:

1. All stock solutions prepared in steps 1 and 2 should be autoclaved, except Fe-EDDHA.
2. Fe-EDDHA is Sequestrene 138 from Novartis. We prepare the 20 mM stock solution by taking 18.7 grams per liter from the material we have, which includes 6% iron. The solution should be well mixed before each use.

3. The use of two separate x20 stocks, and preparation of the final (x1) solution as described, are essential for prevention of CaSO4 precipitation.

MINERAL ANALYSIS

Plant material is harvested, divided into the different tissues, and washed two times in distilled water and two times in double-distilled water. Roots are washed once in tap water prior to these washes. The material is immediately transferred into a 70oC incubator and dried for three days. Dry plant material is crushed into a fine powder. Samples of 250 mg are ashed by incubation for 10 h at 500oC, and dissolved in 10 ml of 0.1 M nitric acid ("Suprapur" grade, Merck). After 30 min at 80oC, samples are brought to 25 ml with H_2O. When the available amount of dry plant material is less than 250 mg, the samples are brought to a final volume of less than 25 ml, but not less than 10 ml. Each sample is then filtered through 1 mm Whatman paper. Mineral content was determined by inductively coupled plasma atomic emission spectrometry (using an ICP-AES model `Spectroflame' from Spectro, Kleve, Germany).

INTER-AND INTRA-EXPERIMENTAL VARIATIONS IN THE MINERAL COMPOSITION OF ROOTS, LEAVES, AND STEMS OF COL-0 PLANTS

The mineral content of roots, leaves, and stems of the Col-0 accession of *A. thaliana* was examined in the hydroponic growth system. We and others found that there is variation in the mineral content observed in different experiments, even when the experimental conditions are kept as constant as possible. This is due to inevitable variations in the environmental conditions that alter the physiological state of the plants. It is, therefore, interesting to find if there are minerals that exhibit relatively low inter- and/or intra- experimental variations. We present here the results of five independent experiments, carried out in different seasons and years, in which the mineral composition of Col-0 plants was determined. Table 7.1 shows the levels in roots, leaves, and stems of the major minerals present in the hydroponic solution: the macronutrients potassium (K), calcium (Ca), magnesium (Mg), phosphorus (P), and sulfur (S); the micronutrients iron (Fe), zinc (Zn), manganese (Mn), copper (Cu), molybdenum (Mo), and cobalt (Co); and the mineral sodium (Na). It should be noted that Na is not classified as an essential macro- or micronutrient, but as a beneficial mineral element (Marschner 1995). The extent of inter-experimental variation (the variation between different experiments) is indicated by the coefficient of variation (CV) (see the legends of Table 7.1 for details how the values presented were calculated). To visualize the differences in CV values, values larger that 50% (an arbitrary rank) were indicated in bold letters in Table 7.1. The CV values of the macronutrients were generally lower than those of the micronutrients. Particularly high inter-experimental variations were observed in the root levels of all micronutrients analyzed. This suggests that the root levels of these micronutrients are more prone to variations due to unintended differences in the precise physiological state of the plants in different experiments compared to the leaf or stem levels of the same minerals, or compared to both the root and shoot levels of the macronutrients. It is not surprising that the intra-experimental variations (i.e., the variations between different samples of the same experiment) in the root and leaf levels of most minerals analyzed were lower compared to the inter-experimental variations (Tables 7.1 and 7.2, and data not shown). Particularly low CV values are highlighted in gray in Tables 7.1 and 7.2. The levels of the corresponding minerals can be used as internal standards in experimental designs that allow this manipulation (that is, when the levels of the utilized minerals are not supposed to vary).

Table 7.3 shows a comparison between the mineral concentrations observed by us in leaves of hydroponically grown Col-0 plants and the concentrations observed by Buescher et al. (2010) in shoots of soil-grown Col-0 plants. This comparison is valid since the shoot material harvested by Buescher et al. (2010) included only leaves. The data of Buescher et al. (2010) were based on 3-6 independent experiments, and our data were based on 5 independent experiments. Table 7.3 shows that there is a high similarity in the concentrations of the macronutrients analyzed in the two sets of experiments [K, Ca, Mg, and P; the S content was not reported by Buescher et al. (2010)]. This similarity is notable due to the large differences between the experimental conditions in the two sets of experiments. It is, therefore, possible that the concentrations (or at least the average concentrations observed in several experiments) of some of these macronutrients are relatively stable in leaves of Col-0 plants, at least under non-extreme physiological conditions.

Most ionomic studies include examination of leaves (besides other tissues). The concentrations of K showed relatively low inter- as well as intra-experimental variations (14% and 4%, respectively) (Tables 7.1 and 7.2). The leaf levels of K also showed high similarity between plants grown in hydroponics and those grown in soil under the experimental conditions used by us and by Buescher et al. (2010), respectively (Table 7.3). This suggests that leaf K levels are particularly suitable to serve as internal standards in experimental designs that allow such manipulation (i.e., when the levels of K are not supposed to vary).

DISTRIBUTION OF MINERAL IN ROOTS, LEAVES, AND STEMS OF COL-0 PLANTS

In most ionomic studies of *A. thaliana*, the mineral composition of stems was not determined. Figure 7.2 presents the ratio between the mineral content of roots, leaves, and stemsof hydroponically grown Col-0 plants. It should be noted that these ratios reflect the total amounts of each mineral in the indicated tissues. These values can be considerably different from the free levels of the same minerals in different cellular compartments. For example, a major proportion of the minerals in roots is deposited in complexes in the root apoplast. Figure 7.2 shows that the amounts of Ca and Mg are, by far, higher in leaves than in roots or stems. The distribution between roots and leaves seen here for Ca is close to the distribution of this element in most dicot plants [reviewed by Conn & Gilliham (2010)]. The fact that the highest concentration of Ca is found in the leaves is related to the fact that this elementis phloem-immobile and is, therefore, not readily redistributed after deposition in the shoot [reviewed by Conn & Gilliham (2010)]. The K levels are about 50% higher in stems than in leaves (Fig. 7.2B). The mineral P shows slight, but statistically significant, higher levels in roots than in leaves. The mineral S is unique among the macronutrients with respect to its concentrations that are two times higher in roots than in leaves (Fig. 7.2).

Table 7.1. The elemental composition of roots, leaves, and stems of Col-0 plants as determined in five experiments

		Roots			Leaves			Stems		
	Mineral	Average	SD	CV	Average	SD	CV	Average	SD	CV
Macro-nutrients	K	62.938	17.684	28	49.2695	7.0722	14	68.4901	18.381	27
	Ca	6.701	0.7227	11	43.2389	14.698	34	14.5582	1.2711	9
	Mg	2.4165	0.3782	16	11.8031	2.7514	23	5.1593	0.545	11
	P	10.494	1.7523	17	8.6381	1.6471	19	7.6944	2.4785	32
	S	19.2347	4.6952	24	10.4239	1.662	16	9.1496	4.3863	48
	Na	6.0529	4.2422	70	6.842	3.1237	46	4.0972	4.1012	100
Micro-nutrients	Fe	1.0366	1.1379	110	0.0614	0.0219	36	0.0327	0.019	58
	Zn	1.1301	0.7509	66	0.1882	0.0811	43	0.0955	0.0195	20
	Mn	0.2753	0.379	138	0.053	0.0287	54	0.0192	0.0049	25
	Cu	0.017	0.0104	61	0.0054	0.0011	21	0.0063	0.0004	7
	Mo	0.1016	0.088	87	0.0101	0.0043	42	0.0084	0.005	60
	Co	0.0097	0.0046	48	0.003	0.0024	78	0.0002	0.0001	35

This table presents the results of five independent experiments. In each experiment, *A. thaliana* Col-0 plants were germinated in Jiffy pellets, transferred to sand and perlite, and then to hydroponics, and harvested when they were about six-weeks old. Plants were maintained in a greenhouse at 24°C with a photoperiod of 16 h light and 8 h darkness. In each experiment, plants were grown in four separate 5- liter hydroponic containers, each containing about 25 plants. The organs (roots, leaves or stems) of all plants grown in the same container were harvested together and treated as one sample. For each experiment, we calculated the average concentration (g · kg^{-1} dry weight) of each mineral in the four samples of each organ (derived from the four containers).

Table 7.2. The elemental composition of roots and leaves of Col-0 plants as determined in one experiment

		Roots			Leaves		
	Mineral	Average	SD	CV	Average	SD	CV
Macro-nutrients	K	59.2833	7.1657	12	55.0859	2.1145	4
	Ca	6.2584	0.7591	12	31.6133	1.1190	4
	Mg	2.2233	0.1298	6	8.7049	0.3059	4
	P	10.9846	0.9259	8	9.3453	1.1018	12
	S	19.4929	1.4356	7	11.0383	1.0447	9
	Na	13.1094	2.3329	18	6.1646	1.3165	21
Micro-nutrients	Fe	3.0601	0.3414	11	0.0985	0.0138	14
	Zn	2.1601	0.1500	7	0.2411	0.0224	9
	Mn	0.9521	0.3136	33	0.0609	0.0039	6
	Cu	0.0322	0.0124	39	0.0066	0.0031	48
	Mo	0.2165	0.0449	21	0.0107	0.0032	30
	Co	0.0163	0.0103	63	0.0058	0.0011	18

The five values obtained for the content of each mineral in each organ (derived from the five independent experiments) were used to calculate the average and standard deviation (SD) values presented in the table. The coefficient of variation (CV) is the percentage of the SD from the average. CV values larger than 50% (an arbitrary rank) are indicated in bold letters. The lowest CV value among those obtained for the macronutrients analyzed in each organ is highlighted in gray.

The table presents the results of one experiment in which *A. thaliana* Col-0 plants were germinated in Jiffy pellets, transferred to sand and perlite, and then grown in four separate 5-liter containers, each containing about 25 plants. Plants were maintained in a greenhouse at 24ºC with a photoperiod of 16 h light and 8 h darkness. The organs (roots or leaves) of all plants grown in the same container were harvested together and treated as one sample. The table shows the average concentration ($g \cdot kg^{-1}$ dry weight) and the SD of each mineral in the four samples (derived from the four containers) of each organ. The CV is the percentage of the SD from the average. CV values larger than %50 (an arbitrary rank) are indicated in bold letters. The lowest CV values among those obtained for the different macronutrients analyzed in each organ are highlighted in gray.

Table 7.3. Comparison between the data obtained in the present study and in soil-grown plants

	Mineral	Leaf content in this work	Leaf content in the work of Buescher et al. (2010)	Leaf content in this work relative to the data of Buescher et al. (2010)
Macro-nutrients	K	49.2695	46.1000	1.0688
	Ca	43.2389	45.0000	0.9609
	Mg	11.8031	12.9000	0.9150
	P	8.6381	9.7000	0.8905
	Na	6.8420	0.8600	7.9559
Micro-nutrients	Fe	0.0614	0.1000	0.6140
	Zn	0.1882	0.0610	3.0847
	Mn	0.0530	0.0636	0.8336
	Cu	0.0054	0.0018	2.9524
	Mo	0.0101	0.0055	1.8384
	Co	0.0030	0.0019	1.6353

This table shows the levels (g · kg-1 dry weight) of minerals in Col0- leaves obtained in the present study (average values of the five experiments reported in Table 7.1) and in the ionomic study of Buescher et al. (2010). The right column presents the ratio between the mineral content of leaves in the present study and in the study of Buescher et al. (2010). Ratios close to 1 are indicated in bold letters.

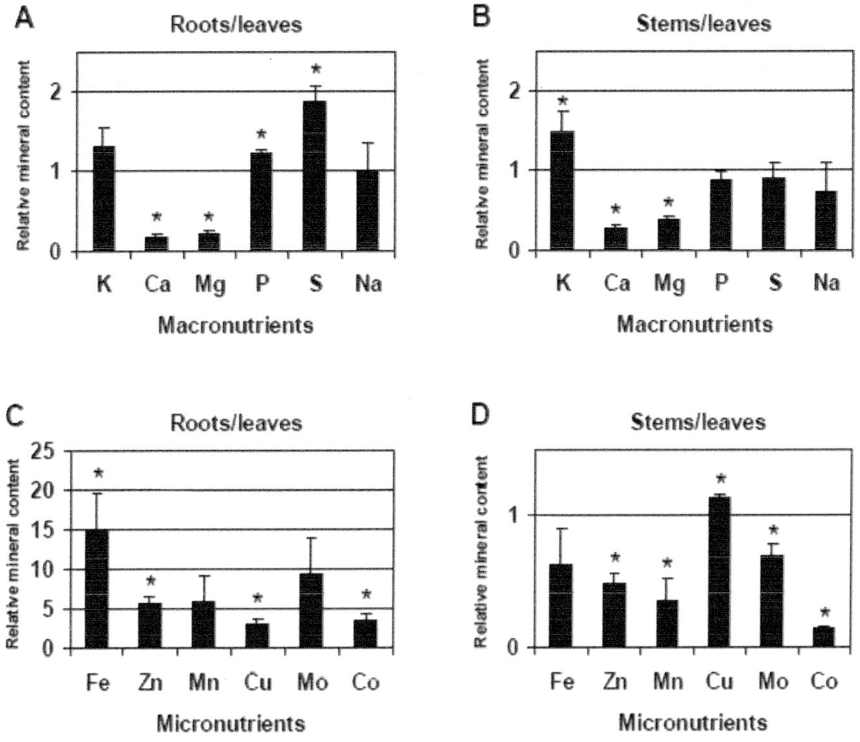

Fig. 7.2. The ratio between the mineral content of roots, leaves and stems of Col-0 plants grown hydroponically

The ratios between the mineral content (that was determined in terms of g · kg-1 dry weight) of roots and leaves, or stems and leaves, were separately calculated for each of the five experiments described in Table 7.1. Each column shows the average and the standard error (SE) of the ratios obtained in each of the independent experiments. The graphs present the root-to-leaf (A, C) and stem-to-leaf (B, D) levels of the macronutrients (A, B) and micronutrients (C, D). The mineral Na is shown together with the macronutrients. The asterisks indicate statistically significant differences between the content of each mineral in roots and leaves (A, C) or stems and leaves (B, D).

The concentrations of most micronutrients analyzed (Fe, Zn, Cu, and Co) are higher in roots than in leaves (Fig. 7.2C). Among the four indicated minerals, the Zn and Co levels are also significantly lower in stems than in leaves (Fig. 7.2D). This suggests that mobilization of Znand Co in the vascular system is particularly low, at least in Col-0 plants grown in the physiological conditions described here. The data presented in Figure 7.2C and D suggest that this is also the case for Fe, Mn, and Mo, although apparently with higher inter-experimental variations since the differences in the content of these minerals in the different organs were not always statistically significant. One outcome of this situation is a restriction in the levels of heavy metals that reach the sensitive reproductive organs. The restricted mobility of the micronutrients in the vascular system is related to processes occurring at both the xylem and phloem. These processes include (among others) accumulation in the xylem parenchyma of roots and shoots, and control of mineral loading into the xylem and phloem (Marschner 1995). The mobility of most micronutrients in the phloem is lower than that of most macronutrients (Marschner 1995).

In some ionomic studies of *A. thaliana*, the stems, when present, were harvested as part of the shoots. The data presented here show that the leaf and stem contents of many minerals, including K, Ca, Mg, Zn, Mn, Cu, Mo, and Co, are significantly different (Fig. 7.2B, D). It is, therefore, important that shoot material harvested for ionomic analysis will be divided into separate leaf and stem samples or will include, as much as possible, constant proportions of leaves and stems. However, the study of the stem ionome *per se* is an essential step towards achieving a better understanding of long-distance transport processes in plants.

COMPARISION BETWEEN THE ROOT AND LEAF IONOME OF THE COL-0 AND C24 ACCESSIONS OF *A. THALIANA*

Ionomic analysis of different *A. thaliana* accessions can provide valuable information. Comparing the shoot ionome of different *A. thaliana* accessions was utilized for identification of over 100 QTLs for mineral accumulation (Buescher et al., 2010). Identification of differences in the root ionome between different accessions can contribute to the discovery of novel genes that control mineral uptake and transport in plants. Towards this goal, we present here the differences between the root and leaf ionome of the Col-0 and C24 accessions of *A. thaliana* [C24 was not included among the 12 accessions whose shoot ionome was analyzed by Buescher et al. (2010)]. Figure 7.3 shows both the absolute and relative levels of each mineral in roots and leaves of plants grown hydroponically. The levels of Ca, P, and Zn were almost identical in both roots and leaves of the two accessions. Both roots and leaves of the C24 accession exhibited significantly lower levels of Mo and Fe compared to Col-0. A statistically significant reduction was observed in the levels of K, Mg, Mn, and Cu in leaves of C24 compared to Col-0. However, the differences in the leaf content of K and Mg were minor. The only mineral that showed a significant reduction in its root but not leaf content in C24 plants was S.

Among the minerals that differed between the two accessions, a particularly big difference was seen in the root and leaf content of Mo. A strong reduction in the content of this mineral in shoots of various *A. thaliana* accessions, as compared to Col-0, was previously observed by Buescher et al. (2010). In a subsequent study it was shown, using hydroponic growth systems, that the Ler accession had a reduced content of Mo not only in shoots but also in roots, when compared to Col-0 (Baxter et al., 2008). This suggested that the low Mo content in shoots was not due to enhanced accumulation of Mo in the roots, but rather to reduced Mo uptake by the roots (Baxter et al., 2008). Grafting experiments demonstrated that the differences in Mo content in shoots were driven solely by the roots. It was found that the mitochondrial Mo transporter (MOT1) is responsible for the variation in Mo content in both*thaliana* plants (accessions Col-0 and C24) were germinated in Jiffy pellets, transferred to sand and perlite, and then to hydroponics. For each accession, plants were grown in four separate 5-liter containers containing about 25 plants each. Plants were maintained in a greenhouse at 24°C with a photoperiod of 16 h light and 8 h darkness. The organs (roots or leaves) of all plants grown in the same container were harvested together and treated as one sample. Two graphs are presented for each mineral. One graph presents the averages and standard errors of the mineral concentrations (g · kg^{-1} dry weight) obtained in the four containers of each accession. The second graph presents the relative levels of each mineral (C24 relative to Col0-) in each organ. The asterisks, which are presented in both graphs, indicate a statistically significant difference (p<0.05 in Student's *t*-test) between the levels of minerals in the two accessions. R, roots; L, leaves.

Interestingly, Mn content is significantly lower in leaves, but not roots, of C24 as compared to Col-0 (Fig. 7.3). This suggests that the trait(s) that governs the difference in Mn content between Col-0 and C24 is not related to uptake in roots but to transport and/or accumulation in leaves. The mechanisms for uptake, transport, and internal sequestration of Mn in plants are now starting to be understood (Cailliatte et al., 2010; reviewed by Pittman, 2005). It will be necessary to determine if the indicated difference between Col-0 and C24 is related to a novel trait(s) or to some of the currently identified proteins involved in internal sequestration of Mn. The latter proteins include CAX2 (Hirschi et al., 2000), the ER Ca/Mn pump ECA1 (Wu et al., 2002), and putative homologs of the *Stylosanthes hamata* Mn transporter MTP1 (Delhaize et al., 2003).

Co levels were apparently higher in leaves of C24 compared to Col-0 (Fig. 7.3). The difference was not statistically significant (p=0.09 in Student's *t*-test), possibly due to the difficulty to accurately determine Co amounts, which were close to the detection limit of the ICP machine. The Fe and Co transporter FPN2, which sequesters Co into the root vacuoles and lowers its mobilization to the shoot, was identified as the protein responsible for the increased levels of Co in the Ts-1 and Se-0 accessions of *A. thaliana* as compared to Col-0 (Morrissey et al., 2009). The Ts-1 and Se-0 accessions have a truncated version of the *FPN2* gene, and it is possible that the C24 accession shares this property.

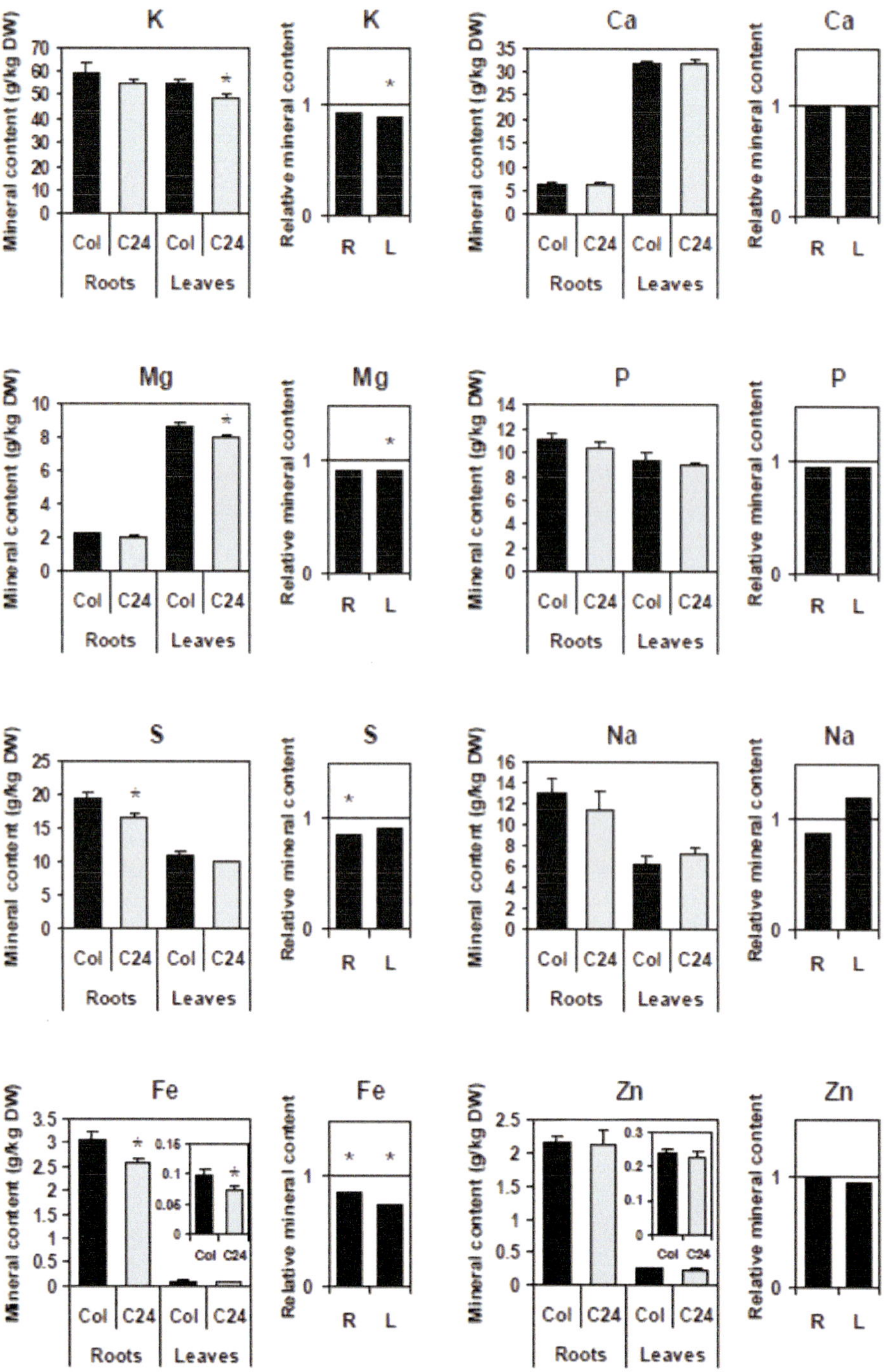

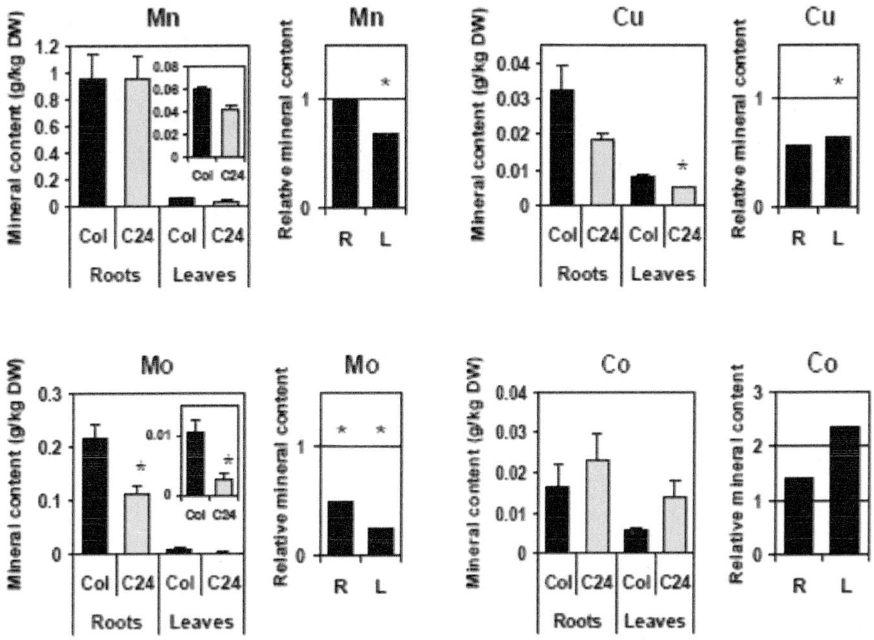

Fig. 7.3. The differences between the root and leaf ionome of the Col-0 and C24 accessionsshoots and roots of the various accessions (Baxter et al., 2008). This study demonstrated the significance of investigating the mineral content of roots in order to enhance understanding of the mechanisms involved.

An interesting finding reported here is the lower content of Cu in leaves, and perhaps also roots, of the C24 compared to the Col-0 accession (the *p* values in Student's *t*-test were 0.009 and 0.076 for leaves and roots, respectively) (Fig. 7.3). Among the *A. thaliana* accessions studied by Buescher et al. (2010), no accession showed a reduction in its Cu content relative to Col-0. The only accession whose Cu content differed from that of Col-0 was Ler-2, which showed a 69% increase in the Cu content of shoots. This makes Ler-2 and C24 interesting candidates for the search for putative novel genes that control the Cu content of plants. It is, of course, possible that the difference in the Cu content of the indicated accessions is related to one of the currently identified genes that control Cu homeostasis in plants [reviewed by Palmer & Guerinot (2009) and by Pilon (2011)]. Our data suggest, although not with complete certainty, that the reduction in the Cu content of C24 is related to processes occurring also, or possibly solely, in roots. The currently identified transport proteins that affect the Cu content of roots include COPT1, which is involved in Cu uptake into roots (Sancenon et al., 2004), HMA5, whose knock-out increased the Cu content of roots but not shoots (Andres-Colas et al., 2006), and COPT5, which functions in the inter-organ allocation of Cu (Klaumann et al., 2011).

A statistically significant reduction in Fe content was observed in both leaves and roots of C24 as compared to Col-0 (Fig. 7.3). Among the *A. thaliana* accessions investigated by Buescher et al. (2010), no accession showed a reduction in its Fe content compared to Col-0, whereas two accessions showed increased Fe content. It will be interesting to determine whether the gene(s) responsible for the differences in the Fe content of the indicated accessions is among the many Fe-homeostasis genes identified thus far [reviewed by Conte & Walker (2011), Jeong & Guerinot (2009), and by Palmer et al. (2009)]. Our data suggest that the reduction in the Fe content of C24 compared to Col-0 is related to processes occurring also, or possibly solely, in roots.

AT A GALNCE

- This chapter emphasizes the practical and potential contribution of hydroponic growth systems to the identification of genes and processes governing the plant ionome and, in particular, to the study of processes occurring in roots. Hydroponic systems are a useful tool for obtaining sufficient amounts of clean roots for mineral analyses. These systems also enable the application of accurate concentrations of minerals. Nevertheless, the absolute levels of minerals in plants can vary between different experiments due to unavoidable differences in the precise physiological conditions of the plants. The data presented here indicate that the root levels of the micronutrients are more prone to variations due to unintended differences in the precise physiological state of the plants in different experiments compared to the leaf or stem levels of the same minerals, or compared to both the root and shoot levels of the macronutrients.
- Interesting conclusions were reached by comparing the leaf ionome reported by Buescher et al. (2010) with that observed by us in Col-0 plants grown in soil or hydroponics, respectively. Despite the large differences in the experimental conditions, the concentrations of all macronutrients analyzed (K, Ca, Mg, and P) were very similar in the two sets of experiments. In contrast, the concentrations of all micronutrients analyzed in the two systems (Fe, Zn, Cu, Mo, Co, and – to a lesser extent – Mn) were relatively different (Table 7.3). This suggests that the macronutrient content in *A. thaliana* leaves varies to a lower extent due to altered environmental conditions compared to the micronutrient content. In particular, the leaf content of K showed low inter- and intra-experimental variation in the hydroponic system, as well as similarity with the values obtained in plants grown in soil in the experimental system described by Buescher et al. (2010). This suggests that the leaf content of K can be utilized as an

internal standard in experimental designs that allow this manipulation (that is, when the levels of K are not supposed to vary).

- The data presented here also showed that for many minerals, the ion composition of the *A. thaliana* stem is considerably different from that of the leaf. This point should be taken into consideration in the design of ionomic studies. The data reported here about the differences between the root and leaf ionomes of the Col-0 and C24 accessions may contribute to the efforts to utilize the natural variations between different *A. thaliana* accessions for identification of novel genes that control mineral uptake and transport in plants. Since minerals are absorbed into plants through roots, it is also important to gain knowledge of the root ionome, and hydroponic systems can act as a useful tool in such a study.

REFERENCES

Andres-Colas N.; Sancenon V.; Rodriguez-Navarro S.; Mayo S.; Thiele D.J.; Ecker J.R.; Puig S. & Penarrubia L. (2006). The *Arabidopsis* heavy metal P-type ATPase HMA5 interacts with metallochaperones and functions in copper detoxification of roots. *Plant Journal*, Vol. 45, pp. 225-236.

Baxter I. (2009). Ionomics: studying the social network of mineral nutrients. *Current Opinion in Plant Biology*, Vol. 12, pp. 381-386.

Baxter I.; Muthukumar B.; Park H.C.; Buchner P.; Lahner B.; Danku J.; Zhao K.; Lee J.; Hawkesford M.J.; Guerinot M.L. & Salt D.E. (2008). Variation in molybdenum content across broadly distributed populations of *Arabidopsis thaliana* is controlled by a mitochondrial molybdenum transporter (*MOT1*). *Plos Genetics*, Vol. 4.

Buescher E.; Achberger T.; Amusan I.; Giannini A.; Ochsenfeld C.; Rus A.; Lahner B.; Hoekenga O.; Yakubova E.; Harper J.F.; Guerinot M.L.; Zhang M.; Salt D.E. & Baxter I.R. (2010). Natural genetic variation in selected populations of *Arabidopsis thaliana* is associated with ionomic differences. *Plos One*, Vol. 5.

Cailliatte R.; Schikora A.; Briat J.F.; Mari S. & Curie C. (2010). High-affinity manganese uptake by the metal transporter NRAMP1 is essential for *Arabidopsis* growth in low manganese conditions. *The Plant Cell*, Vol. 22, pp. 904-917.

Chen Z.; Watanabe T.; Shinano T.; Okazaki K. & Osaki M. (2009). Rapid characterization of plant mutants with an altered ion-profile: a case study using *Lotus japonicus*. *New Phytologist*, Vol. 181, pp. 795-801.

Conn S. & Gilliham M. (2010). Comparative physiology of elemental distributions in plants. *Annals of Botany*, Vol. 105, pp. 1081-1102.

Conte S.S. & Walker E.L. (2011). Transporters contributing to iron trafficking in plants. *Molecular Plant*, Vol. 4, pp. 464-476.

David-Assael O.; Saul H.; Saul V.; Mizrachy-Dagri T.; Berezin I.; Brook E. & Shaul O. (2005). Expression of AtMHX, an *Arabidopsis* vacuolar metal transporter, is repressed by the 5' untranslated region of its gene. *Journal of Experimental Botany*, Vol. 56, pp. 1039-1047.

Delhaize E.; Kataoka T.; Hebb D.M.; White R.G. & Ryan P.R. (2003). Genes encoding proteins of the cation diffusion facilitator family that confer manganese tolerance. *The Plant Cell*, Vol. 15, pp. 1131-1142.

Ghandilyan A.; Ilk N.; Hanhart C.; Mbengue M.; Barboza L.; Schat H.; Koornneef M.; El Lithy M.; Vreugdenhil D.; Reymond M. & Aarts M.G. (2009). A strong effect of growth medium and organ type on the identification of QTLs for phytate and mineral concentrations in three *Arabidopsis thaliana* RIL populations. *Journal of Experimental Botany*, Vol. 60, pp. 1409-1425.

Gibeaut D.M.; Hulett J.; Cramer G.R. & Seemann J.R. (1997). Maximal biomass of *Arabidopsis thaliana* using a simple, low-maintenance hydroponic method and favorable environmental conditions. *Plant Physiology*, Vol. 115, pp. 317-319.

Hermans C. & Verbruggen N. (2005). Physiological characterization of Mg deficiency in *Arabidopsis thaliana*. *Journal of Experimental Botany*, Vol. 56, pp. 2153-2161.

Hirschi K.D.; Korenkov V.D.; Wilganowski N.L. & Wagner G.J. (2000). Expression of *Arabidopsis CAX2* in tobacco. Altered metal accumulation and increased manganese tolerance. *Plant Physiology*, Vol. 124, pp. 125-133.

Hoagland D.R. & Arnon D.I. (1938). The water culture method for growing plants without soil. *California Agricultural* Experiment *Station Circulars*, Vol. 347, pp. 1-39.

Jeong J. & Guerinot M.L. (2009). Homing in on iron homeostasis in plants. *Trends in Plant Science*, Vol. 14, pp. 280-285.

Johnson C.M.; Stout P.R.; Broyer T.C. & Carlton A.B. (1957). Comparative chlorine requirements of different plant species. *Plant and Soil*, Vol. 8, pp. 337-353.

Klaumann S.; Nickolaus S.D.; Fuerst S.H.; Starck S.; Schneider S.; Neuhaus H. & Trentmann O. (2011). The tonoplast copper transporter COPT5 acts as an exporter and is required for interorgan allocation of copper in *Arabidopsis thaliana*. *New Phytologist*, Vol. 192, pp. 393-404.

Laganowsky A.; Gomez S.M.; Whitelegge J.P. & Nishio J.N. (2009). Hydroponics on a chip: Analysis of the Fe deficient *Arabidopsis* thylakoid membrane proteome. *Journal of Proteomics*, Vol. 72, pp. 397-415.

Lahner B.; Gong J.M.; Mahmoudian M.; Smith E.L.; Abid K.B.; Rogers E.E.; Guerinot M.L.; Harper J.F.; Ward J.M.; McIntyre L.; Schroeder J.I. & Salt D.E. (2003). Genomic scale profiling of nutrient and trace elements in *Arabidopsis thaliana*. *Nature Biotechnology*, Vol. 21, pp. 1215-1221.

Liu J.; Yang J.; Li R.; Shi L.; Zhang C.; Long Y.; Xu F. & Meng J. (2009). Analysis of genetic factors that control shoot mineral concentrations in rapeseed (*Brassica napus*) in different boron environments. *Plant and Soil*, Vol. 320, pp. 255-266.

Marschner, H. (1995). Mineral Nutrition of Higher Plants, Academic Press, London, San Diego.

Morrissey J.; Baxter I.R.; Lee J.; Li L.; Lahner B.; Grotz N.; Kaplan J.; Salt D.E. & Guerinot M.L. (2009). The ferroportin metal efflux proteins function in iron and cobalt homeostasis in *Arabidopsis*. *The Plant Cell*, Vol. 21, pp. 3326-3338.

Norton G.J.; Deacon C.M.; Xiong L.; Huang S.; Meharg A.A. & Price A.H. (2010). Genetic mapping of the rice ionome in leaves and grain: identification of QTLs for 17 elements including arsenic, cadmium, iron and selenium. *Plant and Soil*, Vol. 329, pp. 139-153.

Palmer C.M. & Guerinot M.L. (2009). Facing the challenges of Cu, Fe and Zn homeostasis in plants. *Nature Chemical Biology*, Vol. 5, pp. 333-340.

Pilon M. (2011). Moving copper in plants. *New Phytologist*, Vol. 192, pp. 305-307.

Pittman J.K. (2005). Managing the manganese: molecular mechanisms of manganese transport and homeostasis. *New Phytologist*, Vol. 167, pp. 733-742.

Prinzenberg A.E.; Barbier H.; Salt D.E.; Stich B. & Reymond M. (2010). Relationships between growth, growth response to nutrient supply, and ion content using a recombinant inbred line population in *Arabidopsis*. *Plant Physiology*, Vol. 154, pp. 1361-1371.

Salt, D.E.; Baxter, I. & Lahner, B. (2008). Ionomics and the study of the plant ionome, *Annual Review of Plant Biology*, Vol. 59, pp. 709-733.

Sancenon V.; Puig S.; Mateu-Andres I.; Dorcey E.; Thiele D.J. & Penarrubia L. (2004). The *Arabidopsis* copper transporter COPT1 functions in root elongation and pollen development. *Journal of Biological Chemistry*, Vol. 279, pp. 15348-15355.

White P.J. & Broadley M.R. (2009). Biofortification of crops with seven mineral elements often lacking in human diets - iron, zinc, copper, calcium, magnesium, selenium and iodine. *New Phytologist*, Vol. 182, pp. 49-84.

Williams L. & Salt D.E. (2009). The plant ionome coming into focus. *Current Opinion in Plant Biology*, Vol. 12, pp. 247-249.

Wu Z.Y.; Liang F.; Hong B.M.; Young J.C.; Sussman M.R.; Harper J.F. & Sze H. (2002). An endoplasmic reticulum-bound Ca^{2+}/Mn^{2+} pump, ECA1, supports plant growth and confers tolerance to Mn^{2+} stress. *Plant Physiology*, Vol. 130, pp. 128-137.

CHAPTER-8

COMPREHENDING THE ABSORPTION OF NUTRIENTS, HARMFUL SUBSTANCES, AND POLLUTANTS BY PLANT ROOTS IN HYDROPONIC CULTIVATION

INTRODUCTION: AN OVERVIEW

In the realm of hydroponic cultivation, a cutting-edge and soilless method of growing plants, a fundamental aspect of plant health and growth lies in the understanding of how plant roots uptake essential nutrients, as well as how they interact with potentially harmful substances and pollutants present in the hydroponic environment. This knowledge serves as the cornerstone for optimizing crop yields, ensuring plant vitality, and mitigating the risk of contamination.

The understanding of plant uptake (nutrients, toxic and polluting elements) is crucial for the future food needs of humanity given the explosive growth of the world population and the anthropogenic pressure on the environment which significantly modify the homeostasis of the balanced global cycles. Rice and banana are of fundamental interest for development policy since they are two major foods for the world population. The understanding of the mechanisms and the optimal conditions of nutrient uptake by these plants is thus important to ensure biomass production. Furthermore, the transfer of toxic and polluting elements in the soil-plant system can influence the nutrient uptake and plant growth, and thus has strong agronomic consequences, in addition to the large environmental consequences.

Cultivation of plant in mineral nutrient solution rather than in soil allow scientists to study the relationship between nutritional status and plant growth, but also the impact of biotic and a biotic factors on the plant development. Besides its usefulness for agricultural industry, hydroponic is an ideal culture device to isolate factors affecting plant growth and better understand the essential, beneficial, toxic and polluting effects of the chemical elements, and the chemical transfer of the elements from hydro/pedosphere to biosphere. The understanding of nutrient transfer is indeed crucial for the productivity of the Earth's biosphere in order to (i) maintain the homeostasis of elements between lithosphere, atmosphere and biosphere, (ii) produce sufficient food well-allocated between north and south hemisphere, and (iii) constrain the mobility of toxic and polluting elements in the soil- plant systems and their transfer in the hydrosphere. The chemical elements present in the soluble phase of soil are very reactive with the interfaces of roots in the rhizosphere, and their transfer in the biosphere clearly depends on the activity of other elements in the aqueous phase. Furthermore, in soil-plant systems, the activity of one element in soil solution depends also on many other environmental factors than its own activity in the interfaces of roots: acidity and redox potential of solution, the equilibrium between aqueous and solid phases (neoformation, adsorption, complexation, co-precipitation...), temperature, humidity, enzymatic activity, type of microorganisms, ...

Aluminium toxicity is a primary factor limiting the growth and yield of the majority of plants grown in mineral acid soils of tropical and subtropical areas (Horst, 1995; von Uexküll & Mutert, 1995). This toxicity occurs when soil acidification causes Al solubilization (Lindsay, 1979) and thereby increases the concentration of Al^{3+} ion, the most phytotoxic Alform in the soil solution (Kochian, 1995). Numerous studies have demonstrated the marked depressive effects of soluble aluminium (Al) on water and nutrient uptake, root and shoot growth, and mineral content in different plant species (Quang et al., 1996; Voigt et al., 1999). In addition, since 1945, the toxic radioisotope ^{137}Cs has been released in the environment by nuclear weapons testing, controlled discharge of waste effluents from nuclear plants, and accidental release (Avery, 1996). A general consensus emerged that radiocesium exhibits a biogeochemical behavior rather similar to that of potassium. In this regard, radiocesium displays a relative mobility in various soil-plant systems through uptake by plants and organisms (Carter, 1993), threatening the food chain. The close link between soil solution and roots strongly controls the solubility of nutrients but also of toxic metals, which can cause severe losses in agricultural yield. It is thus crucial to better constrain the root-solution interface through hydroponic culture experiment in order to offer the current better practices for land use management. The dynamic of elements in soil-plant systems depends on ion exchange reactions between soil matrix, solution phase and root. These processes can be assessed by hydroponic studies (Dufey et al., 2001), which demonstrate that the release of protons and organic substances in the rhizosphere strongly influence the mobilization of nutrients and toxic elements (Hinsinger, 1998).

In this chapter, we develop hydroponic devices to study both **root-solution interface**, and **soil-solution-root reactions.** We report hydroponic solutions developed for banana, rice and tree seedlings (Douglas fir and Black pine) as close as possible to the actual concentrations of the soil solution. We also describe a culture medium including solid minerals as a source of nutrients and polluting elements. We focus on the use of hydroponic studies to better understand the dynamics of nutrients (Ca, Mg, K, Na), beneficial (Si), toxic (Al) and polluting (^{137}Cs) elements in the solution-root interfaces, but also the uptake mechanisms, assimilation and translocation of these elements by plants and their subsequent role on plant growth. Finally, we develop the use of geochemical tracers such as stable silicon isotopes and Ge/Si ratios as promising tools to understand plant physiological mechanisms through the quantification of Si isotope and Ge-Si fractionations between dissolved Si source and banana plant, between roots and shoots, and within shoots.

Combining hydroponic culture studies and knowledge in soil science and biogeochemistry, we then point the importance of hydroponic culture to suggest agronomic and environmental advices for plant growth in Earth's Critical Zones characterized by a very poor nutrient stock in old and weathered tropical soils, and largely influenced by global change processes.

DEVICES FOR HYDROPONIC CULTIVATION

Numerous devices are reported in literature to conduct plant growth experiments in hydroponics. In this section, we describe the principles of the devices that we regularly used in our laboratory in the last forty years.

Growth Conditions and Composition of Nutrient Solutions

Prior to hydroponic culture in controlled conditions, seeds are surface-sterilized with 5% H_2O_2 and rinsed five times with demineralized water. The seeds are germinated, and then weaned in an adequate nutrient solution tanks before selecting uniform seedlings for theexperiments. Batches of seedlings are grown in separate pots and placed on a perforated plate of expanded polystyrene which limits water loss by evaporation (Fig. 8.1).

The hydroponic experiments are conducted in growth chambers at a specific photon flux density, relative humidity, day/night temperature for each specific plants and environments (Table 8.1). Depending on genus, plants are supplied with a solution obtained by mixing salts, boric acid and FeEDTA to reach realistic concentrations and match common plant nutrient requirements. In table 8.1, we describe hydroponic solutions as close as possible to the actual concentrations of the soil solution for banana, rice and coniferous tree seedlings. For banana, the nutrient solution composition is determined from (i) the nutrient requirements and the mineral equilibria as established for young bananas, and (ii) the realistic ion concentrations in solutions of tropical acid soils (Rufyikiri et al., 2000b). The composition of the nutrient solution for rice mimicks soil conditions, *i.e.* is very dilute with respect to most of the nutrient solutions used in plant physiological work (Tang Van Hai et al., 1989). The nutrient solution for the cultivation of Douglas fir and Black pine (Cornelis et al., 2010b) matches common tree seedling nutrient requirements (Ingestad, 1971), using the optimum NH4–N/NO_3–N ratio of 40/60.

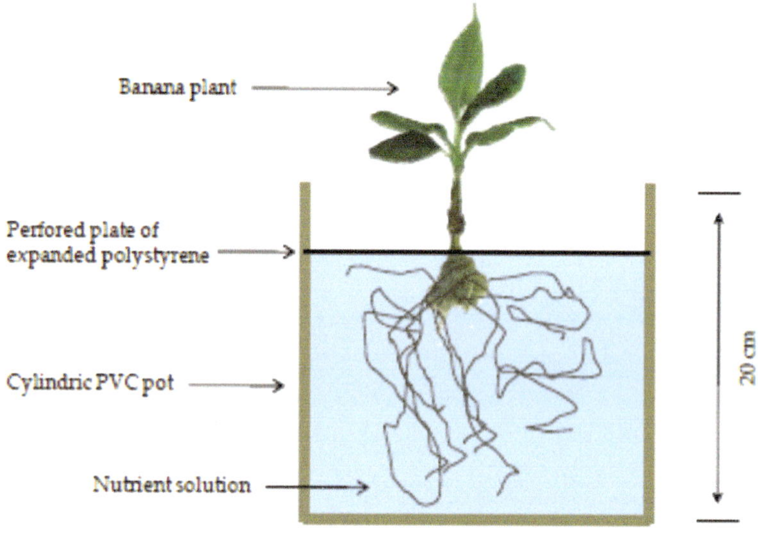

Fig. 8.1. Hydroponic culture device with the main compartments (only aqueous medium).

In most hydroponic devices, there is no support for the plant roots (only aqueous medium; Fig. 8.1), while some hydroponic systems have a solid medium of support. With an aqueous medium, the hydroponic culture device can be (i) static where the nutrient solution is not renewed, or (ii) dynamic where the nutrient solution is continuously renewed at a specific rate with a peristaltic pump. A solid medium in the device (Fig. 8.2) allows studying the origin of the mobilization of elements.

The choice of the devices is thus adapted to the purpose of the research. We describe here the principles of the hydroponic devices that we regularly used in the last forty years to study the soil-plant relations. They will be refer to as Device 1, 2 and 3, respectively, throughout the Chapter.

	Banana	Rice	Douglas fir Black pine		
	Nutrient solution			Hoagland solution	Yoshida solution
macroelements (mM)					
$Ca(NO_3)_2$	0.9	/	4.8	6.8	/
$CaSO_4$	0.05	/	1.6	/	/
$CaCl_2$	0.05	0.25	1.6	/	1
KCl	0.5	0.2	4.0		/
K_2SO_4	0.25	/	4.0	/	0.35
KH_2PO_4	/	0.2	/	0.5	/
KNO_3	/	/	/	5	/
$MgCl_2$	0.05	/	0.4	/	/
$MgSO_4$	0.05	0.4	0.4	3.3	1.6
NH_4Cl	0.1	/	3.2	/	/
$(NH_4)2SO_4$	0.05	/	3.2	/	/
NH_4NO_3	/	0.7	/	1	2.8
NaH_2PO_4	0.05	/	0.2	/	0.3
NaOH	/	/	/	/	/
microelements (µM)					
H_3BO_3	80	46	90	46	18.5
FeEDTANa	80	65	80	65	/
$FeCl_2.6H_2O$	/	/	/	/	35.8
$MnCl_2$	8	12	8	12	9
$ZnSO_4$	0.8	1	0.8	1	0.2
$CuSO_4$	0.8	0.3	0.8	0.3	0.2
$(NH_4)_6Mo_7O_{24}$	5.6	/	5.6	/	0.5
$Na_2MoO_4.2H_2O$	/	0.5	/	0.5	/

Growth chamber conditions

Luminosity	446 µE/m^{-2} s^{-1} for 12h day light	250 µE/m^{-2} s^{-1} for 12h day light	448 µE/m^{-2} s^{-1} for 8h day light
Temperature	28/25 °C day/night	30/22°C day/night	20/18 °C day/night
Relative humidity	90%	80% day 90% night	75%

Table 8.1. Nutrient solution and phytotron chambers conditions used for banana (Rufyikiri et al. 2000), rice (Tang Van Hai et al. 1989) and tree seedlings (Cornelis et al. 2010b) cultivation. Hoagland solution (1950) and Yoshida solution (1976) are given for comparison.

Static Nutrient Solution Device (Device 1)

When the nutrient solution is not renewed (Fig. 8.1), the ion concentrations are calculated to maintain an optimum nutrient status during the experiment. Twice a week, the volumes of the remaining solutions are weighed to estimate water loss, which is immediately balanced by adding demineralized water. Water loss through evaporation is measured in six control pots. Cumulative water uptake is calculated as the difference between water loss and evaporation. The potential nutrient uptake by mass flow (MFU) is defined by the product of water uptake and nutrient concentration in the hydroponic solution.

Continuous Nutrient Flow Device (Device 2)

Nutrient solution can be continuously renewed at a rate of 104 ml h^{-1} pot^{-1}, corresponding to an expected residence time of the solution in the pot close to 24 h. The replacement solution ("input solution") is supplied at the bottom of the pot and the solution in excess ("output solution") is released and collected via an overflow pipe. Water uptake is assessed by weighing input and output solutions of each pot over 1 day. Nutrient uptake is calculated as the difference between daily input and output quantities for each pot (*i.e.*, the difference between the product of concentration and solution volume at input and output).

Culture Device with Substrate (Device 3)

A special culture vessel device developed by Hinsinger et al. (1991, 1992) allows to simulate a macroscopic rhizosphere in one-dimensional geometry. Thanks to this experiment, the sources of nutrients for plants can be identified and quantified. Three experimental designs were developed with differing substrate constitution and differing type of contact between roots and substrate:

- Agarose-clay substrate in contact with a linear rhizosphere through a planar surface (Fig. 8.2a).
- Agarose-soil mixture in contact with a linear rhizosphere through a planar surface (Delvaux et al. 2000)
- Quartz-clay substrate in contact with the entire roots to simulate a real situation in soils (Fig. 8.2b).

UPTAKE OF NUTRIENT AND TOXIC ELEMENTS BY BANANA AND RICE

Crop yields in very acidic soils are often limited by increasing aluminium concentrations in the soil solution. However, in field conditions, we are not able to vary the Al concentration in soil solution since the dynamic equilibrium between solid and aqueous phases imply modifications of soluble and exchangeable Al concentrations, and then modifications of pH and availability of other nutrients.

The interpretation of the specific Al effect on the crop yields is thus impossible. This explains why most studies on the toxicity/resistance to Al have been conducted on nutrient solutions, which allow varying the total concentration of Al regardless of the concentration of other elements, provided that they remain undersaturated with respect to solid phases involving Al (Tang Van Hai, 1993).

Banana

The production of bananas and plantains in the tropics is of prime importance both for local consumption and for export market. Despite the fact that acid soils are common in thetropics and that bananas have large nutrient requirements, little is known on the tolerance of bananas to soil acidity, Al toxicity and mineral deficiencies caused by Al stress. In this regard, hydroponic studies are very useful to understand the mineral nutrition of different banana cultivars.

Free Al is directly toxic to plant roots and, in most cases, is little absorbed or translocated to the aerial plant parts (Bernal & Clark, 1997; Rufyikiri et al., 2000b). Rufyikiri et al. (2000b, 2001) performed a hydroponic experiment using a continuous nutrient flow device (device2) to study the effect of Al on young banana plants. Vitroplants were supplied by Vitropic (Montpellier, France) for Grande Naine. The other cultivars (Agbagba, Obino l'Ewaï, Igitsiri and Kayinja) were produced in the *Musa* Germplasm Transit Centre (Heverlee, Belgium) of the International Network for Improvement of Banana and Plantain (INIBAP). They were weaned for 3 weeks in an aerated nutrient solution. Among 50 vitroplants per cultivar, the six tallest but homogeneous individuals were selected. The vitroplants were then transferred to 2.5 l pots (one plant per pot). Nutrient solution described above was supplied continuously by peristaltic pumps at a rate of 104 $ml.h^{-1}$ per pot. The mean residence time of solution in pots was 1 day. The experiments were conducted in growth chambers at 448$\mu E.m^{-2}.s^{-1}$ photon flux density for 12h per day, %90 relative humidity, °25/28C day/night temperature for the low altitude cultivars, and °20/24C day/night temperature for the banana plants cultivated in the highlands (Igitsirir and Kayinja). Half of the plants were supplied with the same nutrient solution described above but with 78.5 µM Al (added as $Al2(SO4)3.18H_2O$.

For most parameters (appearance of new leaves, total biomass, pseudostem height, leaf surface area, growth of lateral roots, number and diameter of root axes), the two plantains Agbagba and Obino l'Ewaï appeared more Al-resistant and Kayinja more Al-sensitive than both Grande Naine and Igitsiri (Rufyikiri et al. 2000b). In all plants parts for all cultivars, Al reduced Ca and Mg contents, increased K and P contents, and did not change significantly N content. Since Mg content in plant parts was lower than the deficiency threshold concentrations reported in literature, leaf yellowing and marginal necrosis developing on Al-treated bananas is attributed to Mg deficiency (Rufyikiri et al., 2000b). The same hydroponic experiment was established to measure daily water and nutrient uptake (Rufyikiri et al., 2001). Water and nutrient absorption measurement were carried out twice a week during 40 days. Water uptake was assessed by weighing input and output solutions of each pot over 1 day. Aluminium reduced plant water and nutrient uptake and cumulative detrimental effects were observed. The water uptake was only 30-40% of the control and nutrient uptake rates were reduced by more than 50% relatively to the control. It is also demonstrated thanks to this hydroponic experiment that plantain bananas were more resistant to Al.

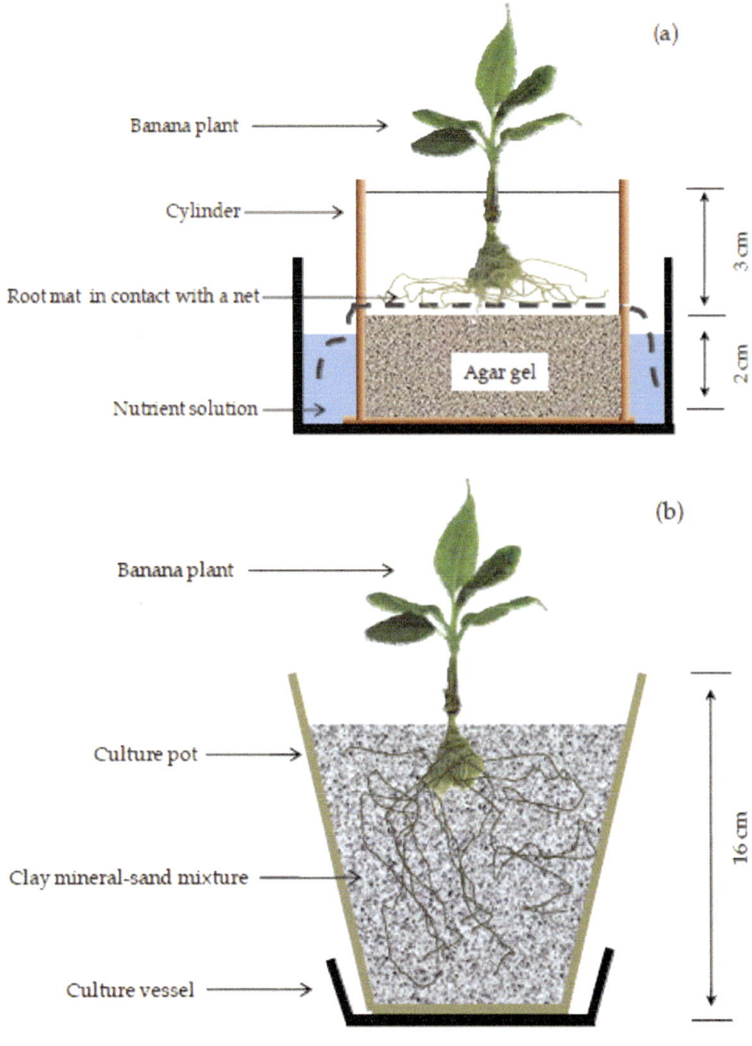

Fig. 8.2. (a) Culture device "agarose-clay substrate" and "agarose-soil mixture", adapted from Hinsinger et al. (1992) – (b) Culture device "quartz-clay substrate", adapted from Rufyikiri et al. (2004)

Aluminium ions are tightly fixed to root exchange sites (Dalghren et al., 1991; Tice et al., 1992), which originate from carboxylate groups of uronic acids of pectin and hemicelluloses in the root cell-walls (Horst et al., 1995). The subsequent changes in (i) physical properties which become less plastic and more resistant to elongation and (ii) in plasma-membrane properties and associated proteins and lipids are evoked, among others, in the literature to explain Al toxicity (Horst et al., 1995; Kochian, 1995). For instance, it has been demonstrated in hydroponic culture that bananas (*Musa* spp.), as many other crop species, accumulate aluminum (Al) in roots when grown in nutrient solution containing Al ions. Aluminum cancompete with calcium (Ca) and magnesium (Mg) on the root exchange sites, which has been reported as a possible cause for Al toxicity to the plant (Rufyikiri et al., 2003). The fixation and accumulation of Al in roots is accompanied by changes in the concentration of other cations, the most obvious effect being the noticeable decrease in Mg content. Competition between Al and other cations on the charged constituents in the apoplast can thus explain the detrimental effects of Al on plants (Horst, 1995; Keltjens, 1995). In non-acid soils, ionic Al activity in solution is low so that Ca, that has also high affinity for carboxylate groups, is dominating the exchange complex roots (Dufey et al., 2001). Despite this fact and thanks to hydroponic studies, Rufyikiri et al. (2002) suggest that the Al/Mg ratio on roots could be a better indicator of Al toxicity that the Al/Ca ratio. Indeed, among the Cu-extractable cations (Dufey & Braun, 1986), Ca was occupying about 50-60% of the exchange sites even when Al was present in the growth medium (78.5 µM), while adsorbed Mg was drastically reduced on root exchange sites (Rufyikiri et al., 2002).

The hydroponic systems are ideal devices to study the resistance of mycorrhizal plants to Al toxicity compared with non mycorrhizal plants (Rufyikiri et al., 2005). In monoxenic cultivation, Rufyikiri et al. (2000a) show that arbuscular mycorrhizal (AM) fungi (*Glomus intraradices*) can be effective in alleviating the aluminium toxicity to banana plants. In that study, both banana plants and AM fungi grew very well in sand under continuous nutrient flow. During the 5 weeks preceding Al application, an average of one leaf was produced per week. The nutrient uptake rates by banana plants decreased with the increase of the addition of Al in the nutrient solution. The Mg uptake rate decreased by 4 to 18 times in the nutritive solution concentrated in Al (78 and 180 µM) for banana plants without arbuscular mycorrhizal fungi. A significant positive effect of arbuscular mycorrhizal fungi on plant growth was observed with aluminium treatment, and was most pronounced at the highest concentration. The benefits, compared with nonmycorrhizal plants, included: increase in shoot dry weight, uptake of water and of most nutrients (Ca, Mg, NH4-N and P), and in calcium, magnesium and phosphorus content, particularly in roots; decrease in aluminium content in root and shoot; and delay in the appearance of aluminium-induced leaf symptoms. The higher nutrient uptake rates in the mycorrhizal plants suggest that mycorrhizal roots were less affected than nonmycorrhizal roots by the the presence of Al.

The banana plant is a model case study since this plant is very sensitive to Al stress (Rufyikiri 2000a, 2000b) and it exhibits very large K requirements and growth rate (Lahav, 1995), as well as selective root uptake of NH4 (Rufyikiri et al., 2001). Thus banana plant has a high potential for acidifying its own rhizosphere, and thereby in solubilizing toxic forms of Al. The differences in symptoms observed between banana plants grown without and with Al is mainly related to Al toxicity, rather than to low pH caused by Al or by proton excretion by roots (excess of cations over anions absorbed). As explained above, banana plants (*Musa* spp.) are very sensitive to Al, which is mobilized in acid soil conditions. These plants may, however, contribute to their own intoxication because their roots can excrete protons in large quantities, implying nutrient mobilization from silicate minerals. A net proton excretion by roots can largely influence Al solubilization, and results from excess root+uptake of cations relatively to anions (Marschner, 1995), particularly with NH4uptake. Theprotons flux generated by roots can be neutralized by the weathering of aluminosilicates (Hinsinger et al., 1992; Hinsinger, 1998), which release Si and plant nutrients (Hinsinger et al., 2001), and can lead to an increase of soluble Al up to toxic levels for plants. As suggested by Rufyikiri et al. (2004), the rhizospheric weathering of aluminosilicates can thus be considered as a source not only for nutrient acquisition by plant roots (Hinsinger, 1998), but also for plant intoxication by trace metals (Lombi et al., 2001).

In this respect, the culture devices with substrate (device 3) allow to study the impact of the chemical environment of rhizosphere on the mobilization of elements from the solid phases in soil (Hinsinger et al., 1992; Delvaux et al., 2000; Rufyikiri et al., 2004). In the study of Rufyikiri (2004) (Fig. 8.2b), the sole source of Al is montmorillonite or kaolinite. The root- induced acidification and subsequent weathering of the minerals involved a preferential release of Al in the aqueous phase of the quartz: clay substrates, an increase of both ammonium-extractable Al from kaolinite and Mg from smectite. These respective root- induced mobilizations led to relatively large root uptake of Al in quartz: kaolinite and Mg in quartz: smectite. These results support that the mobilization of, respectively, Al and Mg are the limiting steps in the dissolution of the kaolinite and montmorillonite, just as demonstrated in previous chemical weathering studies (Nagy, 1995).

From a mica-agar substrate (Fig. 8.2a), Hinsinger et al. (1992) demonstrated that K is released and concomitantly the mica lattice expanded after 3 days of continuous exposure to the root mat, in response to K uptake by plants. The vermiculitization of mica (expansion of the mica lattice through replacement of

K by more hydratable cations Ca and Mg) is detectable up to1.5 mm from the root mat of ryegrass after 4 days. The trioctahedral micas such as phlogopite can thus contribute significantly to the supply of K around the most active parts of the roots.

Rice

The toxicity of Al present in the soil solution is also one of the major causes of the poor performance of rice grown on acid soils. The presence of Al is the unavoidable consequence of the spontaneous acidification of the soil either by the respiration of micro-organisms or the exchange hydrolysis of the cation adsorption sites of clay minerals (Eickman & Laudelout, 1961; Gilbert & Laudelout, 1965). Under certain environmental conditions, the normal acidification process may be greatly enhanced as in poorly buffered soils subject to acid precipitation or in acid sulphate soils where the oxidation of sulphur compounds involves the liberation of huge amounts of sulphuric acid. This is the case in the Mekong Delta in Vietnam where about 2 million ha are characterized by important environmental constraints. A study of the physiological response to Al toxicity of the rice plant in controlled hydroponic conditions is thus of great importance (Tang Van Hai et al., 1989).

A study of the effect of increasing aluminium concentration in a nutrient solution described above for rice (Table 8.1) has shown that Al exerts stimulation on dry matter production and nutrient uptake until a concentration threshold, which is influenced by nutrient solution+composition and cultivar (Tang Van Hai et al., 1989). Nitrogen uptake either as NH4 or-NO_3 was clearly influenced by aluminium concentration when its instantaneous value was+measured by the technique of the continuously flowing culture solution. The NH4uptakerate of two cultivars was such that the more sensitive variety to aluminium took up less+NH4and acidified less the culture solution flowing through the root system with aresidence time of a fewhours.

On the other hand, for plants like cereals, it is useful to identify criteria for early diagnosis of Al resistance with respect to grain yield before reaching maturity. Therefore, Tang Van Hai et al. (1993) evaluate whether the grain yield of 11 rice cultivars grown in nutrient solution with high Al concentration can be related to plant characteristics at the vegetative stage. Rice cultivars were supplied with a solution characterized by the composition described in Table 8.1. To test the toxicity of Al, a concentration of 150 and 400 μM Al was chosen since a concentration of Al between 15-45 μM stimulates the metabolic processes. In contrast to biomass production and mineral concentrations in shoots and roots, the tilering capacity (total number of tillers per plant) is a remarkable characteristic for early assessment of the effect of Al on grain production, as maximum tillering occurs 35-45 days after planting.

In waterlogged fields, the yield losses of rice in acid soils can be explained by the toxicity of aluminium and iron (Fageria et al., 1988; Genon et al 1994; Moore et al. 1990; Tang Van Hai et al., 1989, 1993). The increase of pH following reduction reactions can decrease the solubility of aluminium but induce the solubilization of ferrous ions well known for its toxicity (Ponnamperuma, 1972). Since the variations of pH and redox potential are interdependent in waterlogged soils, we are not able to fix a concentration of Al and Fe in aqueous phase of soil in field or pot experiments. The complex physicochemical behaviour of Al and Fe in soils justifies therefore the interest of controlled hydroponic studies in nutritive solution in order to better constrain the consequences of Al and Fe excess on the rice growth. In hydroponic experiment using rice-solution described in Table 8.1, the ncrease of Al (from 74 to 370 μM) and Fe (from 18 to 357 μM) concentrations decrease the grain yield and biomass. The toxicity threshold of Fe (125-357 μM) is relatively low relative to the field threshold of 5000 μM (van Breemen & Moorman, 1978). The difference is certainly due to chemical speciation of Fe2+ in soil solution, influenced by the chemical equilibrium between solid and aqueous phases, and the complexing of organic anions (Quang et al., 1996). Moreover a high concentration of Fe2+ in soil solution does not necessarily coincide with a high concentration of Fe2+ in the aerated rhizosphere in which the formation of oxyhydroxides occur as a mechanism of protection against Fe toxicity (Becker & Ash, 2005).

Phosphate uptake rate did not affect plant biomass, but increase grain yield, i.e. it alleviated the yield losses due to increase of Al and Fe concentrations (Fig 8.3a). The increase of P uptake rate could result in the formation of Al-P complexes ($AlHPO_{4+}$) that reduce the Al3+ concentration (Quang et al., 1996) (Fig. 8.3b). When P concentration exceed Al concentration, uncomplexed P is in the form of $H_2PO_4^{2-}$. Conversely, the uncomplexed Al is mainly Al3+ and AlSO4+ (ratio of 3-4:1) (Quang et al., 1996). Iron is not involved in ionic associations and is mainly as Fe2+. Thus Fe toxicity is not reduced through a physicochemical process with P.

Some authors suggest that P could enhance oxidizing potential of the rhizosphere decreasing the availability of ferrous iron (van Breemen & Moorman, 1978). The effect of $AlHPO_{4+}$ complexes and formation of Fe oxides on the surface of roots are described in Fig. 8.4. Based on the statistical analysis performed by Quang et al. (1996), grain yield (Y) is related to the concentrations of the elements in nutritive solution (expressed in μM) by thefollowing equation:

$$Y = 5.47 - 0.0079\ Al - 0.0058\ Fe + 0.0098\ P\ (r^2=0.86)$$

These results cannot be directly extrapolated to the field because the soil solution can be re- alimented in Al and Fe through dissolution of solid phases, counteracting the positive effect of ionic complexation. However, controlled studies in hydroponics allow to understanding the mechanisms of mineral nutrition and can be extrapolated to the field by using equilibrium program between solid and aqueous phase, such as "Visual Minteq" and "PHREEQC".

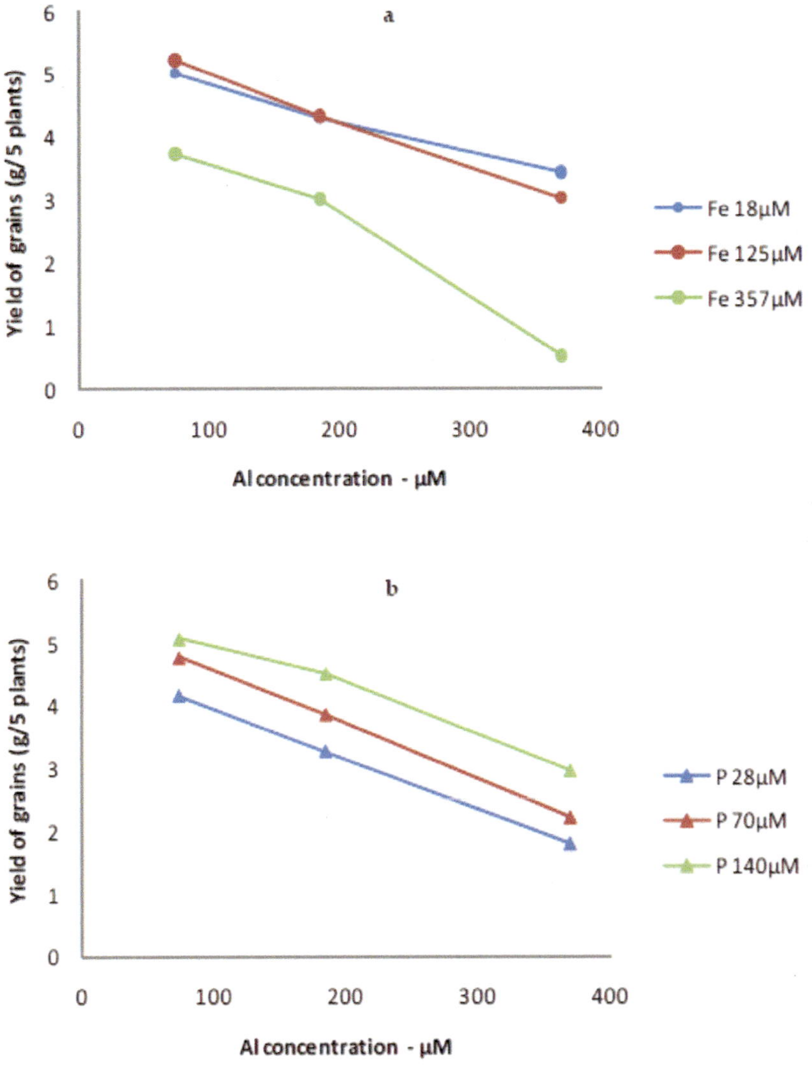

Fig. 8.3. Impact of the concentration of Al in nutritive solution on the grain yield: (a) for the three concentrations of Fe and (b) for the three concentrations of P (adapted from Quang et al., 1996)

In another study, Quang et al. (1995) aimed to study the impact of temperature gradient on rice growth in acid sulphate soils thanks to a hydroponic experiment with rice-nutrient solution (Table 8.1). Two varieties of rice (IR64 and X2) were cultivated in controlledconditions in phytotrons at 19/21°C and 28/32°C (day/night) for 56 days. After 7 days, the young plants were transferred to a rice-nutrient solution. Two series of 60 plants per variety tightened with cotton on plastic disks were grown on 20L tanks, and the solutions were renewed every 2 days for 2 weeks. From those plants, eight series of 5 plants per variety were transferred on 450-ml vessels to which nutrient solution was continuously supplied by peristaltic pumps at a rate of about 100 $ml.h^{-1}$. The P concentration was increased up to 2 $mg.l^{-1}$ to ensure this element was not limiting for plant growth. Besides the tillering retardation at lower temperature (°20C), the hydroponic experiment demonstrated that the translocation of Fe from roots to shoots was stimulated upon rising temperature leading to plant death on the most acid soil at °30C. Indeed severe reducing conditions were created at °30C: redox potential (Eh) dropped rapidly down to about 0 V which is a value out of the range of Fe^{3+}/Fe^{2+} buffering. Parallel to Eh drop, pH increased up to about 6.5-6 at °30C, which prevented plants from Al toxicity commonly observed at pH below 4.5.

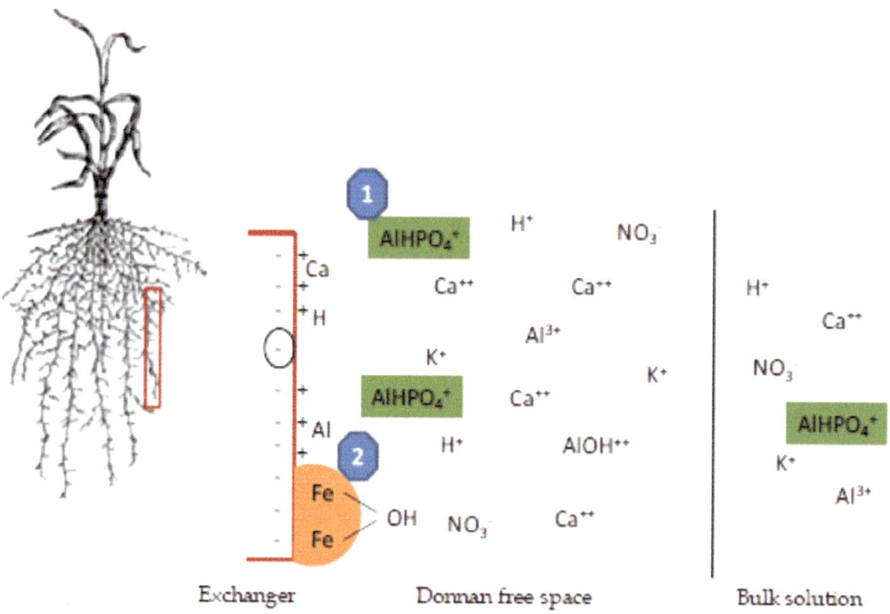

Fig. 8.4. Diagram of the ion exchange model adapted from Dufey et al. (2001). The two positive mechanisms of P to decrease the toxicity of Al and Fe are represented by (1) $AlHPO_{4+}$ complex and (2) formation of oxy-hydroxides layer on the surface of the roots.

UPTAKE OF POLLUTING ^{137}CS ELEMENT BY PLANTS

Since the radioactive fallout from nuclear weapons tests and the Chernobyl (1986) and Fukushima (2011) nuclear accidents, radioactive air plumes contaminated large territories. The ^{137}Cs is one of the major radionuclides and implies considerable environmental issues because of its relatively long half-life (30.17 years). Soil acts as a major sink-source compartment in ^{137}Cs fluxes through plant and food chains (Delvaux et al., 2001).

The uptake of radiocesium by plants is proportional to its concentration in the solution around roots. However, root uptake of ^{137}Cs is largely influenced by K status of the solution. Above 1 mM K, varying K concentration has no significant effect on trace Cs uptake (Shaw et al., 1992). Nonetheless, some studies in hydroponic conditions showed that root uptake of ^{137}Cs is significantly increased with decreasing K concentration in solution, at K concentration below 1 mM (Cline & Hungate, 1960; Smolders et al., 1996). Smolders et al. (1996) have used ^{137}Cs spiked nutrient solution experiment to grow wheat at four concentrations of K added as KNO_3 (250 ,50 ,25 and 1000 μM). They have shown that ^{137}Cs activity concentrations in 18-day old plants drastically increased (123-fold in the shoot and 300-fold in the root) when K concentration decreased from 1000 μM to 25 μM. Such low K concentration around the roots are not abnormal in soil solution. So, amongst the different fertilizers, potassium fertilizers are generally most efficient in reducing ^{137}Cs transfer (Nisbet et al., 1993).

The behavior of ^{137}Cs in soils is largely affected by the rhizosphere solution composition. This latter differs from the bulk soil solution since many factors (pH, ionic strength, redox potential, presence of root exudates and specific rhizospheric microflora) affect the physicochemical processes controlling trace element mobility and availability, making the rhizosphere a unique environment where plant roots and soil minerals strongly interact (McLaughlin et al., 1998). In this regard, the bioavailability of radiocesium varies extensively between soils because of differences in (i) ^{137}Cs retention, affecting ^{137}Cs concentration in solution, and hence its supply to plant roots, and (ii) K availability, affecting the ^{137}Cs root uptake process. Based on these concepts, Absalom et al. (1999) developed a soil-plant ^{137}Cs transfer model in which the soil solution concentrations of ^{137}Cs and K were estimated from (i) the clay content and (ii) the exchangeable K content, and also from the total ^{137}Cs content in soil and the time after the contamination. More precisely, radiocesium is specifically retained on vermiculitic sites neighboring micaceous wedge zones (Delvaux et al. 2001; Maes et al., 1999). In acid soils, hydroxyl interlayered vermiculite weakly retain radiocesium. Organic matter can therefore influence the retention of trace Cs through Al complexation which deplete the concentration of Al in the soil solution and hence impedes Al-interlayering and maintains the interlayer sites accessible for Cs retention (Delvaux et al., 2001). Organic matter can also increase ^{137}Cs soil-to-plant transfer by a dilution effect of highly fixing sites born by mica-like minerals (illite, vermiculite, micaceous mixed-layered clays), particularly in upper soil horizons (Kruyts & Delvaux, 2002).

In addition, the rhizospheric mobilization of ^{137}Cs (Delvaux et al., 2000; Kruyts et al., 2000) has been characterized through experimental cropping device adapted from Hinsinger et al. (1992) (Fig. 8.2a). In this device, the soil samples were pre-treated to be homoionic with Ca^{2+} ions, whereas the plantlets of ryegrass previously germinated in a K-free nutrient medium. The soil artificially ^{137}Cs contaminated (~1000 Bq cm^{-3}) was mixed with a 10 $g.l^{-1}$ agar gel prepared by using a solution free of potassium (Hinsinger et al. 1992). The nutrient concentrations of solution were in macroelements (mM): 3.5 $Ca(NO_3)2.4H_2O$, 1 $MgSO_4$, 1 NaH_2PO_4; in microelements (μM), 10 H3BO3, 100 FeEDTANa, 20 MnCl2, 0.2 $ZnSO_{4,}$ 0.2 $CuSO_{4,}$ 0.2 $Na2MoO4.2H_2O$. The close contact between soil and root consists of a macroscopic rhizosphere intensifying the soil-root interaction. Using this device with a collection of 47 soil horizons from 17 pedons of widely varying soil properties from semi- natural environments (Delvaux et al., 2000), the ^{137}Cs soil-plant transfer factor varied almost200-fold between soil materials and was strongly negatively correlated to soil vermiculite content. Kruyts et al (2000) investigated the relative contributions of the horizons of forest soil in the rhizospheric mobilization of radiocesium. Assuming negligible horizon to horizon transfer and equivalent root exploration in each soil horizon, the respective contributions of the soil horizons to the ^{137}Cs soil-to-plant transfer were 96.7% in Of (organic fragmented horizon), 0.13% in OAh (transitional organo-mineral horizon), 1.34% in Ah (organo-mineral horizon), and 1.84% in Bw (illuvial horizon). Such contributions are quite similar to those measured by Thiry et al. (2000) in the same forest soil, but using a pot experiment approach involving young spruce plants. This suggests that short experimental measurements are well designed to provide reproducible constrain on the ^{137}Cs mobilization, which provides a way to avoid time consuming pot experiments with longer procedures.

The soil-root interaction has been presented in terms of chemical processes occurring in the rhizosphere. The role of fungi, particularly the mycorrhizal fungal species in this interaction must be developed, from a mechanistic point of view. In this regard, monoxenic culture systems on a synthetic medium rapidly became a powerful system for studies of physiological and element transport processes in mycorrhizal symbioses (Rufyikiri et al., 2005). Among the microbial species, arbuscular mycorrhizal (AM) fungi are obligate symbionts that colonize the root cortex developing extraradical mycelia (ERM) that ramifies in the soil. It was shown that the ERM of an AM fungus can take up, possibly accumulate and unambiguously translocate ^{137}Cs to the roots (Declerck et al., 2003).

UPTAKE AND TRANSPORT OF SILICON IN PLANT

Silicon (Si) is the second most mass-abundant element (28.8%) of the Earth's crust (Wedepohl 1995) and it occurs in a large range of minerals at the Earth's surface, ranging from <0.5 wt% to ~47 wt% in the pedosphere (McKeague & Cline, 1963). As ultimate source, chemical weathering of silicate minerals liberates dissolved Si as monosilicic acid. The H_4SiO_4 contributes to soil formation through biogeochemical reactions such as neoformation of secondary minerals, adsorption onto Fe and Al (hydr)oxides and uptake by plants. The H_4SiO_4 concentration in the soil solution is controlled by biogeochemical processes and ranges from 0.01 to 1.99 mM (Karathanasis, 2002), with most common concentrations of about 0.6–0.1 mM (Faure, 1991).

In terrestrial ecosystems, Si is distributed between plant and soil before its land-to-rivers transfer since it is taken up as aqueous monosilicic acid (H4SiO4) and translocated to transpiration sites (where it polymerizes as amorphous biogenic opal called phytolith), which returns to the soil within organic residues (Smithson, 1956). By accumulating in terrestrial plants to a similar extent as some major macronutrients (%10-0.1 Si dry weight), Si becomes largely mobile in the soil-plant system. Understanding the soil-plant Si cycle in surface environments and the dissolved Si export from soils into rivers is crucial (Cornelis et al., 2011) given that the marine primary bio-productivity depends on the availability of H_4SiO_4 for phytoplankton CO2-consumers that requires Si. In the terrestrial Si cycle, higher plants largely contribute to the global Si cycle since their annual Si production ranges from 1.7 to $5.6*10^{12}$ kg Si yr^{-1} (Conley, 2002), which rivals Si production of phytoplankton-diatoms in the oceans ($10*6.7^{12}$ kg Si yr^{-1}) (Tréguer et al., 1995).

The biomineralization of amorphous silica seems to be restricted to some plant families (Epstein, 1999; Hodson et al., 2005) and appear in various shapes depending on the location of Si deposits and plant species (Carnelli et al., 2001, 2004). Based on active, passive or exclusive mechanisms of Si uptake, plant species are classified as high-, intermediate- or non-accumulator, respectively (Takahashi et al., 1990). A classification of the plant kingdom shows that the majority of Si high-accumulators (1–10 wt% Si in shoots) belong to the monocotyledons (e.g. banana, bamboo, sugar cane, soybean, rice, wheat, barley, sorghum and oat), while most dicotyledons absorb Si passively (0.5–1.0 wt% Si in shoots), and some dicots such as legumes have limited Si uptake (<0.5 wt% Si in shoots) (Liang et al., 2007; Ma et al., 2001; Ma & Takahashi, 2002).

SILICON NUTRITION IN PLANTS

The essentiality of Si for terrestrial plants is still extensively debated (Richmond & Sussman, 2003; Takahashi et al., 1990). So far, only two groups of plants are known to have an absolute and quantitatively major requirement for Si: the diatoms and other members of the yellow-brown or golden algae, the Chrysophyceae and the Equisitaceae (Epstein, 1999). Silicon is not considered as an essential element for higher plants, but its beneficial effects on growth have been reported in a wide variety of crops, including rice, wheat, barley, and cucumber (Korndörfer and Lepsch, 2001; Ma et al., 2001; Ma & Takahashi, 2002; Ma, 2004). There is a general consensus that Si improves the plant resistance to various biotic and abiotic stresses. Silica deposition in leaves is a resistant structural component (Raven, 1983), providing a more upright position, which favours light interception (Epstein, 1994; Marschner, 1995), and thus can play a role in the C cycle promoting photosynthesis. Moreover, biogenic silica in plant tissues creates a hard outer layer that serves as a defense against fungal and insect attacks (Bélanger et al., 2003; Kablan et al., 2011; Sangster et al., 2001; Vermeire et al., 2011). Finally, it is widely accepted that Si alleviates the toxicity of Al and other metal ions such as Mn in higher plants (Liang et al., 2007). Currently, there is no evidence of Si involved in plant metabolism (Ma et al., 2001) since no Si bearing organic compound has been identified in higher plants to date (Knight & Kinrade, 2001). The beneficial effects of Si remain reduced under optimum growth conditions, but are more obvious under stress conditions (Epstein, 1994; Bélanger et al., 1995). The Si accumulators wheat (*Triticum aestivum*) and rice (*Oryza sativa*), the premier crops for the nutrition of mankind, are susceptible to a variety of diseases if the Si supply is low (Epstein, 2001). Understanding the mechanisms that control Si uptakes by terrestrial plants (both for high and non-accumulator Si plants) in controlled hydroponics conditions is thus essential.

In **temperate forest ecosystems**, previous field studies show that the Si uptake varies considerably among tree species, ranging between 2 and 44 kg Si ha^{-1} yr^{-1} (Bartoli 1983; Gérard et al. 2008; Cornelis et al. 2010a). The different accumulation of Si in aerial parts may be due to the varying abilities of Si root uptake, soil mineralogical composition, transpiration rate and availability of silicic acid in the forest soil solution. Furthermore, such a difference of Si concentration in forest vegetation could affect tree seedling growth. The beneficial effect of Si supply on pine seedling growth has been demonstrated under water- stress conditions (Emadian & Newton, 1989), but remains poorly explored in optimal conditions.

With a hydroponic study, Cornelis et al. (2010b) aimed to understand the contrasting Si uptake between Douglas fir and Black pine and quantify the effect of Si on the growth of tree seedlings. For this purpose, *Pseudostsuga menziesii* and *Pinus nigra* seedlings were grown for 11 weeks in hydroponics (device 1) with a wide range of Si concentration in the nutrient solution (0.2-1.6 mM). The seeds were germinated for 15 days in the dark at day/night temperatures of 20/18°C. They were then weaned for 30 days in nutrient solution tanks before uniform seedlings were selected for the experiments. Batches of six seedlings were grown in separate cylindrical PVC pots containing 2.5 l of nutrient solution and placed on a perforated plate of expanded polystyrene which limits water loss by evaporation (Fig. 8.5). The experiment was conducted over 11 weeks in growth chambers at conditions described above (Table 8.1).

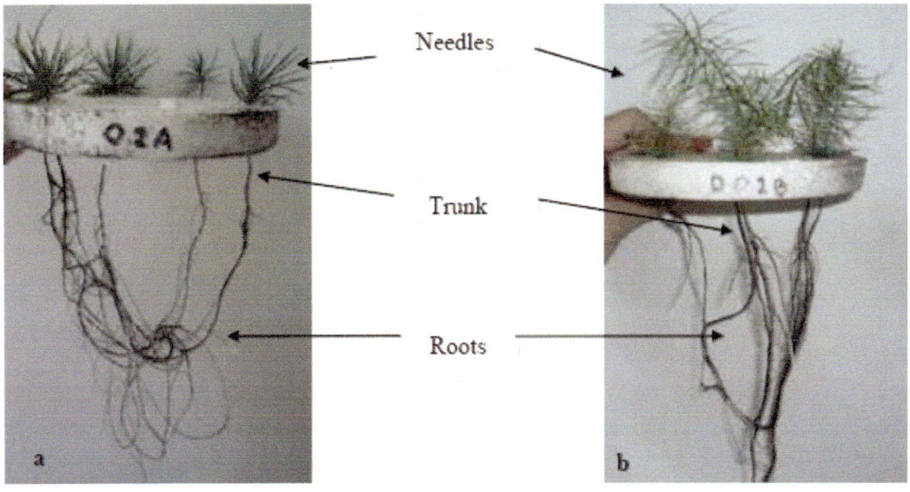

Fig. 8.5. Different parts of tree seedlings cultivated in hydroponics (from Cornelis et al. 2010b)

This experimental study using tree seedlings in hydroponic conditions confirms the in-situ observation of higher Si accumulation in Douglas fir leaves as compared with those of Black pine. Cornelis et al. (2010b) demonstrate that the mechanisms of Si uptake in hydroponics are identical between tree species: passive (mass-flow driven) uptake at realistic Si concentrations (0.2 mM) and rejective uptake at higher Si concentrations in nutrient solution. The contrasting Si accumulation in the leaves of coniferous tree seedlings could be attributed to the significant difference in the transpiration rates. The higher Si concentrations in nutrient solutions (0.8 and 1.6 mM) do not affect the growth of tree seedlings, probably because they are not subjected to stress conditions. This finding helps us to understand the mechanisms controlling the different Si uptakes by forest vegetations in order to better predict Si pathways in soil-tree systems.

In **banana plantations**, there are some evidences that fungic resistance is associated with Vertisols characterized by high concentration of Si in soil solution, whereas pathogens incidence seems to be high on largely desilicicated ferrallitic soil (reviewed by Delvaux,1995). Banana roots are able to induce silicate dissolution thereby increasing silicon availability in the rhizosphere (Hinsinger et al., 2001; Rufyikiri et al., 2004). In this respect, a study was conducted in hydroponic by Henriet et al. (2006) (device 2) to investigate the effect of Si supply on banana growth, water uptake and nutrient uptake under optimal and sub- or non-optimal conditions (Fig. 8.6). An experimental hydroponic device was designed to maintain the plants in optimal nutritional conditions and to allow the measurement of the daily water and nutrient uptake. The experiment was conducted in a growth chamber at conditions defined above (Table 8.1). Banana plants were grown in cylindrical PVC pots containing 2.5 l of nutrient solution which was continuously renewed at a rate of 104 ml h^{-1} pot^{-1}, corresponding to an expected residence time of the solution in the pot close to 24 h. The composition of the nutrient solution is described in Table 8.1. Four treatments were applied in addition to a control, defined by concentrations of silicon in the nutrient solution of 0.0 (control), 0.08, 0.42, 0.83 and 1.66 mM, and to cover as much as possible the range of Si concentration that can be found in soil solutions (0.01-1.99mM). As described by Henriet et al. (2006), Si was supplied as H_4SiO_4 obtained by dissolving sodium metasilicate in demineralized water, followed by leaching on a protonated cation-exchange resin (Amberlite® IR-120). Neither Si precipitation nor H_4SiO_4 deprotonation was expected because Si concentration was below the solubility limit (<1.79 mM Si) and pH ranged between 5 and 6.5 (Stumm & Morgan 1996).

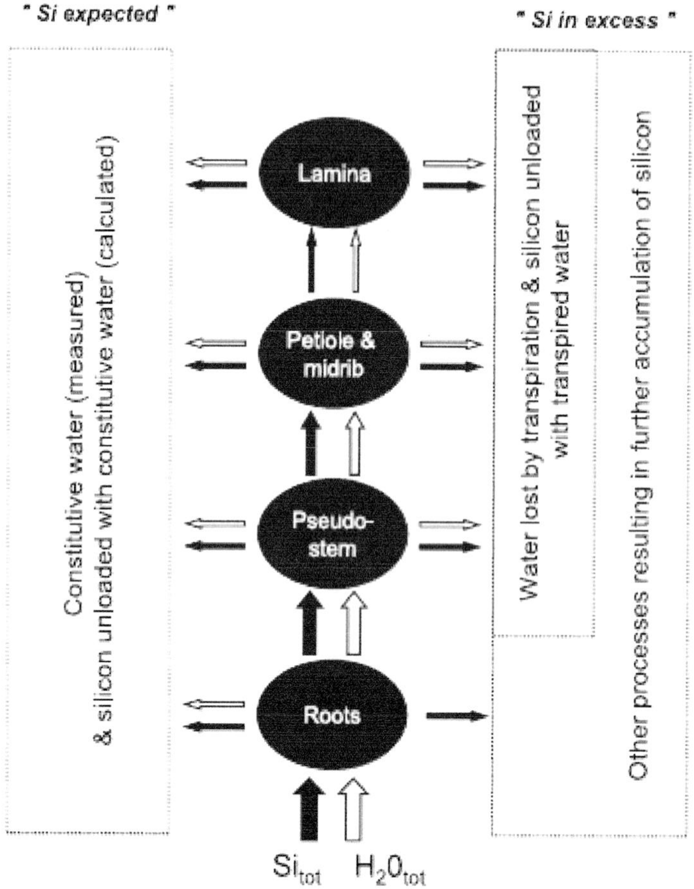

Fig. 8.6. Quantitative model of water and silicon movement and distribution in the plant (from Henriet et al., 2006)

Henriet et al. (2006) demonstrate that the level of Si supply did not affect plant growth, nor the rate of water and nutrient uptake. The rate of Si uptake and the

Si concentration in plant tissues increased markedly with the Si supply. The field studies confirm these results, since the leaf Si concentration is positively correlated with plant-available Si content ($CaCl_2$- extractable), soil Si content and total reserve in weatherable minerals (Henriet et al. 2008a). Furthermore, soil weathering stage directly impacts the soil-to-plant transfer of silicon, and thereby the stock of biogenic Si in a soil–plant system involving a Si-accumulating plant (Henriet et al., 2008b). At the highest Si concentrations (1.66 mM), silicon absorption was essentially driven by mass flow of water (passive transport). However, at lower Si concentrations (0.02–0.83 mM), it was higher than its uptake by mass flow and caused the depletion of silicon in the nutrient solution, suggesting the existence of active processes of silicon transport in banana plant.

Recent studies demonstrate the existence of active Si transporters in rice roots (*Lsi1* and *Lsi2*) and shoots (*Lsi6*), responsible for the high Si uptake capacity or rice (Ma et al. 2006, 2007; Yamaji et al. 2008). *Lsi1* is an influx transporter of silicic acid, while *Lsi2* is an active efflux transporter of the same chemical compound. These findings imply that the active transport process operates in some places along the Si trajectory from the root to xylem loading sites. For banana plants, the increasing Si content in shoot organs (pseudostem < petiole and midrib < young lamina < old leaf) supports the major role of transpiration in silicon accumulation and was not dependent on silicon supply (Fig. 8.6).

GEOCHEMICAL TRACING OF SILICON UPTAKE

An additional approach to better understand Si uptake in plant is to use geochemical tracers such as **stable silicon isotopes** or **Ge/Si ratios**.

Silicon has three stable isotopes of atomic mass units ^{28}Si (27.976927), ^{29}Si (28.976495), and ^{30}Si (29.973770) with respective abundance 92.23%, 4.67%, 3.10% (Faure & Mensing, 2005). Those isotopes can be discriminated by chemical, physical or biological processes, inducing an isotope fractionation between two compartments. Phytoliths were shown to display large Si isotope variations (Douthitt, 1982) supporting an impact of plant biological uptake on Si isotopes. Therefore, studying silicon stable isotope fractionation within plants is promising to better understand plant physiological mechanism. Banana plants represent an ideal case study to quantify plant-induced Si isotopic fractionation since banana is accumulating silicon in the sheath cells of vascular bundles in leaves and pseudostem. Through hydroponic experiment in controlled conditions, Opfergelt et al. (2006a, 2006b) aimed to quantify the Si isotope fractionation between dissolved Si source to banana plant, and between plant parts. The banana plants were cultivated in the same conditions than for the study of Henriet et al. (2006) (device 2). The Si isotope compositions of the plant parts and of the nutrient solution were determined using a multicollector plasma source mass spectrometer (MC-ICP-MS). The results are expressed as $\delta^{29}Si$ relative to the NBS28 standard, with an average precision of ±0.08‰, with $\delta^{29}Si$ = [($^{29}Si/^{28}Si$*sample*)/($^{29}Si/^{28}Si$*NBS28*)- 1]×1000. A positive $\delta^{29}Si$ value stands for a heavy Si isotope composition relative to the standard, i.e. relatively enriched in heavy Si isotopes, whereas a negative $\delta^{29}Si$ value represents enrichment in light Si isotopes. The results of this study performed in hydroponics confirmed that plants fractionate Si isotopes by depleting the source solution in light Si isotopes ($\delta^{29}Si$ = +0.06‰ in nutrient solution and -0.34‰ in root cortex) (Fig. 8.7) with a plant-source fractionation factor ($^{29}\varepsilon$) of ‰0.11±0.4-.

More specifically, Si isotope compositions of the various plant parts indicate that a similar mechanism of heavy isotopes discrimination should occur at three levels in the plant: at the root epidermis, for xylem loading and for xylem unloading. At each step, a preferential passage of light Si isotopes contributes to a progressive isotopic fractionation of the solution moving from the uptake sites in roots to the transpiration termini in lamina, and results in reproducible Si isotope compositions of the plant organs consistent with their position along the trajectory of the solution. On mature banana (*Musa acuminata* Colla, cv Grande Naine) from Cameroon, $\delta^{29}Si$ values range from -0.11‰ in the pseudostem to +0.51‰ in the lamina, with a net increase towards heavier isotopic composition in the upper parts of the plant (Opfergelt et al. 2006b, 2008). This strongly accords with results obtained in vitro on banana plantlets cultivated in hydroponics, where the Si isotope composition was increasingly heavier from pseudostems to lamina (Fig. 8.7).

Another useful tracer to understand Si uptake in plant is the **Ge/Si ratio.** Germanium (Ge) is a trace element which behaves as a chemical analog to Si, but is fractionated relative to Si in plants. Therefore, the Ge/Si ratio is a promising tool to trace the plant impact on surface processes (Kurtz et al., 2002; Derry et al., 2005). A study was conducted in hydroponic by Delvigne et al. (2009) on banana plant to better understand the Ge/Si fractionation in plants. The growth chamber conditions and the nutrient solution are described in Table 8.1. Initial nutrient solutions were doped both in Si (13.5 mg/l) and Ge (0.32 mg/l), and no additional Si or Ge was provided during the experiment (closed system for Si and Ge; as in device 1). In contrast with the finite pool kept for Si and Ge, the nutrient requirements were adapted to plant need and continuously supplied as dilute nutrient solutions with peristaltic pumps at a rate of 104 ml h^{-1} pot^{-1} (device 2). Ge was provided as Germanic acid Ge(OH)4 from GeCl4 reaction with NaOH (Azam, 1974). Ge concentrations were higher than those observed in natural waters (20–1 ng/l Ge; Mortlock & Froelich, 1996), but toxical evidence in barley appears at much higher Ge content in solution (>1.44 mg/l; Halperin et al., 1995).

Delvigne et al. (2009) demonstrate that Ge uptake by plants is similar to the Si uptake, and no discrimination against Ge appears to occur at the uptake step. This is in agreement with what is reported for rice roots (Takahashi et al., 1976; Ma et al., 2001). This contrasts with the Si isotopic observations (Opfergelt et al., 2006a) that demonstrate a preferential incorporation of the light Si isotopes at the root interface. The Si transporters likely responsible for the isotope fractionation may not differentiate between Ge and Si. This might explain why Si-rich plants like rice show also a tendency to accumulate more Ge. There is a large Ge/Si fractionation between roots and shoots, with Ge being trapped in roots as a probable response to the toxicity of this element for the plant. This selective Ge entrapment in roots is likely controlled by the relative higher affinity of Ge to form organic complexes than Si. Owing to Ge trapping in roots, the xylem sap entering the shoots is depleted in Ge explaining the low Ge/Si reported in phytoliths (Derry et al., 2005; Blecker et al., 2007). In shoots Ge closely follows Si throughout the transpiration pathway with no evidence of any discrimination whereas Si isotopes are significantly fractionated along the transpiration pathway (Opfergelt et al., 2006a, 2006b, 2008).

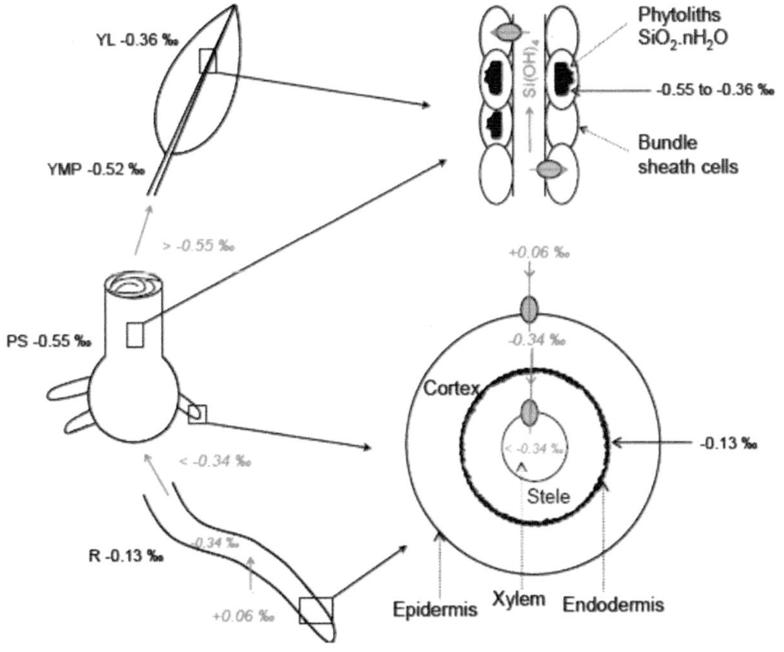

Fig. 8.7. Proposed schematic model of global Si isotope fractionation in banana plant. Black = Si isotope composition ($\delta^{29}Si$) of solid precipitated silica (phytoliths); grey and italic = Si isotope composition ($\delta^{29}Si$) of aqueous H4SiO4. R = roots, PS = pseudostem, YMP = young midribs and petioles, YL = young lamina (from Opfergelt et al., 2006a).

This supports that combining hydroponic studies on stable Si isotopes and Ge/Si ratio in plants relative to a nutrient solution is useful to investigate Si uptake, transport and deposition in plant, and can provide new insights on Si and Ge mobility through plants.

AT A GLANCE

- In field experiments, it is difficult to isolate factors affecting the nutrition of plants and mechanisms of element uptake. Indeed, the dynamic of nutrient (Ca, Mg, Na, K) beneficial (Si), toxic (Al) and polluting (^{137}Cs) elements largely depends on the physico-chemistry of the bulk soil solution and more particularly of the rhizospheric solution. We have illustrated in this Chapter that hydroponic culture was a suitable device to better constrain each of the external factors influencing the transfer of elements in the soil-plant system (pH, redox potential, activity of elements, dissolution/neoformation of minerals ...). Moreover, hydroponic devices allow to simulate environmental changes and their impact on the plant growth and nutritional status (humidity, temperature, luminosity ...).
- An accurate knowledge of the dynamic of nutrient, toxic and polluting elements in soil- plant systems is vital to preserve the Earth's critical zone defined as "the external terrestrial layer extending from the outer limits of vegetation to the lower boundary of groundwater, inclusive of all liquid, gas, mineral and biotic components" (Brantley et al., 2007). This terrestrial system is permanently subject to environmental constraints which disrupt the homeostasis of the biogeochemical cycle of elements since the extent of human activities. The agronomic and environmental management of the soil-plant system is crucial to mitigate global change processes, while ensuring biomass production, which is related to the future needs of humanity.

REFERENCES

Absalom, J.P., Young, S.D., Crout, N.M.J., Nisbet, A.F., Woodman, R.F.M., Smolders, E. & Gillet, A.G. (1999). Predicting soil to plant transfer of radiocesium using soil characteristics. *Environmental Science & Technology*, Vol.33, No.8, pp. 1218-1223

Avery, S.V. (1996). Fate of cesium in the environment: distribution between the abiotic and biotic components of aquatic and terrestrial ecosystems. Journal of Environmental Radioactivity, Vol. 30, No.2, pp. 139-171

Azam, F. (1974). Silicic-acid uptake in diatoms studied with [68Ge] Germanic acid tracer. *Planta*, Vol. 121, pp. 205–212

Bartoli, F. (1983) The biogeochemical cycle of silicon in two temperate forest ecosystems. *Environmental Biogeochemistry Ecol. Bull.*, Vol.35, pp. 469– 476

Becker, M. & Asch, F. (2005). Iron toxicity in rice-conditions and management concepts. *Journal of Plant Nutrition and Soil Science*, Vol.168, pp. 558-573

Bélanger, R.R., Benhamou, N., & Menzies, J. G. (2003) Cytological evidence of an active role of silicon in wheat resistance to powdery mildew (Blumeria graminis f. sp tritici). *Phytopathology*, Vol.93, No.4, pp. 402–412

Bélanger, R.R., Bowen, P.A., Ehret, D.L., & Menzies, J. G. (1995). Soluble silicon: its role in crop and disease management of greenhouse crops. *Plant Disease*, Vol.79, pp. 329– 336

Bernal, J.H. & Clark, R.B. (1997). Mineral acquisition of aluminium-tolerant and –sensitive sorghum genotypes grown with varied aluminium. *Communications in soil science and plant analysis*, Vol.28, pp. 49-62

Blecker, S.W., King S.L., Derry L.A., Chadwick O.A., Ippolito J.A. & Kelly E.F. (2007). The ratio of germanium to silicon in plant phytoliths: Quantification of biological discrimination under controlled experimental conditions. *Biogeochemistry*, Vol.86, pp. 189– 199

Brantley S., Goldhaber M.B. & Vala Ragnarsdottir K. (2007). Crossing disciplines and scales to understand the critical zone. *Elements*, Vol. 3, pp. 307-314

Carter, M.W. (1993). *Radionuclides in the Food Chain*, Springler Verlag, New York

Carnelli, A.L., Madella, M. & Theurillat, J.-P. (2001). Biogenic silica production in selected alpine plant species and plant communities. *Annals of Botany*, Vol.87, pp. 425–434

Carnelli, A.L., Theurillat, J.-P. & Madella, M. (2004). Phytolith types and type-frequencies in subalpine-alpine plant species of the European Alps. *Review of Palaeobotany Palynology*, Vol.129, pp. 39–65

Cline, J.F. & Hungate, F.P. (1960). Accumulation of potassium, caesium-137 and rubidium- 86 in bean plants grown in nutrient solutions. *Plant Physiology*, Vol.35, pp. 826–829

Conley, D.J. (2002). Terrestrial ecosystems and the global biogeochemical silica cycle. Global Biogeochemical Cycles, Vol.16, No.4, pp. 1121, doi:10.1029/2002GB001894

Cornelis, J.-T., Ranger, J., Iserentant, A. & Delvaux, B. (2010a) Tree species impact the terrestrial cycle of silicon through various uptakes. *Biogeochemistry*, Vol.97, pp. 231– 245

Cornelis, J.-T., Delvaux, B. & Titeux, H. (2010b). The contrasting silicon uptakes by coniferous trees: a hydroponic experiment on young seedlings. *Plant and Soil*, Vol.336, pp. 99–106

Cornelis, J.-T., Delvaux, B., Georg, R.G., Lucas, Y., Ranger, J. & Opfergelt, S. (2011). Tracing the origin of dissolved silicon transferred from various soil-plant systems towards rivers: a review. *Biogeosciences*, Vol.8, pp. 89-112

Dahlgren, R.A., Vogt, K.A. & Ugolini, F.C. (1991). The influence of soil chemistry on fine root aluminium concentrations and root dynamics in a subalpine spodosol, Washington State, USA. *Plant and Soil*, Vol.133, pp. 117-129

Declerck, S., Dupré de Boulois, H., Bivort, C. & Delvaux, B. (2003). Extraradical mycelium of the arbuscularmycorrhizal fungus*Glomus lamellosum*can take up, accumulate and translocate radiocaesium under root-organ culture conditions. Environmental Microbiology, Vol.5, pp. 510–516

Delvaux, B. (1995). Soils, In: *Bananas and plantains*, S. Gowen, (Ed), 230-257, Chapman & Hall, London, UK

Delvaux, B., Kruyts, N. & Cremers, A. (2000) Rhizospheric mobilization of radiocaesium in soils. *Environmental Science & Technology*, Vol.34, pp. 1489–1493

Delvaux, B., Kruyts, N., Maes, E. & Smolders, E. (2001). Fate of radiocesium in soil and rhizosphere. In: *Trace elements in the rhizosphere*, G.R. Gobran, W.W. Wenzel, E. Lombi, (Eds), 61-91, CRC Press LLC, UK

Delvigne, C., Opfergelt, S., Cardinal, D., Delvaux, B. & André, L. (2009) Distinct silicon and germanium pathways in the soil-plant system: evidence from banana and horsetail. *Journal of Geophysical Research – Biogeosciences*, Vol.114, G02013, doi:10.1029/2008JG000899.

Derry, L.A., Kurtz A.C., Ziegler K. & Chadwick O.A. (2005). Biological control of terrestrial silica cycling and export fluxes to watersheds. *Nature*, Vol.433, pp. 728– 731

Douthitt, C.B. (1982). The geochemistry of the stable isotopes of silicon. Geochimica et Cosmochimica Acta, Vol.46, pp. 1449–1458.

Dufey, J., Genon, J., Jaillard, B., Calba, H., Rufyikiri, G. & Delvaux B. (2001). Cation exchange on plant roots involving aluminium: experimental data and modeling, In: Trace Elements in the Rhizosphere, R. George, W.W. Wenzel, E. Lombi (Eds.), 227-252, CRC Press LLC

Dufey, J. & Braun, R. (1986). Cation-exchange capacity of roots – titration, sum of exchangeable cations, copper adsorption.*Journal of Plant Nutrition*, Vol.9, No.9, pp. 1147-1155

Eeckman,J.P.&Laudelout,H.(1961).Chemicalstabilityofhydrogen-montmorillonite suspensions. *Kolloid Z.*, Vol.178, pp. 99- 107

Emadian, S.F. & Newton, R.J. (1989). Growth enhancement of loblolly pine (Pinus taeda L.) seedlings by silicon. Journal of Plant Physiology, Vol.134, pp. 98–103

Epstein, E. (1994). The Anomaly of silicon in plant biology. *Proceedings of the National Academy of Sciences of United States of America*, Vol.91, pp. 11–17

Epstein, E. (1999). Silicon. *Annual Review of Plant Physiology*, Vol.50, pp. 641–664

Epstein, E. (2001). Silicon in plants: Facts vs. Concepts, In: *Silicon in agriculture*, L. E. Datnoff, G. H. Snyder & G. H. Korndörfer (Eds), 1-15, Elsevier, The Netherlands

Fageria N.K., Baligar, V.C. & Wright, R.J. (1988) Aluminum toxicity in crop plants. *Journal of Plant Nutrition*, Vol.11, pp. 309-319

Faure, G. (1991). *Principles and application of inorganic geochemistry*, New York, MacMillan Faure, F. & Mensing, T. M. (2005). *Isotopes: Principles and Applications.* Wiley and sons

Genon, J., De Hepce, N., Dufey, J.E., Delvaux, B. & Hennebert, P.A. (1994). Iron toxicity and other chemical constraints to rice in highland swamps of Burundi. *Plant and Soil* Vol.166, pp. 109-115

Gérard, F., Mayer, K. U., Hodson, M. J. & Ranger, J. (2008). Modelling the biogeochemical cycle of silicon in soils: Application to a temperate forest ecosystem. *Geochimica et Cosmochimica Acta*, Vol.72, No.3, pp. 741–758

Gilbert, M. & Laudelout, H. (1965) Exchange of hydrogen ions in clays. *Soil Science*, Vol.100, pp. 157-162

Halperin, S.J., Barzilay, A., Carson, M., Roberts C., Lynch, J. & Komarneni, S. (1995). Germanium accumulation and toxicity in barley. *Journal of Plant Nutrition*, Vol.18, pp. 1417– 1426

Henriet, C., Draye, X., Oppitz, I., Swennen, R. & Delvaux, B. (2006). Effects, distribution and uptake of silicon in banana (Musa spp.) under controlled conditions. *Plant and Soil*, Vol.287, pp. 359–374

Henriet, C., Draye, X., Dorel, M., Bodarwe, L. & Delvaux, B. (2008a). Leaf silicon content in banana (Musa spp.) reveals the weathering stage of volcanic ash soils in Guadeloupe. *Plant and Soil*, Vol.313, pp. 71–82

Henriet, C., De Jaeger, N., Dorel, M., Opfergelt, S. & Delvaux, B. (2008b) The reserve of weatherable primary silicates impacts the accumulation of biogenic silicon in volcanic ash soils. *Biogeochemistry*, Vol.90, pp. 209–223

Hinsinger, P. (1998). How do plant roots acquire mineral nutrients? Chemical processes involved in the rhizosphere. *Advances in Agronomy*, Vol.64, pp. 225–265

Hinsinger, P., Jaillard, B. & Dufey, J.E. (1992). Rapid weathering of a trioctahedral mica by the roots of ryegrass. *Soil Science Society of America Journal*, Vol.56, pp. 977–982

Hinsinger, P., Dufey, J.E. & Jaillard, B. (1991). Biological weathering of micas in the rhizosphere as related to potassium absorption by plant roots, *Proceedings of the Int. Soc. Roots Res. Congress,* pp 98-105, In: Plant Roots and their Environment, McMichael, B.L., Persson, H. (Eds.), Uppsala, Sweden. 22–26 August 1988, Elsevier, Amsterdam, The Netherlands

Hinsinger, P., Barros, O.N.F., Benedetti, M.F., Noack, Y. & Callot, G. (2001). Plant-induced weathering of a basaltic rock: experimental evidence. *Geochimica et Cosmochimica Acta*, Vol. 5, pp. 137–152

Hoagland, D.R. & Arnon A.I. (1950). The Water-Culture Method for Growing Plants Without Soil, California Agricultural Experiment Station Circular 347, Berkeley

Hodson, M. J., White, P. J., Mead, A. & Broadley, M. R. (2005). Phylogenetic variation in the silicon composition of plants. Annals of Botany, Vol.96, pp. 1027–1046

Horst, WJ. (1995). The role of the apoplast in aluminium toxicity and resistance of higher plants: a review. *Zeitschrift fur Pflanzenernahrung und Bodenkunde*, Vol.158, pp. 419- 428

Ingestad, T. (1971). A definition of optimum nutrient requirements in birch seedlings. II. *Physiol Plant*, Vol.24, pp. 118–125

Kablan, L., Lagauche, A., Delvaux, B. & Legrève A. (2011). Reduction of Black Sigatoka disease on banana in response to silicon nutrition. *Plant Disease*, accepted.

Karathanasis, A.D. (2002). Mineral equilibria in environmental soil systems, In: *Soil Mineralogy with environmental applications*, J.B. Dixon & D.G. Schulze (Eds.), 109-151, Soil Science Society of America, Madison, USA

Keltjens, W.G. (1995). Magnesium uptake by Al-stressed maize plants with special emphasis on cation interactions at root exchange sites. *Plant and Soil*, Vol.171, pp. 141-146

Knight, C.T.G. & Kinrade, S.D. (2001). A primer on the aqueous chemistry of silicon, In: *Silicon in agriculture*, L. E. Datnoff, G. H. Snyder & G. H. Korndörfer (Eds), 57-84, Elsevier, The Netherlands

Kochian, LV. (1995). Cellular mechanisms of aluminium toxicity and resistance in plants. *Annual Review of Plant Physiology and Plant Molecular Biology,* Vol.46, pp 237-260.

Korndörfer, G.H. & Lepsch, I. (2001) Effect of silicon on plant growth and crop yield, In: *Silicon in agriculture*, L. E. Datnoff, G. H. Snyder & G. H. Korndörfer (Eds), 133-147, Elsevier, The Netherlands

Kruyts, N., Thiry, Y., & Delvaux, B. (2000). Respective horizon contributions to 137Cs soil to- plant transfer: a rhizospheric experimental approach. *Journal of Environmental Quality*, Vol.29, pp. 1180–1185

Kruyts, N. & Delvaux, B. (2002). Soil organic horizons as a major source for radiocesium biorecycling in forest ecosystems. *Journal of Environmental Radioactivity,* Vol.58, pp 175 –190

Kurtz, A., Derry, L.A & Chadwick, O.A. (2002). Germanium-silicon fractionation in the weathering environment. Geochimica et Cosmochimica Acta, Vol.66, pp 1525– 1537

Lahav, E. (1995). Banana nutrition. In: *Bananas and Plantains*, S. Gowen (Ed.), 258-316, Chapman & Hall, London, UK

Liang, Y.C., Sun, W., Zhu, Y.G. & Christie, P. (2007). Mechanisms of silicon-mediated alleviation of abiotic stresses in higher plants: A review. *Environmental Pollution*, Vol.147, pp. 422–428

Lindsay, W.L. (1979). *Chemical equilibria in soils*, John Wiley, New York

Lombi, E., Wenzel, W.W., Gobran, G.R. & Adriano, D.C. (2001). Dependency of phytoavailability of metals on indigenous and induced rhizosphere processes: a review. In: *Trace Elements in the Rhizosphere,* Gobran, G.R., Wenzel, W.W., Lombi, E. (Eds.), 3-24, CRC Press, London, UK

Ma, J. F., Miyake, Y. & Takahashi, E. (2001). Silicon as a beneficial element for crop plants, In: *Silicon in Agriculture*, L. E. Datnoff, G. H. Snyder & G. H. Korndörfer (Eds), 17– 39, Elsevier, Amsterdam, Netherlands.

Ma, J. F. & Takahashi, E. (2002) *Soil, fertilizer, and plant silicon research in Japan*, Elsevier, The Netherlands

Ma, J. F. (2004). Role of silicon in enhancing the resistance of plants to biotic and abiotic stresses, *Soil Science and Plant Nutrition*, Vol.50, pp. 11–18

Ma, J. F., Tamai, K., Yamaji, N., Mitani, N., Konishi, S., Katsuhara, M., Ishiguro, M., Murata, Y. & Yano, M. (2006) A silicon transporter in rice. *Nature*, Vol.440, pp. 688–691

Ma, J. F., Yamaji, N., Mitani, N., Tamai, K., Konishi, S., Fujiwara, T., Katsuhara, M. & Yano, M. (2007). An efflux transporter of silicon in rice. *Nature*, Vol.448, pp. 209–212

Maes, E., Vielvoye, L., Stone, W. & Delvaux, B. (1999). Fixation of radiocaesium traces in a weathering sequence mica → vermiculite → hydroxy interlayered vermiculite. *European Journal of Soil Science*, Vol.50, pp. 107–115

Marschner, H. (1995). *Mineral Nutrition of Higher Plants*, second ed., Academic Press, London, UK

McKeague, J. A. & Cline, M. G. (1963). Silica in soils. *Advances in Agronomy*, Vol.15, pp. 339– 396

McLaughlin, M.J.M., Smolders, E. & Merckx R. (1998). Soil-root interface: Physico-chemical processes, In: *Soil Chemistry and Ecosystem Health*, P.M. Huang (Ed.), chap. 8, SSSA Special Publication, American Society of Agronomy, Madison, WI

Moore, P.A.J., Attanandana, T., Patrick, W.H.Jr. (1990). Factors affecting rice growth on acid sulfate soils. *Soil Science Society of America Journal*, Vol.54, pp. 1651-1656

Mortlock, R.A. & Froelich, P.N. (1996). Determination of germanium by isotope dilution- hybride generation inductively coupled plasma mass spectrometry. *Analytica Chimica Acta,* Vol.332, pp. 277–284

Nagy, K.L. (1995). Dissolution and precipitation kinetics of sheet silicates, In: *Chemical Weathering Rates of Silicates Minerals*, A.F. White & S.L. Brantlev (Eds.), 173-234, Mineralogical Society of America, BookCrafters, Inc, Chelsa, Michigan

Nisbet, A.F., Konoplev, A.V., Shaw, G., Lembrechts, J.F., Merckx, R., Smolders, E., Vandecasteele, C.M., Lonsjo, H., Carini, F. & Burton, O. (1993). Application of fertilizers and ameliorants to reduce soil to plant transfer of radiocesium and radiostrontium in the medium to long-term – summary. *Science of the Total Environment*, Vol.137, pp. 173-182

Opfergelt, S., Delvaux, B., André, L. & Cardinal, D. (2008). Plant silicon isotopic signature might reflect soil weathering degree. *Biogeochemistry*, Vol.91, pp. 163-175

Opfergelt, S., Cardinal, D., Henriet, C., Draye, X., André, L. & Delvaux, B. (2006a). Silicon isotopic fractionation by banana (Musa spp.) grown in a continuous flow device. *Plant and Soil*, Vol.285, pp. 333– 345

Opfergelt, S., Cardinal, D., Henriet, C., André, L. & Delvaux, B. (2006b). Silicon isotopic fractionation between plant parts in banana: In situ vs. in vitro. *Journal of Geochemical Exploration*, Vol.88, pp. 224– 22

Ponnamperuma, F.N. (1972). The chemistry of submerged soils. *Advances in Agronomy*, Vol.24, pp. 26-96

Quang, V.D., Tang Van Hai, Kanyama, T.E. & Dufey, J.E. (1996). Effets combinés de l'aluminium, du fer, et du phosphore sur l'absorption d'ions et le rendement du riz (*Oryza sativa* L.) en solution nutritive. *Agronomie*, Vol.16, pp. 175-186

Quang, V.D., Tang Van Hai & Dufey, J.E. (1995). Effect of temperature on rice growth in nutrient solution and in acid sulphate soils from Vietnam. *Plant and Soil*, Vol.177, pp. 73-83

Raven, J. A. (1983). The transport and function of silicon in plants. *Biological Reviews*, Vol.58, pp. 179–207

Richmond, K. E. & Sussman, M. (2003). Got silicon? The non-essential beneficial plant nutrient. *Current Opinion in Plant Biology*, Vol.6, No.3, pp. 268–272

Rufyikiri, G., Declerck, S., Dufey, J.E. & Delvaux B. (2000a). Arbuscular mycorrhizal fungi might alleviate aluminium toxicity in banana plants. *New Phytologist*, Vol.148, pp. 343-352

Rufyikiri, G., Nootens, D., Dufey, J.E. & Delvaux, B. (2000b). Effect of aluminium on bananas (Musa spp.) cultivated in acid solutions. I. Plant growth and chemical composition. *Fruits* Vol.55, pp. 367–379

Rufyikiri, G., Dufey, J.E., Nootens, D. & Delvaux, B. (2001). Effect of aluminium on bananas (Musa spp.) cultivated in acid solutions. II. Water and nutrient uptake. *Fruits,* Vol.56, pp. 3–14

Rufyikiri, G., Dufey, J.E., Achard, R. & Delvaux, B. (2002). Cation exchange capacity and aluminium-calcium-magnesium binding in roots of bananas cultivated in soils and in nutrient solutions. *Communications in soil science and plant analysis*, Vol.33, pp. 991-1009

Rufyikiri, G., Genon J.G., Dufey, J.E. & Delvaux, B. (2003) Competitive adsorption of hydrogen, calcium, potassium, magnesium and aluminium on banana roots: experimental data and modeling. *Journal of Plant Nutrition*, Vol.26, No.2, pp. 351- 368

Rufyikiri, G., Nootens, D., Dufey, J.E. & Delvaux, B. (2004). Mobilization of aluminium and magnesium by roots of banana (*Musa* spp.) from kaolinite and smectite clay minerals. *Applied Geochemistry*, Vol.19, pp. 633-643

Rufyikiri, G., Kruyts, N., Declerck, S., Thiry, Y., Delvaux, B., Dupré de Boulois, H. & Joner, E. (2005). Uptake, assimilation and translocation of mineral elements in monoxenic cultivation systems, In: Soil Biology, S. Declerck, D.G. Strullu & J.A. Fortin (Eds.), 201-215, Springer-Verlag, Berlin, Germany

Sangster, A.G., Hodson, M.J. & Tubb, H.J. (2001). Silicon deposition in higher plants, In: *Silicon in Agriculture*, L. E. Datnoff, G. H. Snyder & G. H. Korndörfer (Eds), 85-113, Elsevier, The Netherlands

Shaw, G., Hewamanna, R., Lillywhite, J. & Bell, J.N.B. (1992). Radiocesium uptake and translocation in wheat with reference to the transfer factor concept and ion competition effects. *Journal of Environmental Radioactivity*, Vol.16, pp. 167-180

Smithson, F. (1956). Plant opal in soil. *Nature*, Vol.178, pp. 107

Smolders, E., Kiebooms, L., Buysse, J., & Merckx, R. (1996). 137Cs uptake in spring wheat (Triticum aestivum L. cv. Tonic) at varying K supply: I. The effect in soil solution culture. *Plant and Soil*, Vol.178, pp. 265–271

Stumm, W & Morgan J.J. (1996). *Aquatic chemistry-chemical equilibria and rates in natural waters.* Wiley, New York, p 1022

Takahashi, E., Syo, S. & Miyake, Y. (1976). Comparative studies on silica nutrition in plants.

Effect of germanium on the growth of plants with special reference to silicon nutrition (in Japanese). *Journal of the Science of Soil and Manure Japan*, Vol.47, pp. 183– 190

Takahashi, E., Ma, J. F. & Miyake, Y. (1990). The possibility of silicon as an essential element for higher plants. *Comments on Agricultural and Food Chemi*stry, Vol.2, pp. 99–122

Tang Van Hai, Truong Thi Nga & Laudelout, H. (1989). Effect of Aluminium on the mineral nutrition of rice. *Plant and Soil*, Vol.114, pp. 173-185

Tang Van Hai, Houben, V., Nzok Mbouti, C. & Dufey, J.E. (1993). Diagnositc précoce de la résistance de cultivars de riz (Oryza sativa L) à la toxicité aluminique. *Agronomie*, Vol.13, pp. 853-860

Tice, K.R., Parker, D.R. & DeMason, D.A. (1992). Operationally defined apoplastic and symplastic aluminium fractions in root tips of Al-intoxicated wheat. Plant Physiology, Vol.100, pp. 309-318

Thiry, Y., Kruyts, N. & Delvaux, B. (2000). Respective horizon contributions to cesium-137 soil-to-plant transfer: a pot experiment approach. *Journal of Environmental Radioactivity*, Vol.45, pp. 1194 –1199

Tréguer, P., Nelson, D.M., Van Bennekom, A.J., De Master, D.J., Leynaert, A. & Quéguiner, B. (1995). The silica balance in the world ocean: a reestimate. *Science*, Vol.268, pp. 375–379

van Breemen, N. & Moorman, F.R. (1978). Iron-toxic soils, In: *Soils and Rice*, 781-799, The International Rice Research Institute IRRI, Philippines, 781-799

Vermeire, M.L., Kablan, L., Dorel, M., Delvaux, B., Risède, J.M. & Legrève, A. (2011). Protective role of silicon in the banana-Cylindrocladiumspathiphylli pathosystem. *European Journal of Plant Pathology*, in press, DOI 10.1007/s10658-011-9835-x.

Voigt, P.W., Godwin, H.W. & Morris, D.R. (1999). Effect of four acid soils on root growth of clover seedlings using a soil-on-agar procedure. *Plant and Soil*, Vol.205, pp. 51-56

von Uexkull, H.R. & Mutert, E. (1995). Global extent, development and economic impact of acid soils. *Plant and Soil*, Vol.171, pp. 1-15

Yamaji, N., Mitani, N. & Ma, J. F. (2008). A transporter regulating silicon distribution in rice shoots. *Plant Cell*, Vol.20, pp. 1381–1389

Yoshida, S., Forno, D.A., Cook, J.H. & Gomez, K.A. (1976). *Laboratory Manual for Physiological Studies of Rice*, International Rice Research Institute, Philippines, pp. 61

CHAPTER-9

HYDROPONIC SYSTEMS AND ECO-FRIENDLY REMEDIATION

INTRODUCTION: AN OVERVIEW

Hydroponic systems offer a sustainable approach to agriculture by enabling the cultivation of plants without soil, using nutrient-rich water solutions instead. This innovative method not only conserves water but also reduces the need for harmful pesticides and fertilizers. Beyond agriculture, hydroponic systems also find application in environmental cleanup initiatives.

They can be employed in eco-friendly remediation projects to purify contaminated water sources, aiding in the restoration of ecosystems and contributing to a greener, cleaner environment. These versatile systems exemplify the potential of technology to address both food production challenges and environmental concerns simultaneously.

Water pollution refers to any chemical, physical or biological change in the quality of water that is detrimental to human, plant, or animal health. Water pollution affects all the major water bodies of the world such as lakes, rivers, oceans and groundwater. Polluted water is unfit for drinking and for other consumption processes. It may also be not suitable for agricultural and industrial use.

TYPES OF WATER POLLUTION

Water pollution takes various forms, with each type having unique sources and consequences. One common type is chemical pollution, resulting from the release of industrial and agricultural chemicals into water bodies, leading to toxicity and ecological imbalances. Another type is microbial pollution, caused by the presence of harmful pathogens like bacteria and viruses, which can lead to waterborne diseases. Physical pollution includes the introduction of solid materials such as plastics and debris, which not only degrades water quality but also harms aquatic life. Lastly, nutrient pollution, often caused by excess nutrients from fertilizers and sewage, can lead to harmful algal blooms and oxygen depletion, disrupting the aquatic ecosystem. Understanding these types of water pollution is crucial for effective mitigation and conservation efforts to protect our vital water resources.

Toxic Substances

The greatest contributors to toxic pollution are oil spills, herbicides, pesticides, industrial compounds and heavy metals such as mercury (Hg), cadmium (Cd), chromium (Cr), nickel (Ni), lead (Pb) arsenic (As), copper (Cu), zinc (Zn), among others. Organic pollution occurs when an excess of organic matter, such as manure or sewage, enters the water. When organic matter increases in a water body, the number of decomposers will increase. These microorganisms grow rapidly and use a great deal of oxygen during their growth. This leads to a depletion of oxygen as the decomposition process occurs and consequently limits oxygen availability to aquatic organisms.

As the aquatic organisms die, they are broken down by decomposers which lead to further depletion of the oxygen levels. A type of organic pollution can occur when inorganic pollutants such as nitrogen and phosphates accumulate in aquatic ecosystems. High levels of these nutrients cause an overgrowth of plants and algae. The enormous decay of this algae and plant matter become organic material in the water and lowers the oxygen level causing suffocation of fish and other organism in a water body. This overall process is known as eutrophication.

Thermal Pollution

Thermal pollution can occur when water is used as a coolant near a power or industrial plant and then is returned to the aquatic environment at a higher/cooler temperature than it was originally. Thermal pollution can have a disastrous effect on life in an aquatic ecosystem as temperature increases/decrease the amount of oxygen in the water, thereby reducing the aquatic life presence.

Natural Pollution

Also called Ecological pollution, takes place when chemical pollution, organic pollution or thermal pollution is caused by nature rather than by human activity. An example of natural pollution would be an increased rate of siltation of a waterway after a landslide which would increase the amount of sediments in runoff water. Another example would be when a large animal, such as a deer, drowns in a flood and a large amount of organic material is added to the water as a result. Major geological events such as a volcano eruption might also be sources of ecological pollution.

Important Sources of Water Pollution

The most important sources of water pollution are domestic wastes, industrial effluents and agricultural wastes. Other sources include oil spills, atmospheric deposition, marine dumping, radioactive waste and eutrophication.

- *Domestic sewage:* is wastewater generated from the household activities. It contains organic and inorganic materials such as phosphates, nitrates, heavy metal-containing wastes. Organic materials are food and vegetable waste, whereas inorganic materials come from soaps and detergents.
- *Industrial Effluents:* Manufacturing and processing industry wastes contain organic pollutants and other toxic chemicals. Some of the pollutants from industrial source include Cd, Pb, Hg, As, asbestos, nitrates, phosphates, oils, etc. Wastewater from food and chemical processing industries contribute more to water pollution than the other industries such as distilleries, leather processing industries and thermal power plants. Also dye industries generate wastewater which changes the water quality especially water color. Many of the big industries have come up with wastewater treatment plants.
- However, it is not the case with small-scale industries. Water can also become contaminated with toxic or radioactive materials from industry, mine sites and abandoned hazardous waste sites. For instance, in 1932, the Minamata disease in which nearly 1,800 people died and many more suffered occurred due to consumption of fish containing high amounts of methyl mercury. It was caused by release of methyl mercury from Chisso Corporation's chemical factory.
- Additionally, an indirect effect by industrial activity is when acid precipitation is caused as burning fossil fuels emit sulfur dioxide into the atmosphere. The sulfur dioxide reacts with the water in the atmosphere, creating rainfall which contains sulfuric acid. As acid precipitation falls into lakes, streams and ponds it can lower the overall pH of the waterway, affecting plant life, and subsequently the whole food chain. It can also leach heavy metals from the soil into the water, killing fish and other aquatic organisms. Because of this, air pollution is potentially one of the most threatening forms of pollution to aquatic ecosystems.
- *Agricultural Waste:* includes manure, slurries and runoffs. Farms often use large amounts of herbicides and pesticides, both of which are toxic pollutants. These substances are particularly dangerous to life in rivers, streams and lakes, where toxic substances can build up over a period of time. The runoffs from these agricultural fields cause water pollution to the nearby water sources. The seepage of fertilizers and pesticides

causes groundwater pollution, which is commonly known as leaching. Fertilizers can increase the amounts of nitrates and phosphates in the water, which can lead to eutrophication. Allowing livestock to graze near water sources often results in organic waste products being washed into the waterways and can also lead to eutrophication.

Heavy Metal Pollution in Water

Heavy metal pollution of freshwater environments is a serious environmental problem in the industrial areas. Water pollution by heavy metals (elements with an atomic density greater than 6 g cm^{-3}) has become therefore a global issue that need considerable attention towards combating. The common heavy metals that have been identified in polluted water include As, Cu, Cd, Pb, Cr, Ni, Hg and Zn. The release of these metals without proper treatment poses a significant threat to public health because of their persistence, *biomagnification* and accumulation in food chain. Their presence in water is due to discharges from residential dwellings, groundwater infiltration and industrial discharges. Their occurrence and accumulation in the environment is a result of direct or indirect human activities, such as rapid industrialization, urbanization and anthropogenic sources.

Severe toxic effects of heavy metal intake include reduced growth and development, cancer, organ damage, nervous system damage, and in extreme cases, death. The danger of heavy metal pollutants in water lies in two aspects of their impact. Firstly, heavy metals have the ability to persist in natural ecosystems for an extended period. Secondly, they have the ability to accumulate in successive levels of the biological chain, thereby causing acute and chronic diseases.

Water Treatment for Heavy Metal Removal

Several methods of removing heavy metals from water based on chemical and microbiological processes have been developed with a degree of success. Control over the quality and composition of industrial waste, including the removal of heavy metals, may take advantage principally of physicochemical methods based on chemical precipitation and coagulation (flocculation) followed by sedimentation, flotation, ionic exchange, reverse osmosis, extraction, microfiltration, adsorption on activated carbon, etc. However, these techniques are associated with high costs if large volumes, low metal concentrations, and high clean-up standards are involved. The insufficient effectiveness of the traditional heavy metal removal from water techniques has led to the search for more economical and simple procedures for the primary and (or) final removal of heavy metals from wastewater (Salt et al. 1995). Among these promising techniques is the phytoremediation of industrial wastewater, which involves the removal of heavy metals by adsorption, accumulation, or precipitation using higher aquatic and terrestrial plants, and the subsequent processing, utilization or burial of the contaminated biomass in special areas.

PHYTOREMEDIATION

The term phytoremediation ("*phyto*" meaning plant, and the Latin suffix "*remedium*" meaning to restore) actually refers to a diverse collection of emerging plant-based technologies that use either naturally occurring or genetically engineered plants for cleaning contaminated environments (Sarma, 2011). The primary motivation behind the development of phytoremediative technologies is the potential for low-cost remediation (Garbisu & Alkorta, 2001). Phytoremediation use plants to remove, reduce, degrade, or immobilize environmental contaminants, primarily those of anthropogenic origin, with the objective of restoring area sites to functional conditions for private or public applications. Research on phytoremediation has focused on the use of plants to: 1) accelerate degradation of organic contaminants, usually in concert with root rhizosphere microorganisms, or 2) remove/extract hazardous heavy metals from soils or water. Phytoremediation of contaminated sites is appealing because it is relatively inexpensive and aesthetically pleasing to the public compared to traditional remediation strategies.

Response of Plants to Metal Pollution

The general response of plants growing on a metal contaminated soil is categorized into the following:

1. *Hyperaccumulators:* These are species of plants that absorb and concentrate high levels of heavy metals either in their roots, shoots and/or leaves. By definition, a hyperaccumulator must accumulate at least 100 mg g^{-1} (0.01 % dry weight), Cd, As and some other trace metals, 1,000 mg g^{-1} (0.1 dry weight.) cobalt (Co), Cu, Cr, Ni and Pb and 10,000 mg g^{-1} (1 % dry weight.) (Reeves & Baker, 2000). Hyperaccumulators take up high amounts of a toxic substance, usually a metal or metalloids in their shoots during normal growth and reproduction (Baker & Whiting, 2002). Hyperaccumulators are found in 45 different families with the highest among the Brassicaceae (Reeves & Baker, 2000). One such hyperaccumulator, *Thlaspi caerulescens* is a well-known Zn hyperaccumulator able to accumulate close to 30,000 and 10,000 mg kg^{-1} Zn and Cd respectively in the shoot dry matter without growth reduction (Milner & Kochian, 2008).

2. *Metal Excluders (tolerant):* This category of plant species can grow on soil with concentration of a particular elements that are toxic to most other plants by means of preventing metals from entering their aerial parts and so maintain constant metal concentration in the soil around their roots (Ghosh & Singh, 2005)

3. *Metal Indicators:* In this plants, the extent of metal accumulation reflects metal concentration in the rhizospheric soil/water. Indicator species have been reported for mine prospecting studies to find new ore bodies (Chaney et al., 2007). In general, shows poor control over metal uptake and transport processes.

Different Types of Phytoremediation

Depending on the underlying processes, polluted matrix, applicability, and nature of the contaminant, phytoremediation can be broadly categorized as:

Phytodegradation

Also called phytotransformation, is the breakdown of organic contaminants taken up by plants through metabolic processes within the plant, or the breakdown of contaminants surrounding the plant through the effect of compounds (such as enzymes) produced by theplants. Complex organic pollutants are degraded into simpler molecules and are incorporated into the plant tissues to help the plant grow faster. Plants contain enzymes (complex chemical proteins) that catalyse and accelerate chemical reactions. Some enzymes break down and convert ammunition wastes, others degrade chlorinated solvents such as trichloroethylene (TCE) and others degrade herbicides.

Phytostimulation

Also called rhizostimulation or plant-assisted bioremediation/degradation, is the breakdown of organic contaminants in the rhizosphere area (water/soil surrounding the roots of plants) through microbial activity and enhanced by the presence of plant roots. This process is generally slower than phytodegradation. Microorganisms (bacteria, fungi) consume and digest organic substances for nutrition and energy. In this way, certain microorganisms can digest organic pollutants such as hydrocarbons, pesticides or solvents and break them down into harmless products in a process called biodegradation. Natural substances released by the plant roots (exudates) – sugars, acids and alcohols – contain organic carbon that provides food, attract microorganisms to the rhizosphere and enhance their activity.

Phytovolatilization

Is the uptake and transpiration of a contaminant by a plant (e.g. poplars), with release of the contaminant (mainly organic) or a modified form of the contaminant to the atmosphere. Phytovolatilization occurs as growing trees and other plants take up water and the contaminants. Some of these contaminants can pass through the plants to the leaves and evaporate, or volatilize, into the atmosphere.

Phytoextraction

Also called phytoaccumulation, refers to the uptake of metals from soil by plant roots into above-ground portions of plants. Certain plants, i.e. hyperaccumulators, absorb unusually large amounts of metals in comparison to other plants. After the plants are allowed to grow for some time, they are harvested and either incinerated or composted to recycle the metals. This procedure may be repeated as necessary to bring soil contaminant levels down to allowable limits. If plants are incinerated, the ash must be disposed of in a hazardous waste landfill, but the volume of ash will be less than 10% of the volume that would be created if the contaminated soil itself were dug up for treatment. Metals such as Ni, Zn and Cu are the best candidates for removal by phytoextraction because the majority of the approximately 400 known plants that absorb unusually large amounts of metals have a high affinity for accumulating these metals (Reeves & Baker, 2000).

Phytostabilization

Is the use of certain plant species to immobilize inorganic contaminants in the groundwater and soil through adsorption onto roots, absorption and accumulation by roots, or rhizosphere-mediated precipitation. This process is intended to reduce the mobility of the contaminant and prevent migration to the groundwater or atmosphere, with a consequent reduction of the pollutant bioavailability (Grimaldo & López-Chuken, 2011). This techniquecan be used to revegetate sites where natural vegetation is lacking due to high metal concentrations in surface soils. Metal-tolerant species can be used to restore vegetation to the sites, thereby decreasing the potential migration of contamination through wind erosion and transport of exposed surface soils and leaching of soil contamination to groundwater.

Phytofiltration

More commonly called rhizofiltration is the use of roots to uptake also store contaminants from an aqueous growth matrix. The description of this hydroponic-based technology is expanded in the next section.

RHIZOFILTRATION

Background

Because heavy metal pollution affects the quality of drinking water supply and wastewater discharge, great efforts have been made in the last two decades to reduce pollution in water resources to reach environmental sustainability (Goal 7 of the Millennium Development Goals). This section is therefore aimed at giving a general overview of rhizofiltration – a hydroponic-based environmental biotechnology - as a cost-effective and sustainable alternative for the remediation of heavy metals pollutants in drinking water and wastewater treatment systems.

As mentioned previously, metal pollutants in wastewater, superficial water and groundwater are most commonly removed by chemical precipitation or flocculation, followed by sedimentation and disposal of the resulting sludge (Ensley, 2000). A promising alternative to these conventional clean-up methods is rhizofiltration ('*rhizo*' means root), a plant-based technique designed for the removal of metals in aquatic environments. In rhizofiltration plant roots grown in water absorb, concentrate and precipitate toxic metals and organic chemicals from polluted effluents (Vallini et al., 2005). The plants can be used as filters in constructed wetlands (Kang, 2011) or in a hydroponic setup (Candelario-Torres et al., 2009).

Rhizofiltration Technology

Rhizofiltration is primarily used to remediate extracted groundwater, surface water, and wastewater with low contaminant concentrations. It is defined as the use of plants, both terrestrial and aquatic, to absorb, concentrate, and precipitate contaminants from polluted aqueous sources in their roots. Rhizofiltration can be used for Pb, Cd, Cu, Ni, Zn, and Cr, which are primarily retained within the roots. The plants to be used for rhizofiltration clean- up are raised in greenhouses with their roots in water. Contaminated water is either collected from a waste site or the plants are planted in the contaminated area allowing an in- situ treatment, minimizing disturbance to the environment, where the roots then take up the water and the metal contaminants dissolved in it. As the roots become saturated with contaminants, they are harvested.

During this hydroponic process, plants can adsorb or precipitate onto roots (or absorb into the roots) the metal contaminants that are in solution surrounding the root zone (Dushenkov et al., 1995). Changes in rhizosphere pH and root exudates may also causemetals to precipitate near root surfaces. As roots become saturated with the metal contaminants, plants can be harvested for disposal or reutilization (Sas-Nowosielska et al., 2004). Dushenkov et al. (1995), suggested that that plants used for rhizofiltration should preferably accumulate metals only in the roots since the translocation of metals from roots to shoots would decrease the cost-effectiveness of rhizofiltration by increasing the amount of contaminated plant residue needing disposal. In contrast, Straczek et al. (2009) suggest that the capacity for rhizofiltration can be increased by using plants with an enhanced ability to translocate metals within the plant. Despite this difference in opinion, it is apparent that proper plant selection should be based in the total amount of metal removed from the polluted water indistinctly whether metal is accumulated in roots or shoots (Figure 9.1).

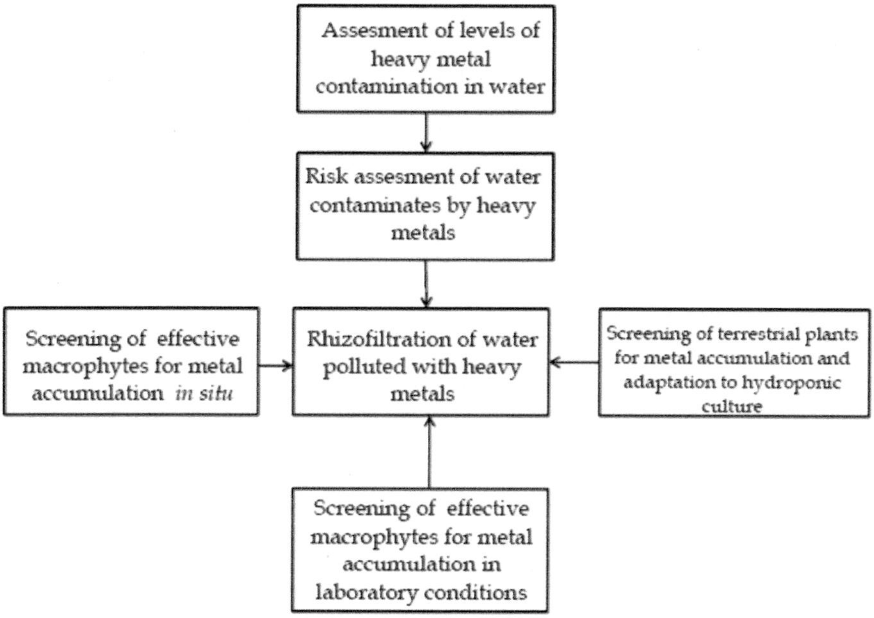

Fig. 9.1. Scheme of phytoremediation process for water contaminated by heavy metals. Modified from Galiulin et al. (2001).

Plant Species for Rhizofiltration

Dushenkov & Kapulnik (2000) described the model characteristics of plants used for rhizofiltration. Plants should be able to tolerate and accumulate significant amounts of the target metals in conjunction with easy handling, low maintenance cost, and a minimum of secondary waste requiring disposal. It is also desirable plants to produce significant amounts of root biomass or root surface area and high evapotranspiration rates. Several aquatic species have shown the capacity to remove heavy metals from water, for instance, water hyacinth (*Eichhornia crassipes*, Mahmood et al., 2010), pennywort (*Hydrocotyle umbellate,* Khilji & Bareen, 2008), and duckweed (*Lemna minor*, Hou et al., 2007). However, these plants have shown limited potential for rhizofiltration, because they are not efficient atmetal removal, a result of their small, slow-growing root system (Dushenkov et al., 1995). These authors also point out that the high water content of aquatic plants make difficult their drying, composting, or incineration.

Despite limitations, Mahmood et al. (2010) indicated that water hyacinth is effective in removing trace elements in waste streams. For example, Mahmood et al. (2010) demonstrated that water hyacinth would remove silver from industrial wastewater for subsequent recovery with high efficiency in a fairly short time. The accumulation of some other heavy metals and trace elements in many species of wetland plants has also been demonstrated (Romero-Núñez et al., 2011). Water hyacinth has been used successfully in wastewater treatment systems to improve the quality of water by reducing the levels of organic and inorganic nutrients, and readily reducing the level of heavy metals in acid mine drainage water.

Trace element removal by wetland vegetation can be greatly enhanced by selection of appropriate wetland plant species. The selection is based on the types of elements to be remediated, the geographic location, microclimate, hydrologic conditions, soil properties, and known accumulation capacities of the species. Knowledge of the capabilities of different wetland plant species to absorb and transport trace elements under different conditions is important to know. One such plant is the vascular aquatic plant water hyacinth which is commonly found in tropical and subtropical regions of the world. Water hyacinth is a fast growing, floating plant with a reasonably well-developed fibrous root system and large biomass and it adapts easily to various aquatic conditions.

Hydroponic system involves aeration and therefore is not limited to aquatic species; it often makes use of terrestrial species with large roots and good capacity to accumulate inorganics (Dushenkov & Kapulnik, 2000).

Rhizofiltration Using Terrestrial Plants

The advantages associated with rhizofiltration are the ability to use both terrestrial and aquatic plants for either *in situ* or *ex situ* applications. Terrestrial plants are thought to be more suitable for rhizofiltration because they produce longer, more substantial, often fibrous root systems with large surface areas for metal sorption (López-Chuken & Young, 2010). Another advantage is that contaminants do not have to be translocated to the shoots. Sunflower (*Helianthus annuus*), Indian mustard (*Brassica juncea*), spinach (*Spinacia oleracea*), corn (*Zea mays*) and tobacco (*Nicotiana tabacum*) are among the most promising terrestrial candidates for metal removal in water. The roots of *B. juncea* are effective in the removal of Cd, Cr, Cu, Ni, Pb, and Zn (Dushenkov et al. 1995), sunflower removes Pb (Dushenkov et al. 1995), U (Dushenkov et al. 1997a), 137Cs, and 90Sr (Dushenkov et al. 1997b) and tobacco removes Cd, Ni, Pb, Cr, Zn, Cu, Hg and As (Candelario-Torres et al., 2009) from hydroponic solutions. Similarly, López-Chuken & Young (2010) and López-Chuken et al. (2010) observed that roots of hydroponically grown terrestrial plants such as *B. juncea* and *Z. mays* effectively removed Cd from aqueous solutions. Candelario-Torres et al. (2009) shown that*N. tabacum* plants effectively remove toxic metals, such as Pb, Cd and Cr from polluted solutions.

Perhaps, the best example of a successful rhizofiltration remediation program occurred at Chernobyl, Ukraine, where, sunflowers were successfully used to remediate radioactiveuranium from pond water. Tobacco has been shown to develop long and hairy root systems when grown in nutrient solution, which create an extremely high surface area.

Rhizofiltration: Current Advances

Arthur et al., (2008) published a remarkable and extensive review of the use of rhizofiltration technologies applied to removal and further recovery of metals. While the extensive review of literature reported by the authors, only a few representative examples will be provided here. Rhizofiltration of metal-contaminated water was early investigated by Schulman et al. (1999). They developed a screening method to look for mutants of *Brassica juncea* that had enhanced Cd or Pb accumulation capabilities. The authors found that cell-wall binding and precipitation are the primary mechanisms of Pb accumulation in plants, thus the authors concluded that the hyperaccumulating characteristic of the mutant was due to the increased cell wall per unit of root weight. The ability to remove and recover heavy metals, including Pb, Cr, Zn, Cu and Ni from aqueous solutions was shown in experiments with *Medicago sativa* (alfalfa). Optimum binding was in aqueous solutions at pH 5, and tests showed that binding to alfalfa shoots occurred within five minutes. Similar results were reported by López-Chuken & Young (2010), where a rapid initial reduction of Cd concentration in a hydroponic solution was observed after an initial 3 hours of exposition to maize plants. This was thought to signify a rapid equilibration between the root system and Cd in solution rather than true absorption into roots. Alfalfa biomass is also reported as an effective species for recovering Au(III) from aqueous solutions (Gamez et al., 2003). Recovery of valuable metals by plants is called "*phytomining*".

Rhizofiltration of uranium (U) by terrestrial plants has been investigated by Dushenkov et al. (1997a). They found that certain sunflower species had a high affinity for U and could concentrate it from water into the roots. Phytofiltration of chromium-contaminated water has been studied (Candelario-Torres et al., 2009), and it has been found that several plant species can uptake the toxic Cr(VI) species and reduce the pollutant to the nontoxic form, Cr(III). Water hyacinth supplied with Cr(VI) in nutrient culture accumulated Cr(III) in root and shoot tissues. Reduction to the nontoxic form appeared to occur in the fine lateral roots (Lytle et al., 1998). Phytofiltration of As from potable water has being evaluated using the brake fern (*Pteris vittata*). This plant species is known to tolerate high concentrations of As (Elless et al., 2003).

While some plants may have desirable characteristics for rhizofiltration of toxic metals from the environment, it is critical to better understand the mechanisms behind these capabilities before they can be exploited to the fullest extent for phytoremediation programs (Arthur et al., 2008). Recently, much research has being conducted to elucidate the chemical and physiological interactions involved in metal adsorption and accumulation.

To improve the potential of candidate plants for rhizofiltration, several research studies have been undertaken to select lines and improve plants by breeding. Recently, research using genetically modified (GM) plants has been spreading. Some transgenic plants are designed to be dwarf species. In this way, most of the biomass goes to leaves, and the rest to produce a short stem, similar approach as in the 1960s 'Green Revolution' for food production. Under this circumstances, when the native plants compete with them, do not allow them to develop (Gressel and Al-Ahmad, 2005).

However, according to Gressel and Al-Ahmad (2005), there are some considerations of the use of GM plants in field studies: The presence of wild relatives in the same area; harvesting time; the possibility that the plant become a weed; the possibility of gene flux and plant surviving on its own. Additionally, consideration to leave a fallow zone of about 8-18 m. There is also recommended that no food production at site be allowed during the next growing season. In general, whether GM plants or not, a problem for species used for phytoextraction should be considered: The risk that they can displace native species.

CHALLENGES OF HYDROPONIC PHYTOTECHNOLOGIES

Plant Development and Variability

One of the main problems encountered by rhizofiltration research at laboratory level is the large variability in heavy metals content measured between plants exposed to the same treatment solutions. López-Chuken (2005) made several attempts to minimise this variation. For example, growing a large excess of seedlings in order to select uniform plants prior to treatment application, increase nutrient solution volume and nutrient concentration to reduce minor variation in conditions during the exposure period, validate digestion and analysis procedures against standard reference materials. In addition, it was shown that for an hydroponic trial using *Z. mays* which showed apparently erratic trends with Cd treatments (López-Chuken et al., 2010) there was, nevertheless, a good correlation between root and shoot analysis for individual plants which seems to suggest that the variability in response observed lay with individual plants rather than being caused by a methodical source of error. In general, plant variability in the hydroponic trials followed the qualitative trend: maize > Indian mustard > tobacco (Candelario-Torres et al., 2009; López-Chuken & Young, 2010; López- Chuken et al., 2010).

To date, hydroponic experiments dealing heavy metal uptake, generally involve the use of plants at early steps of development, for example an experiment used young seedlings (wheat harvested 6 days after sowing) with short exposure times to the treated solution ranging between 0 and 200 minutes (Berkelaar & Hale,

2003). In general, seeds from the Fam. Gramineae normally contain high nutrient reserves. Therefore, for short-term trials at early seedling stage, these plants may be still partially absorbing energy from the endosperm, and not completely reflecting nutrient conditions from solution. It is therefore recommended that hydroponic trials dealing with metal uptake, should preferably use plants as mature as possible (at least 6 weeks), and with enough metal exposure time (≥ 10 days) (López-Chuken, 2005). This would allow the study of more mature plants and also minimised any short-term effects arising from the transition from nutrient to treatment solutions.

Berkelaar & Hale (2003) suggested the use of short-term metal accumulation experiments to maintain aseptic conditions and avoid biodegradation when using organic ligands as a reservoir of chelated metal in solution. However, it has been demonstrated an initial rapid sorption of metals from treatment solutions which does not necessarily reflect the normal uptake rate of the plants growing under steady state conditions in the treatment solutions (López-Chuken & Young, 2010). This effect was concluded to be a rapid approach to a pseudo- equilibrium state between root surface sorption sites and nutrient solutions. Furthermore, even when aseptic conditions are not strictly controlled, little effect in metal speciation was observed due to the presence of organic matter in solution of long-term trials (6-8 weeks) (López-Chuken & Young, 2010).

Can Rhizofiltration Effectiveness be Extrapolated to Soil Pollution?

One of the main advantages of the hydroponic-based experiments is that the solution chemistry in contact with plant root surfaces can be unequivocally designed and controlled during uptake trials. This represents a significant advantage over the uncertainties intrinsic to contaminated soil studies. Furthermore, (arguably) reliable morphological and chemical analysis of roots is only possible when using hydroponic growth media. However, the disadvantages of altered plant physiology and short exposure time in hydroponic studies are also well recognised. The choice of soil or hydroponic systems for the study of metal uptake must be dictated by the intention of the study but will always remain a compromise between the desire for control over experimental conditions and the unrealistic side effects of using any medium other than a naturally contaminated soil with a field sown plant allowed to follow its full span of physiological development. Chaney et al., (2005) suggested that the best media to test accumulator capacity by plants is the naturally contaminated soil (long-term), preferably in situ since it represents realistic conditions.

One problem associated with metal uptake trials with hydroponic systems, and other artificial media, is that such studies often seem to adopt very high 'unrealistic' metal concentrations. Recent research has shown that ordinary plants can even reach the metal hyperaccumulator "definition", for example Cd >100 mg kg^{-1} (Baker et al., 2000) under artificial media conditions (de la Rosa et al., 2004).

In a set of soil-based and hydroponic experiments using maize and Indian mustard plants (López-Chuken, 2006; López-Chuken & Young, 2010; López-Chuken et al., 2010) it has been shown that the ratio of Cd concentrations (Cdshoot:Cdroot) for different maize species ranged between 0.09 and 0.43 (mean = 0.24; SD = 0.11) throughout the trials. These results indicated that for all treatments, whether in soil or nutrient solution and even using different maize varieties (hybrid W23/L317; salt tolerant 2001-196-1; mays PI596543 and Cameron), Cd concentrations in roots were consistently larger than in shoots. However, when *B. juncea* plants were used, the ratio of concentrations (Cdshoot:Cdroot) showed large differences between soil (1.63 - 2.65; mean = 2.18; SD = 0.51) and nutrient solution (0.02 - 0.05; mean = 0.03; SD = 0.01) trials, despite using the same variety of Indian mustard (var. G32192). These results may suggest that perhaps not all plants are suitable for hydroponic experiments because plants show different physiological Cd uptake responses when growing in soil or dissimilar artificial media conditions. In the case of *B. juncea* the rapid root-to-shoot transfer observed for soil trials was virtually suspended in solution culture. Thus, under hydroponic conditions some factor (or a combination of factors) controlling Cd accumulation by roots and translocation to shoots may be affected.

Importance of Root Surface Area in Expressing Metal Uptake

Hydroponic trials also offer some advantages over soil experiments where there is a particular interest in the role of the root morphology as a control over metal uptake rates. Although the morphology of roots grown in nutrient solutions will differ from those generated in soils, the entire root can be extracted without physical damage or contamination from soil particulates.

López-Chuken & Young (2010) in a Cd rhizofiltration trial using *Z. mays* observed that the dataset was effectively 'normalised' when including root morphological parameters toexpress Cd uptake rates by plants. In all cases the root surface area (RSA) was the morphological characteristic that best explained changes in Cd uptake by plants. However, other measured root characteristics, (i.e. volume, length, root projected area) were so strongly correlated (R= 0.88-1.00) with the RSA that the differences between expressing Cd uptake rates using these parameters were minimal. Furthermore, Berkelaar & Hale, (2000) in a nutrient solution trial growing wheat, expressed Cd uptake rates per unit root length or 'number of tips' with similar efficiency to that of RSA. It would be hence interesting to include, as future work, the use of rooting hormones on single varieties to investigate the potential enhancement of metal uptake rates by plants. A recent research (in hydroponics) has shown that adding the growth phytohormone IIA (indole-

3-acetic acid) in combination with EDTA, increased Pb accumulation in leaves by about 2800% and by 600%, as compared to Pb content in leaves of lucerne plants exposed to Pb alone and with Pb/EDTA, respectively (López et al., 2005).

Utilization of Phytoremediation by-Products

One of the main concerns about the use of plants with high metal phytoaccumulation characteristics is their post-harvest disposal. Rhizofiltration technologies applied to metal removal generally involve repeated cropping of plants in contaminated water, until the metal concentration drops to acceptable level. The ability of the plants to account for the decrease in water metal concentrations as a function of metal uptake and biomass production plays an important role in achieving regulatory acceptance (Ghosh & Singh, 2005). Although this may sound simple, several factors make it challenging in the field. One of the difficulties for commercial implementation of rhizofiltration has been the disposal of contaminated plant material (Ghosh & Singh, 2005). After each cropping, the plant is removed from the site; leading to accumulation of huge amounts of hazardous biomass waste that needs to be stored or disposed appropriately so that it does not pose any risk to the environment.

Composting and compaction has been proposed as post-harvest biomass treatment by some authors (Blaylock & Huang, 2000) however, leachates generated by composted or compacted biomass will need to be collected and treated appropriately. It has been also reported that plant material may be dried, burned and disposed of in landfill as ash (Keller et al, 2005). Another promising route to utilize biomass produces by phytoremediation in an integrated manner is through thermochemical conversion process (Keller et al, 2005). If rhizofiltration could be combined with biomass generation and its commercial utilization as an energy source, then it can be turned into profit making operation and the remaining ash can be used as bio-ore, the basic principle of phytomining.

Thermochemical energy conversion best suits the rhizofiltration biomass waste because it cannot be utilized in any other way as fodder and fertilizers. Combustion is a rudimentary method of burning the biomass and should be applied only under controlled conditions, whereby volume is reduced to 2–5 %. This method of plant matter disposal has to be carefully evaluated as burning the metal bearing hazardous waste in open will releases the gases and particulates to the environment.

Another alternative is the process called "*gasification*" through which biomass material can be subjected to series of chemical changes to yield clean and combustive gas at high thermalefficiencies. This mixture of gases called as producer gas and/or pyro-gas that can be combusted for generating thermal and electrical energy. The process of gasification of biomass in a gasifier is a complex phenomenon; it involves drying, heating, thermal decomposition (pyrolysis) and gasification, and combustion chemical reactions, which occurs simultaneously and it may be possible to recycle the metal residue from the ash.

Cost Estimates Using Rhizofiltration

Rhizofiltration is a cost-competitive technology in the treatment of large volumes of water containing low, but environmental significant concentrations of heavy metals such as Cr, Pb, and Zn (Candelario-Torres et al., 2009). The commercialization of rhizofiltration systems need to be driven by cost-effectiveness as well as by such technological advantages as applicability to many real conditions, ability to treat high volumes, lesser needed for chemicals, reduced volume of secondary waste, possibility of recycling and the almost secure likelihood of regulatory and public acceptance (Dushenkov et al. 1995).

This hydroponic-based phytotechnology has worked effectively at test sites near the Chernobyl nuclear plant in Ukraine. It has been estimated that the cost to remove radionuclides from water using sunflowers would be between $2 and $6 per 1,000 gallons, including disposal costs On the other hand, a standard treatment of microfiltration and precipitation would cost nearly $80 per 1,000 gallons (EPA, 2000). Glass (1999) estimated that depending on the pollutant, substrate, and alternative remediation methods available, phytoremediation could be typically 2–10-fold cheaper than conventional remediation methods.

Despite obvious advantages, the application of this plant-based technology may be more challenging and susceptible to failure than other methods of similar cost. The production of hydroponically grown plants and the maintenance of successful hydroponic systems in the field will require the qualified personnel, and the facilities and specialized equipment required could exceed the original estimated cost. Perhaps the fundamental benefit of this remediation method is related to positive public perception. Using plants at a site where contamination exists conveys the idea of cleanliness in an area that would have normally been perceived as polluted.

Rhizofiltration and Sustainable Development

Contaminated water in the urban environments and rural areas represents a major environmental and human health problem in the world. As shown above, some plants possess pronounced capacity and ability for the metabolism and degradation of many contaminants and are regarded as "green livers" acting as a sink for environmentally harmful contaminants. It has been reported that green space programmes are conducted in most countries to check the increasing levels of carbon dioxide which causes global warming. But when properly managed and handled through the use of some plants that have phytoremediation property, green space would not only clean the atmosphere of its excess carbon dioxide but also the soil and water from its contaminants.

Developing cost effective and environmentally friendly technologies for the remediation of soils and wastewaters polluted polluted with toxic substances is a topic of global interest.

The Millennium Development Goals-MDGs agreed by the international community includes "environmental basic sanitation" as a critical target (IRC-International Water and Sanitation Centre, 2004). The necessity in decontaminating polluted sites is recognised worldwide, both socially and politically, because of the increasing importance placed on environmental protection and human health. Based on the success recorded by various studies on phytoremediation, rhizofiltration could represent a good alternative to contribute to achieve the above-mentioned goals.

AT A GLANCE

- Metals and other inorganic contaminants are among the most prevalent forms of contamination found at waste sites and the high cost of existing clean-up technologies has led to the search for new clean-up strategies that have the potential to be low-cost, low- impact, visually benign, and environmentally sound. Rhizofiltration as an emerging new clean-up concept needs to be promoted and emphasized and expanded mainly in developing countries due to its low cost and potential to be applicable to a variety of contaminated sites. Selection of the appropriate plant species is a critical process for the success of this technology. Fast growing plants, adapted to hydroponic conditions, with high biomass and good metal uptake ability are needed.
- An extra important advantage of rhizofiltration it that both terrestrial and aquatic plants can be used (Prasad & Oliveira, 2003). Although terrestrial plants require physical support, they generally remove more contaminants than aquatic plants. This system can be either in situ (floating rafts on ponds) or ex situ (an engineered tank system). However, rhizofiltration has some intrinsic limitations (Prasad & Oliveira, 2003) a) pH of the polluted water has to be continually adjusted to obtain optimum metals uptake. b) chemical speciation and interaction of all metallic species in the influent has to be understood (López-Chuken et al., 2010), c) An engineered system is required to control influent flow rate, d) plants may have to be grown in greenhouse, e) periodic harvesting and plant disposal are needed, and f) metal uptake results from laboratory studies might be overestimated and not be achievable under real conditions.

- In general, phytoremediation is a multi- and inter-disciplinary technology that will benefit from research in many different areas. Much still remains to be discovered about the chemical and biological processes that underlie a plant's ability to detoxify and accumulate pollutants. Better knowledge of the biochemical mechanisms involved may lead to: a) the identification of novel genes and the subsequent development of transgenic plants with enhanced remediation capacities, and b) a better understanding of the ecological interactions involved (e.g. plant–microbe interactions in the rhizosphere) among others. This knowledge will help improve risk assessment during the design of rhizofiltration programs as well as alleviation of the associated risks during remediation. Adapting each rhizofiltration system to the specifics of polluted water will become more feasible as more information becomes available; for example to select a combination of plant species with different remediation capabilities to clean up sites containing a mix of contaminants. Preferentially, native plant species will be used in order to promote ecosystem restoration during the cleanup process.
- An interesting perspective for phytoremediation could be the adoption of an integrated approach both for research and commercial purposes. Currently, most plant-basedremediation research is carried out by scientists with expertise in only certain fields e.g. plant molecular biology, plant biochemistry, plant physiology, plant biochemistry, plant physiology, ecology, toxicology or microbiology but phytoremediation would be benefited more by a team of multidisciplinary researchers. Commercially to improve public acceptance phytoremediation could be integrated with landscape architecture with an attractive design so that the area may be used as a park or some other recreational place by the public after the remediation process (Pilon-Smits, 2005).

REFERENCES

Arthur, E.; Rice, P.; Rice, P.; Anderson, T.; Baladi, S.; Henderson, K. & Coats, J. (2005). Phytoremediation - An Overview. *Critical Reviews in Plant Sciences*, Vol.24, No.2, (March 2005), pp. 109–122, ISSN 0735 2689

Baker, A.; McGrath, S.; Reeves, R. & Smith, J. (2000). Metal hyperaccumulator plants: a review of the ecology and physiology of a biochemical resource for phytoremediation of metal-polluted soils, In: *Phytoremediation of contaminated soil and water,* N. Terry & G. Bañuelos, (Eds.), 85-107, Lewis Publishers, ISBN 1-56670- 450-2, Boca Raton, Florida, USA

Baker, A. & Whiting, S. (2002). In Search for the Holy Grail - another step in understanding metal hyperaccumulation? *New Phytologist*, Vol.155, No.1, (June 2002), pp. 1-7, ISSN 1469 8137

Berkelaar, E. & Hale, B. (2000). The relationship between root morphology and cadmium accumulation in seedlings of two durum wheat cultivars. *Canadian Journal of Botany,* Vol.78, No.3, (April 2000), pp. 381 387, ISSN 1916-2790

Berkelaar, E. & Hale, B. (2003). Cadmium accumulation by durum wheat roots in ligand buffered hydroponic culture: Uptake of Cd ligand complexes or enhanced diffusion? *Canadian Journal of Botany,* Vol.81, No.7, (July 2003), pp. 755-763, ISSN

Blaylock, M. & Huang, J. (2000) Phytoextraction of Metals, In: *Phytoremediation of Toxic Metals: Using plants to clean up the environment,* I. Raskin. & B. Ensley, (Eds.), 53-70, John Wiley & Sons, ISBN 978-0-471-19254-1, New York, USA

Candelario-Torres, M.; Ramírez, E.; Loredo J.; Gracia, Y.; Gracia, S.; Esquivel, P.; Urbina, P.; Verástegui, W.; Campos, M. & López-Chuken, U. (2009). Rizofiltración de cromo por *Nicotiana tabacum* en un efluente contaminado simulado: aplicaciones para la minimización de la contaminación de suelos agrícolas, *Proceedings of the XVIII Latinamerican Congress of Soil Science,* pp. 1-5, San José, Costa Rica, November 16-21, 2009

Chaney, R.; Angle, J.; McIntosh, M.; Reeves, R.; Li, Y.; Brewer, E.; Chen, K.; Roseberg, R.; Perner, H.; Synkowski, E.; Broad, C.; Wang, S. & Baker, A. (2005). Using Hyperaccumulator Plants to Phytoextract Soil Ni and Cd. *Zeitschrift für Naturforschung C,* Vol.60, pp. 190-198, ISSN 09395075

Chaney, R.; Angle, J.; Broadhurst, C.; Peters, C.; Trappero, R. & Sparks, D. (2007). Improved Understanding of Hyperaccumulation Yields Commercial Phytoextraction and Phytomining Technologies. *Journal of Environmental Quality,* Vol.36, (August 2007), pp. 1429-1443, ISSN 0047-2425

de la Rosa, G.; Peralta-Videa J.; Montes, M.; Parsons, J.; Cano-Aguilera, I. & Gardea-Torresdey, J. (2004). Cadmium uptake and translocation in tumbleweed (*Salsola kali*), a potential Cd-hyperaccumulator desert plant species: ICP/OES and XAS studies. *Chemosphere,* Vol.55, No.9, (June 2004), pp. 1159-1168, ISSN 0045-6535

Dushenkov, V.; Kumar, P.; Motto, H. & Raskin, I. (1995). Rhizofiltration: The Use of Plants to Remove Heavy Metals from Aqueous Streams, *Environmental Science and Technology,* Vol.29, No.5, (May 2005), pp. 1239-1245, ISSN 0013-936X

Dushenkov, S.; Vasudev, D., Kapulnik, Y., Gleba, D., Fleisher, D., Ting, K. C., and Ensley, B. (1997a). Removal of uranium from water using terrestrial plants. *Environmental Science and Technology,* Vol.31, No.12, (November 2007), pp. 3468–3474, ISSN 0013- 936X

Dushenkov, S.; Vasudev, D.; Kapulnik, Y.; Gleba, D.; Fleisher, D.; Ting, K. & Ensley, B. (1997b). Phytoremediation: A novel approach to an old problem, In: *Global environmental biotechnology,* D.L. Wise, (Ed.), 563-572, Elsevier Science B.V., ISBN 978-9048148363, Amsterdam, The Netherlands

Dushenkov, S. & Kapulnik, Y. (2000). Phytofilitration of metals, In: *Phytoremediation of Toxic Metals: Using plants to clean up the environment,* I. Raskin. & B. Ensley, (Eds.), 89-106, John Wiley & Sons, ISBN 978-0-471-19254-1, New York, USA

Elless, M.; Poynton, C. & Blaylock, M. (2003). Phytofiltration of arsenic from drinking water, *Proceedings of the 226th ACS National Meeting,* New York, NY USA, September 7-11, 2003

Ensley, B. (2000). Rational for use of phytoremediation, In: *Phytoremediation of Toxic Metals: Using plants to clean up the environment,* I. Raskin. & B. Ensley, (Eds.), 3-12, John Wiley & Sons, ISBN 978-0-471-19254-1, New York, USA

Environmental Protection Agency, United States (USEPA). (2000). Introduction to Phytoremediation. EPA 600/R-99/107. U.S. Environmental Protection Agency, Office of Research and Development, Cincinnati, OH.

Galiulin R.; Bashkin, V.; Galiulina, R. & Birch, P. (2001). A critical review: protection from pollution by heavy metals – phytoremediation of industrial wastewater, *Land Contamination & Reclamation,* Vol.9, No.4, (March 2001), pp. 349-357, ISSN 0967-0513

Gamez, G.; Gardea-Torresdey, J.; Tiemann, K.; Parsons, J.; Dokken, K. & Yacaman, J. (2003). Recovery of gold(III) from multi-elemental solutions by alfalfa biomass. *Advances in Environmental Research,* Vol.7, No.2, (January 2003), pp. 563–571, ISSN 1093-0191

Garbisu, C. & Alkorta, I. (2001). Phytoextraction: a cost-effective plant-based technology for the removal of metals from the environment. *Bioresource Technology,* Vol.77, No.3, (May 2001), pp. 229-36, ISSN 0960-8524

Ghosh, M. & Singh, S. (2005). A Review on Phytoremediation of Heavy Metals and Utilization of It's by Products, *Applied Ecology and Environmental Research,* Vol.3, No.1, (June 2005), pp. 1-18, ISSN 1589 1623

Glass, D. (1999). US and international markets for phytoremediation, 1999–2000. Needham, MA: D Glass Associates.

Gressel, J. & Al-Ahmad, H. (2005). Assessing and Managing Biological Risks of Plants Used for Bioremediation, Including Risks of Transgene Flow, *Zeitschrift für Naturforschung C,* Vol.60, pp. 154-165, ISSN 09395075

Grimaldo, C. & López-Chuken, U. (2011). Evaluación del efecto de enmiendas sobre la fitoestabilización de un suelo contaminado por Pb y Zn. M.Sc. Thesis. School of Chemistry, Universidad Autónoma de Nuevo León. June 15, 2011

Houa, W.; Chen, X.; Song, G.; Wang, Q. & Chang, C. (2007). Effects of copper and cadmium on heavy metal polluted waterbody restoration by duckweed (*Lemna minor*). *Plant Physiology and Biochemistry,* Vol.45, No.1, (January 2007), pp. 62-69, ISSN 0981-9428

IRC- International Water and Sanitation Centre (October 2011). Available from http://www.washdoc.info/docsearch/title/126204

Kang, J. (2011). Technique Study on the Low Level Radioactive Liquid Waste Treatment by Constructed Wetlands. *Environmental Science and Management*, Vol.36, No.4, (April 2011), pp. 85-89, ISSN 0301-4797

Keller, C.; Ludwig, C.; Davoli, F. & Wochele, J. (2005). Thermal Treatment of Metal-Enriche Biomass Produced from Heavy Metal Phytoextraction. *Environmental Science and Technology*, Vol.39, No.9, (May 2005), pp. 3359–3367, ISSN 0013-936X

Khilji, S. & Bareen, F. (2008). Rhizofiltration of heavy metals from the tannery sludge by the anchored hydrophyte, *Hydrocotyle umbellata* L. *African Journal of Biotechnology*, Vol.7, No.20, (October 2008), pp. 2711-3717, ISSN 1684-5315

López, M.; Peralta-Videa, J.; Benítez, T. & Gardea-Torresdey, J. (2005). Enhancement of lead uptake by alfalfa (*Medicago sativa*) using EDTA and a plant growth promoter. *Chemosphere*, Vol.61, No.14, (October 2005), pp. 595-598, ISSN 0045-6535

López-Chuken, U. (2005). The Effect of Chloro-complexation on Cadmium Uptake by Plants. PhD Thesis. *The University of Nottingham*, United Kingdom pp. 171

López-Chuken, U. & Young, S. (2010). Modelling sulphate-enhanced cadmium uptake by *Zea mays* from nutrient solution under conditions of constant free Cd^{2+} ion activity. *Journal of Environmental Sciences*, Vol.22, No.7, (July 2010), pp. 1080-1085, ISSN 1001 0742

López-Chuken, U.; Young, S. & Guzmán-Mar, J. (2010). Evaluating a ′Biotic Ligand Model′ applied to chloride-enhanced Cd uptake by *Brassica juncea* from nutrient solution at constant Cd^{2+} activity. *Environmental Technology*, Vol.31, No.3, (February 2010), pp. 307-318, ISSN 0959 3330

Lytle, C.; Lytle, F.; Yang, N.; Qian, J.; Hansen, D.; Zayed, A. & Terry, N. (1998). Reduction of Cr(VI) to Cr(III) by wetland plants: Potential for in situ heavy metal detoxification. *Environmental Science and Technology*, Vol.32, No.20, (August 2010), pp. 3087-3093, ISSN 0013-936X

Mahmood, T.; Malik, S. & Hussain, S. (2010). Biosorption and recovery of heavy metals from aqueous solutions by *Eichhornia crassipes* (water hyacinth) ash. *Bioresources*, Vol.5, No.2, pp. 1244-1256, ISSN 1930-2126

Milner, M. & Kochian, L. (2008). Investigating Heavy-metal Hyperaccumulation using *Thlaspi caerulescens* as a Model System, *Annals of Botany*, Vol.102, No.1, (March 2008), pp. 3-13, ISSN 0305-7364

Pilon-Smits, E. & Freeman, J. (2006). Environmental cleanup using plants: biotechnological advances and ecological considerations, *Frontiers in Ecology and the Environment*, Vol.4, No.4, (May 2006), pp. 203-210, ISSN 1540-9295

Prasad, M. & Oliveira, M. (2003). Metal hyperaccumulation in plants - Biodiversity prospecting for phytoremediation technology, *Electronic Journal of Biotechnology*, Vol.6, No.3, (December 2003), pp. 285-321, ISSN 0717-3458

Reeves, R. & Baker, A. (2000). Phytoremediation of toxic metals, In: *Phytoremediation of Toxic Metals: Using plants to clean up the environment*, I. Raskin. & B. Ensley, (Eds.), 193- 229, John Wiley & Sons, ISBN 978-0-471-19254-1, New York, USA

Romero-Núñez, S.; Marrugo-Negrete, J.; Arias-Ríos, J.; Hadad, H. & Maine, M. (2011). Hg, Cu, Pb, Cd, and Zn Accumulation in Macrophytes Growing in Tropical Wetlands. *Water, Air, & Soil Pollution*, Vol.216, No.1-4, (May 2011), pp. 361-373, ISSN 0049- 6979

Salt, D.; Blaylock, M.; Kumar, N.; Dushenkov, V.; Ensley, B.; Chet, I. & Raskin, I. (1995). Phytoremediation: A novel Strategy for the Removal of Toxic Metals from the Environment Using Plants, *Nature Biotechnology*, Vol.13, No.5, (May 1995), pp. 468- 474, ISSN 1087-0156

Sarma, H. (2011). Metal Hyperaccumulation in Plants: A Review Focusing on Phytoremediation Technology. *Journal of Environmental Science and Technology*, Vol.4, No.2, pp. 118-138, ISSN 1994 7887

Sas-Nowosielska, A.; Kucharski, R.; Małkowski, E.; Pogrzeba, M.; Kuperberg, J. & Kryński, K. (2004). Phytoextraction crop disposal--an unsolved problem, *Environmenta Pollution,* Vol.128, No.3, (February 2004) pp. 373-379, ISSN 0269-7491

Schulman, R.; Salt, D. & Raskin, I. (1999). Isolation and partial characterization of a lead- accumulating *Brassica juncea* mutant. *Theoretical and Applied Genetics*, Vol.99, No.3-4, (December 1998), pp. 398-404, ISSN 1432-2242

Straczek, A.; Duquene, L.; Wegrzynek, D.; Chinea-Cano, E.; Wannijn, J.; Navez, J. & Vandenhove, H. (2010). Differences in U root-to-shoot translocation between plant species explained by U distribution in roots. *Journal of Environmental Radioactivity*, Vol.101, No.3, (March 2010), pp 258-266, ISSN 0265-931X

Vallini, G.; Di Gregorio, S. & Lampis, S. (2005). Rhizosphere-induced Selenium Precipitation for Possible Applications in Phytoremediation of Se Polluted Effluents. *Zeitschrift für Naturforschung C*, Vol.60, pp. 349-356, ISSN 09395075

CHAPTER-10

HYDROPONIC CULTIVATION OF PRICKLY PEAR CACTI: ENHANCING NOPALITO AND FODDER PRODUCTION, YIELD, AND QUALITY

INTRODUCTION: AN OVERVIEW

In recent years, hydroponic systems have garnered considerable attention for their innovative approach to agricultural practices. One area where this methodology shows significant promise is in the cultivation of prickly pear cacti, a plant known for its versatility and resilience in arid environments. This pioneering hydroponic approach offers the potential to not only increase the production of valuable Nopalito, the young and tender pads of the prickly pear cactus, but also to enhance the yield and quality of cactus-based fodder. In this endeavor, we delve into the exciting realm of hydroponic technology and its application in optimizing the productivity and quality of these cactus-derived products, while addressing the challenges of traditional cultivation methods and contributing to sustainable agriculture.

The use of cactus pear for producing young cladodes (nopalitos) and fodder represent an attractive option to intensify plant production in arid and semi-arid regions. Nopalitos are considered functional food (Sáenz et al., 2004) and are used in Mexico since pre-Columbian times (Anaya, 2001). The adjective functional is due to additionally to the nutrient supplies, it provides health benefits and contributes to the prevention of some diseases (Sáenz et al., 2004) as is evidenced by results of investigations on the control of cholesterol and the prevention of some diseases like diabetes and obesity (Paiz et al., 2010; Sáenz, 2000). Nopalitos are served with meals, similar to green beans (Stintzing, 2005). Nopalitos of *Opuntia* are produced and consumed in temperate and dry regions while nopalitos of *Nopalea* genera are used in warm regions of Mexico (Sánchez-Venegas, 1995). Alternatively, fodder of cactus pear is significant in some regions of the world, mostly during the dry season of the year (Flores & Aguirre, 1979; Gonzaga and Cordeiro, 2005). Fodder nutrient quality studies on *Opuntia* and *Nopalea* cladodes show variations among species, variants, growth stage of the sprouts and agronomic handling (López-García et al., 2001; Nefzaoui & Ben Salem, 2001; Pinos-Rodríguez et al., 2006) and it has been stated that they have high carbohydrate and water content but have low nutrient and fiber content (López-García et al., 2001). Furthermore, nutritional potential of cactus pear to mitigate feed and water shortages in dry areas were demonstrated by Tegegnea et al. (2007).

Plant growth is a continuous process that strongly depends on genetic information and environmental conditions; and both, sprouting and growth of the cladodes of *Opuntia* and *Nopalea* occur independent on chronological time. Growth and development can modify some chemical characteristics of plants as those considered important for fodder and nopalito quality (Collins & Frits, 2003; Rodríguez-Félix & Cantwell, 1988). Nevertheless, several studies, as those of Cordeiro et al. (2000), Flores-Hernández et al. (2004), and Mondragón-Jacobo et al. (2001) evaluated the yield, the nutrimental value and quality ofcactus pear cladodes for fodder harvesting shoots on determinate fixed periods and it did not take account the shoot size or at least it did not specify this criteria. Then it is possible to infer that cutting cladodes considering fixed time periods would result in heterogeneous characteristics of shoots in quality and size at the harvest moment.

Cactus-pear productivity, as in other crops, is related to management practices. This is because plant productivity is an indicator of the cumulative effects of environmental factors that affect the growth (Nobel, 1988). Nutrient supply through the application of manure, fertilizer or both, coupled with irrigation, increases cladode production (Gonzaga & Cordeiro, 2005). In fact, it has been showed that under optimal conditions, cactus pear productivity can be equal to or higher than that obtained with highly productive crops as corn, alfalfa, sorghum and other (Nobel, 1988). The most extreme of the best production conditions is hydroponics, which provides up to 450 t ha^{-1} of green fodder in six months (Mondragón-Jacobo et al., 2001). Hydroponics is an efficient system for the use of water (Sánchez & Escalante, 1993), a scarce resource in arid regions. Moreover, from the point of view of experimentation, hydroponics provides a controlled environment which makes possible to reduce variance and to evaluate more precisely the reactions to diverse variation sources.

The aims of this work was to know the productivity and the changes of chemical characteristics as forage and nopalito quality of three *Opuntia* species and one *Nopalea* species, in a hydroponic system. The effects of maturity on studied characteristics were measured using fixed cladode sizes as harvest criterium and not a chronological criterium. As a consequence of size criterium for harvesting shoots, the harvest frequency was also studied. The information presented here is a summary of several works (López-Palacios, 2008; López-Palacios et al., 2010; Ramírez-Tobías, 2006; Ramírez-Tobías et al., 2007a, 2007b, 2010) that the authors have been developed in the last years. Moreover, comparisons of productivity and quality terms as nopalito and fodder against results obtain by production systems on soil are discussed.

MATERIALS AND METHODS

An experiment in a hydroponic system under a greenhouse was conducted in San José de la Peña, Villa de Guadalupe, San Luis Potosí, México (23º 15' 00" NL, 100º 45' 20" WL and 1735 m asl). The greenhouse was an East-West linear tunnel. The climate in San José de la Peña is semiarid, with extreme daily temperatures oscillation and cold winters (García, 2004).

Table 10.1. Chemical composition of the nutritive solution (Gallegos-Vazquez et al., 2000) and nutrient sources used (Ramírez-Tobías et al., 2007a).

Element	Concentration mg L^{-1}	Source
N	150.0	$Ca(NO_3)_2.4H_2O$
Ca	210.0	
P	40.0	KH_2PO_4
K	225.0	K_2SO_4
Mg	40.0	$MgSO_4.4H_2O$
Fe	5.0	Fe chelate
Mn	2.0	$MnSO_4.4H_2O$
Cu	0.1	$CuSO_4.5H_2O$
Zn	0.2	$ZnSO_4.7H_2O$
B	0.6	H_3BO_3
Mo	0.05	$(NH_4)_6Mo_7O_{24}.4H_2O$

The greenhouse had no additional controls of temperature such as heaters or coolers; the greenhouse temperature was recorded with a hygrothermograph, which averaged a yearly 22 ºC temperature with an annual variation of 8 ºC. The hydroponic system were four bench containers made with fluted fiber glass, measuring 1.80 long, 1.20 wide and 0.30 m high. Tezontle, a red volcanic grave, was used as substrate for the cactus pear plants (Ramirez- Tobías et al., 2007b). The particle diameter of the gravel was 2 to 4 mm; it was washed and disinfected with

0.20 ml L^{-1} of liquid solution of sodium hypochlorite as was stated by Calderón et al. (1997). A nutrient solution for growing cactus pear plants (Table 10.1) proven by Gallegos et al. (2000) was applied through sub-irrigation twice a day (09:00 and 13:00 h) on alternating days (Olmos et al., 1999). Since location where the experiment was conductedlacks piped water service, filtered runoff water was used for the nutritive solution. The nutritive solution was provided to plants through a closed irrigation system which allows recycling it. Then, the same solution was recycled during 15 days and totally replaced afterwards in order to preserve conductivity and nutrients at an adequate level. The pH of the nutritive solution was kept at the level of 5.8 by adding sulfuric acid (Calderón et al., 1997).

The spineless cactus pear species assessed proceeded from different climatic regions and their shoots are traditionally used either as nopalito or fodder in México. *Nopalea cochenillifera* (L.) Salm-Dyck lives in warm and humid regions; this species is considered the least adapted to low temperatures. *Opuntia ficus-indica* (L.) Mill. cv. Tlaconopal grows in temperate and subhumid climates. *Opuntia robusta* ssp. *larreyi* (Weber) Bravo, can be found in areas with temperate semidry climate with warm summer and extreme temperature variations. The hybrid *Opuntia undulata* Griffiths x *Opuntia tomentosa* Salm-Dyck is cultivated in a transitional climate from semi-warm to temperate.

Mature cladodes were planted in individual pots; the lower third of their total length was buried into the substrate. The orientation of the cladode sides was north-south due to space restrictions in the permanent greenhouse facilities. The plantation density on each bench container was 16.7 plants m^{-2}, with a distance of 20 cm between rows and 30 cm between plants (Fig. 10.1).

The caulinar photosynthetic area of the parental cladodes was estimated according to Sánchez-Venegas (1995). Considering the length of mature cladode in each species, fourth growth stages (GS) of vegetative sprouts were designed (Table 10.2). GS were set by dividing the mean mature cladode length by four in each species, this number, when multiplied by 1 corresponded to young nopalito (GS1), middle developed nopalito (GS2) when multiplied by 2, developed nopalito (GS3) when multiplied by 3 and mature cladode (GS4) when multiplied by 4. As was observed, the sprouts used for nopalito coincided with young and middle developed nopalito (GS1 and GS2, respectively), although the developed nopalitos(GS3) is used as nopalito and fodder, and the mature cladode (GS4) only as fodder. Every shoots sprouted were harvested from the parent plant from January 2005 to February 2006. Shoots sprouting and growth in each treatment was checked on alternate days, and those that had already reached the size set for the corresponding treatment were harvested and its weight quantified. Each harvest, the period of time between harvests (partial periods) and the production were recorded for each experimental unit or parental cladode. Productivity for each period (kg of dry matter ha^{-1} day^{-1}) was estimated by dividing the production by the respective number of days in the period, productivity for the whole experimental period was the sum of partial periods.

Fig. 10.1. Mature cladodes planted in individual pots on a hydroponic system and greenhouse (Figure took from Ramírez-Tobías et al., 2007b).

In order to study attributes related to nopalito quality pH, resistant to flexion and penetration, color and mucilage content were assessed. The pH was evaluated *in situ*, in sprouts harvested between 13:00 and 14:00 h by introducing the glass electrode and the temperature probe of a portable potentiometer (Oacton®) through an incision into a piece of nopalito (Flores-Hernández et al., 2004). Resistance to flexion and penetration measurements were carried out the day after harvesting, using a universal testing machine (Instron®). Resistance to flexion was assessed as the strength needed to bend the nopalito lengthwise (longitudinal flexibility) and widthwise (transverse flexibility) by 18.5% (Rodríguez-Félix & Villegas-Ochoa, 1997), and both measurements were averaged for each nopalito. Resistance to penetration was measured at the base, center and apex with a 1.5 mm diameter awl working at a speed of 100 mm min^{-1} (Rodríguez-Félix & Villegas-Ochoa, 1997; Guevara et al., 2003; Anzurez-Santos et al., 2004), and a mean resistance was calculated for each nopalito.

Table 10.2. Size ranges, in centimeters, of the four cladode growth stages (GS) in *Opuntia* and *Nopalea* cladodes grown under hydroponic conditions. In all cases, the standard deviation is± 1.5 cm, n=16 (Ramírez-Tobías, 2006).

Growth stages	Species			
	N. cochenillifera	*O. robusta* ssp. *larreyi*	*O. undulata* x *O. tomentosa*	*O. ficus-indica*
GS1	6	8	8	10
GS2	12	17	16	20
GS3	17	25	24	31
GS4	23	34	32	41

Color was registered the day after harvest using a Hunterlab colorimeter (Accu-ProbeTM model HH06) in the apex, center and base of the nopalito. According to the Hunter system (MacDougall, 2002), lightness values (L^*), hue angle and chroma were calculated (Sáenz, 2000; MacDougall, 2002). In order to establish mucilage content, the nopalitos were chopped into small pieces, kept in separate bags inside an ice cooler with a refrigerant gel during transportation, and then were kept at -20°C in the lab. The small pieces of nopalito were dried in a laboratory freeze-drier (ilShin® model TFD5505) and ground or crushed in a mortar. The extraction and quantification as percent of mucilage in dry matter was conducted according to the procedure described by Peña-Valdivia & Sánchez-Urdaneta (2006). This method includes extraction with hot water as solvent and its concentration by precipitation with ethanol.

Nutrimental fodder quality was assessed through the dry matter percentage, crude protein and ash content, which were obtained according to the AOAC (1990) techniques. The percentages of fiber insoluble in neutral and acid detergents were determined following the Van Soest et al. (1991) methods.

A random full-block experimental design was used, with a factorial arrangement of treatments and four replicates. Factors evaluated were cactus pear species (four) and growth stage; evaluations of nopalito quality attributes were made on growth stage one, two and three, and those of productivity and nutrimental fodder quality were made on the four growth stages. Then the experimental units assessed for nopalito quality attributes were a total of 48 and for productivity and nutrimental fodder quality were 64. The data of productivity were analyzed through a covariance analysis in which the caulinar photosynthetic area of parental cladodes was used as a covariable. Variables characterizing the quality of nopalitos and fodder were analyzed through a variance analysis and a multiple comparison of means. The SAS software was used for conducting statistical analysis through the

GLM and LSMEANS procedures (SAS, 1990). The pattern of biomass accumulation was estimated by plotting the cumulative production of partial harvest periods against time. Then, the model that best fits the data of each treatment against time was selected with SigmaPlot Jandel Scientific software (version 10) for the personal computer, this software was also used for selection of the best-fit models.

RESULTS

Nopalito and Forage Productivity

Plant productivity is defined as the rate of energy stored from photosynthetic activity (Odum, 1972); then it is a continuous process in which growth is affected by environmental factors (Nobel, 1988). Nopalito producers usually take its size as one criterium to cut the shoot intended for marketing. As an example, small nopalitos are considered as a highquality product (Flores & Olvera, 1994) which is confirmed by its excellent chemical and physical characteristics (Cantwell, 1992; Rodríguez-Félix & Cantwell, 1988); also authors observed that this kind of nopalitos takes prices in Mexican markets 50 to 100% higher than big ones. On the other hand, nutrimental quality of the bigger cladodes, usually used as fodder, is diminished with growth as will be indicated forward. Notwithstanding, research works cited along this paper, evaluating yield and nutrimental characteristics as fodder, usually refers that harvest of cladodes was made on fixed equal periods which is distantly linked with natural growth of the cladodes. Then, information corresponds to a wide range of sizes available at the time of the cut. Thereby in this work, productivity was analyzed on specific cladode sizes which are compatible with those of nopalito and fodder (Table 10.2).

Cladode productivity increased directly as harvest was carried out at increasing sizes, and the pattern of increase is different among species (Fig. 10.2). Conversely, the harvest of small cladodes as is made when they are cut to use them as nopalitos limits productivity and the harvest of mature cladodes enhance productivity. The productivity increase due to increasing size of the cladode showed some differences among species, being exponential in*ficus-indica* cv. Tlaconopal, *O. undulata* x *O. tomentosa* and *N. cochenillifera*, and linear in *O. robusta* ssp. *larreyi* (Fig. 10.2). The more pronounced increase in biomass with respect to growth stage at harvest, seem to be due to secondary growth (García de Cortázar & Nobel, 1992), characterized by cell wall lignification (Collins & Fritz, 2002). The linear relationship between productivity and growth stages in *O. robusta* ssp. *larreyi* may be explained by its low lignin content respecting to the other three species (Ramírez-Tobías et al., 2007a).

The productivity when harvest was made on the first growth stage kept some similarity with the second growth stage. Although a tendency to increase productivity was registered only in *O. robusta* ssp. *larreyi* (Fig. 10.2). This fact allows recommending the harvest of younger sprouts since yield did no augment for harvesting some bigger cladodes and savvy consumers might pay higher prices as was indicated before. Moreover, at least in Mexico, nopalitos marketing is more common in small cut pieces, which cause confusion in the consumer's perception of the quality of nopalito. Likewise, productivity of very young nopalitos of *O. robusta* ssp. *larreyi* is nearly doubled by that of bigger tender cactus in the same species, and a demerit occurs on visuals attributes associated with quality (López-Palacios et al., 2010). Then, recommendation as indicated should take into account variation in quality attributes due to maturity.

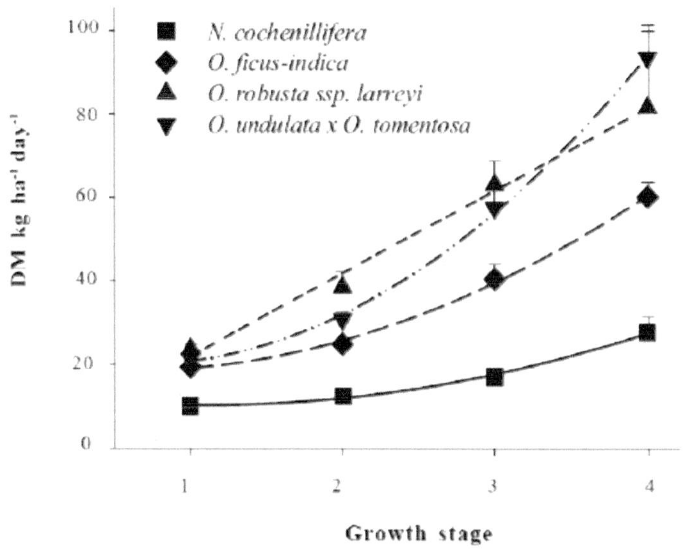

Fig. 10.2. Effects of the development stage on cactus pear productivity grown under hydroponics. Bars indicate standard errors. Shoots were harvested from the parent plant from January 2005 to February 2006 (Figure from Ramírez-Tobías et al., 2010).

Cladode productivity was statistically similar between growth stages three and four, excepting *O. robusta* ssp. *larreyi* which productivity of cladodes in growth stage four was significantly higher than productivity in growth stage three (Fig. 10.2). As will be described forward, cladode maturation both in soil and in hydroponics is characterized by a reduction in its quality as fodder (Gregory & Felker, 1992; Ramírez-Tobías, et al., 2007a). Moreover, in *Opuntia* species, from early mature cladode (growth stage three) to mature cladode (growth stage four) only insoluble fiber in neutral detergent increased and dry matter percent decreased significantly while other attributes as crude protein content, insoluble fiber in neutral detergent and ash kept without significant changes; although in *N. cochenillifera* crude protein, insoluble fiber in neutral detergent kept without changes (Ramírez-Tobías et al., 2007a). The lack of statistical differences in productivity when harvest early mature cladodes and full mature ones allows us to recommend the use of inmature cladodes, although species differences must be taken into account.

With the cladodes of bigger sizes, productivity increased further in *Opuntia* than in *Nopalea* species. From the second growth stage, productivity differences among *Opuntia* species were larger than with *Nopalea*. Two species, *O. robusta* ssp. *larreyi* and *O. undulata* x *O. tomentosa* showed the highest productivity; in contrast, *N. cochenillifera* the lowest (Fig. 10.2). Differences in productivity among *Opuntia* and *Nopalea* species can be due to adaptation to climatic conditions of their respective regions, specifically to temperature. Climatological data of *O. robusta* ssp. *larreyi* and *O. undulata* x *O. tomentosa* areas suggest ample temperature variation in daily and seasonal patterns (García, 2004). Then, the higher productivity of these species might be due to their higher tolerance to broad daily and seasonal temperature variations recorded in the greenhouse, which had not additional controls of temperature (Ramírez-Tobías, 2006). In contrast, climate patterns from regions of *N. cochenillifera* and *O. ficus*-indica cv. Tlaconopal, have less daily and seasonal variations of temperatures; first species has its optimum around a warm temperature regime and second with optimum around a temperate temperature regime (García, 2004).

Probably, temperature variations inside the greenhouse inhibited the growth of the less-tolerant species, resulting in lower productivities. The low productivity of *Nopalea* in relation with *Opuntia* recorded in hydroponics also could be explained by temperature effects. In the north-east semidry region of Brazil, on field conditions, the fodder productivity of *N. cochenillifera* and *O. ficus- indica* cv. Tlaconopal was statistically similar (Gonzaga & Cordeiro, 2005). Temperature reported for some Brazilian regions where *N. cochenillifera* and *O. ficus-indica* growth (Cordeiro et al., 2004) has some correspondence with that of *N. cochenillifera* regions in Mexico (Sánchez, 1995). Thus, differences in productivity between *Nopalea* and *Opuntia* could be indicative of genetic differences between both genera, particularly in low temperature tolerance.

The biomass accumulation pattern of Opuntia and Nopalea from the younger sprouts to the mature ones coincides with the patterns of perennial plants. Along a year of continuous harvest at the four different cladode sizes, the biomass accumulated shows a sigmoid pattern (Fig. 10.3). Along this time, accumulated production showed an initial lag phase lasting 40 days, followed by an exponential rise phase lasting up to 200 days, and a late stabilization phase of accumulating production (Fig. 10.3). Since data referred are from cactus pear growing in hydroponics under a greenhouse lacking of temperature control, it is possible suggest that temperature could be a limiting factor. In fact, inside the greenhouse, the monthly average temperature varied around 8 ⌋C during the registered period

(Fig. 10.4). Furthermore, the raise in monthly average temperature coincided with the accelerated increase on linear phase of the sigmoid trend, and the late stabilization phase seems to be associated to the drop in greenhouse temperature (Figs. 10.3 and 10.4).

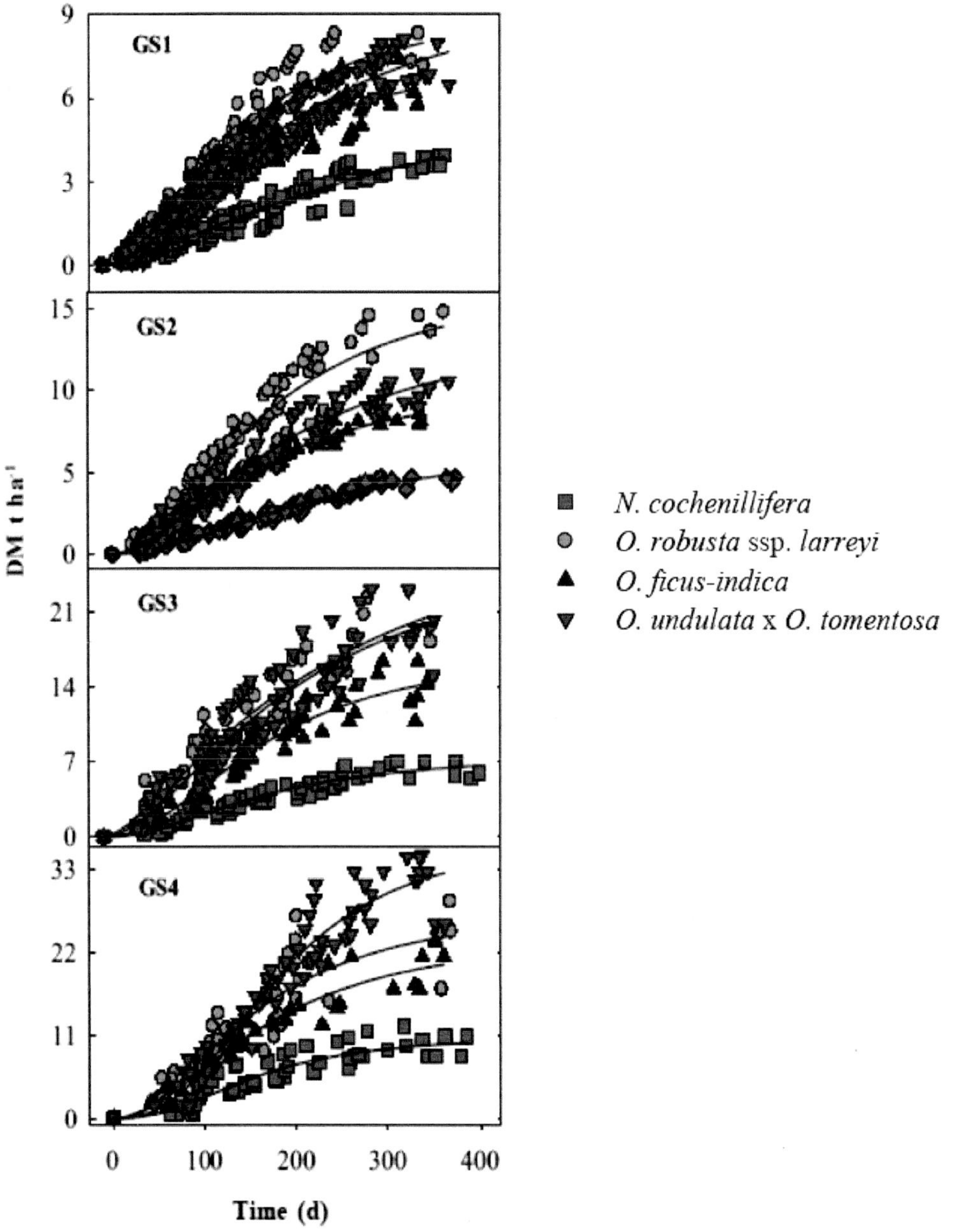

Fig. 10.3. Biomass accumulation pattern in three *Opuntia* and one *Nopalea* species harvested at four different cladode sizes. (From Ramírez-Tobías et al., 2010).

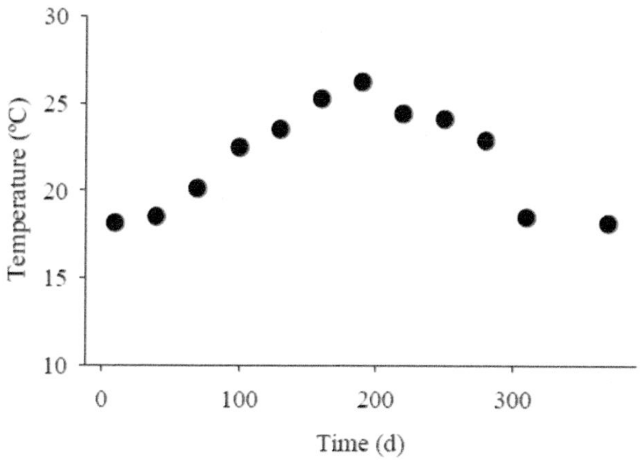

Fig. 10.4. Trend of the monthly average temperature registered inside the greenhouse where *O. ficus-indica, O. robusta* ssp. *larreyi, O. undulata* and *O. tomentosa* and *N. cochenillifera* grew along a full year.

The hydroponics clearly increased productivity of cactus pear. *O. ficus-indica* cv. Tlaconopal is a low productive species that yielded much more in hydroponics than in soil (Table 10.3). In the case of the mature cladode (GS4), the productivity was of 21, 24 and 32 t of dry matter ha^{-1} yr^{-1} in *O. ficus-indica* cv. Tlaconopal, *O. robusta* ssp. *larreyi* and *O. undulata* x *O. tomentosa*, respectively, a value far higher than productivity registered on intensive systems on soil (Table 3). A productivity of 47 t of dry matter ha^{-1} yr^{-1} was estimated for *O. ficus-indica* irrigated daily with a nutrient solution and with special care in promoting high efficiency in the use of solar radiation (Nobel et al., 1992). Hence the productivity of *Opuntia* may be extremely high when optimal conditions are provided (Nobel, 1988), as under hydroponics (Mondragón-Jacobo et al., 2001; Ramírez-Tobías et al., 2010).

Table 10.3. Comparison of productivity of production systems of *O. ficus-indica*. From GS 1 to GS 3 values are tons of cladodes (Fresh matter, FM) ha^{-1} yr^{-1} and in GS4 values are tons of cladodes (Dry matter, DM) ha^{-1} yr^{-1}. Based on Ramírez-Tobías et al., 2010[1], Flores & Olvera, 1994, Blanco-Macías et al., 2004[2] and Gonzaga & Cordeiro, 2005[3]).

Production system	Nopalito (GS 1)	Nopalito (GS 2)	Nopalito / Fodder (GS 3)	Fodder (GS 4)
Hydroponics	142[1]	264[1]	470[1]	21[1]
Traditional row system		40 to 90[2]		
Intensive system		112[2]		10[3]

The criterion used by cactus pear farmers and collectors for harvesting or gathering vegetative shoots, either for nopalitos or fodder, is cladode size, or growth stage as was made in this work. This fact contrasts with criterion generally used where fixed time periods are investigated. As will be indicated, quality of sprouts decays with maturity of the cladode

(López-Palacios et al., 2010; Ramírez-Tobías et al., 2007a). Subsequently, harvest would require a given frequency for best approximation to a cladode size and a desired quality. When cladode size is used to determine the moment of cut, the average period of time between harvests augment; consequently, frequency of harvest is inversely related to the size of cladode at the cut moment. The ideal situation would be achieving high harvest efficiency, it means, a low harvest frequency coupled with a high production every harvest (Ramírez-Tobías et al., 2010). Besides, the harvest frequency depend on the species; *O. ficus- indica* cv. Tlaconopal and *O. robusta* ssp. *larreyi* showed the longest mean periods between harvests, whereas the hybrid variants required the most frequent harvests (Table 10.4).

Table 10.4. Harvest frequency (average period and number of harvests) according to the growth stage (GS) at the cutting moment of four cactus pear species during an experiment evaluation from January 2005 to February 2006. (AIH, average interval time between harvests (days); AH, average number of harvests completed). Values after the sign ± indicate the standard deviation (From Ramírez-Tobías et al., 2010).

Species	GS1		GS2		GS3		GS4	
	AIH	AH	AIH	AH	AIH	AH	AIH	AH
N. cochenillifera	9±1	35±7	13±2	28±5	20±1	20±1	23±9	16±4
O. robusta ssp. *larreyi*	11±2	28±5	14±2	23±3	19±5	15±2	28±9	11±2
O. ficus-indica	12±1	27±3	18±2	19±3	21±3	16±2	35±3	10±1
O. undulata x *O. tomentosa*	9±1	40±1	13±1	27±1	15±2	22±2	21±2	17±2

NUTRIMENTAL FODDER QUALITY

Nutrimental quality was affected significantly by the maturation of each single cladode (Ramírez-Tobías et al., 2007a). As cladode grows and matures, crude protein diminished; protein content is considered a strong indicator of the quality of forages, but in cactus pear cladodes it diminished significantly from the smaller sizes of the cladodes (young and middle development nopalitos, GS1 and GS2) to the bigger ones (developing nopalito and mature cladode, GS3 and GS4) (Fig. 10.5). Similar findings have been observed in cladodes of *Opuntia* and *Nopalea* (Gregory

& Felker, 1992; Nefzaoui & Ben Salem, 2001) and in other forages (Collins & Fritz, 2003). Reductions in the protein content occur due to translocation of amino acids to younger organs or tissues which present higher metabolic activity than older ones (Nobel, 1983). Calculations considering crude protein content and production show that absolute protein content harvested in the mature cladodes did not reduce because increase in biomass compensates the decrease in protein content percentage (Ramírez-Tobías, 2006).

Insoluble fibers in neutral and acid detergents augment significantly from the younger growth stages to the older ones in *Opuntia* species while in *Nopalea* insoluble fiber in neutral detergent decreased and insoluble fiber in acid detergent slightly augmented (Fig. 10.5). Although, the significant increase of both fibers in *Opuntia* species with maturity, the increases in insoluble fibers in neutral detergent were greater than in insoluble fibers in acid detergent. The increase of both types of fibers is common in plants since is a result of lignifications processes due to maturation (Collinz & Fritz, 2003; De Alba, 1971), but smaller increase of insoluble fibers in acid detergent could evidence incipient lignifications. In fact, very young *Opuntia* cladodes lack lignin (Peña-Valdivia & Sánchez-Urdaneta, 2004). Conversely, the fiber increase of cladodes produced in hydroponics probably does not affect digestibility since its dry matter digestibility averaged 90% (Mondragón-Jacobo et al., 2001), while dry matter digestibility of cladodes produced on soil averaged 70 to 75% (Gonzaga & Cordeiro, 2005; Pinos-Rodríguez et al., 2006). Then, the increase of fibers in cladodes might involve mainly cellulose and hemicelluloses, which do not diminish nutrimental quality of cladode produced in hydroponics.

Ash of cladodes produced in hydroponics increased significantly with the maturity (Fig. 10.6). Similar findings were documented in cladodes over different cactus pear plants (Flores- Hernández et al., 2004) and in young and mature cladodes of *O. ficus-indica* cv. Tlaconopal (Rodríguez-Garcia et al., 2008). The augment on ash content is due to mineral accumulation, mainly calcium (Nobel, 1983; Rodríguez-Garcia et al., 2008). The increasing ash content due to maturity might reduce quality of the cladode as fodder since ash content is negatively correlated to digestibility (Pinos-Rodríguez et al., 2006). Notwithstanding, ash is a synonym of minerals (De Alba, 1971) and its input to the feeding of cattle and humans is relevant (Mcconn & Nakata, 2004).

Accumulation of ash through growth took different patterns among species. In the three*Opuntia* species, the ash content augmented with age of the cladode until developingnopalito (growth stage three); but from this point, ash did not change through fourth stage in *O. ficus-indica* cv. Tlaconopal while in *O. robusta* ssp. *larreyi* and in the hybrid species it reduced significantly.

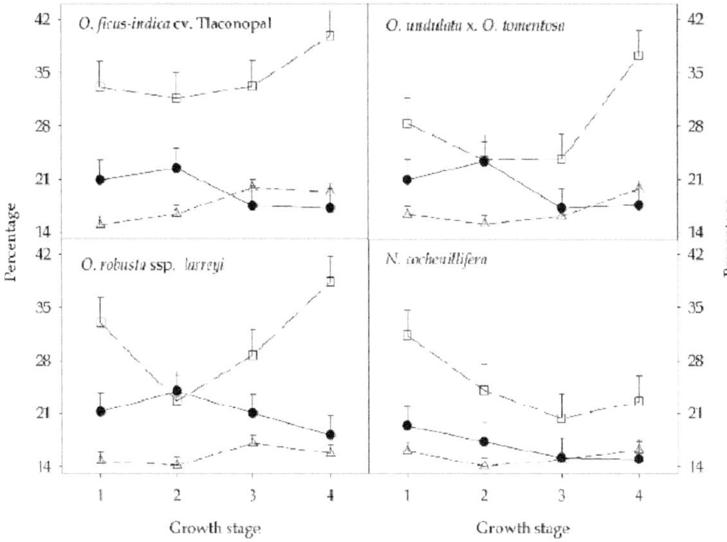

Fig. 10.5. Changes on the nutritional content due to maturity of cladodes of three Opuntia species and one of Nopalea grown in hydroponics under a greenhouse. (Insoluble fiber in acid detergent, insoluble fiber in neutral detergent and crude protein). Vertical bar indicate standard error. (Modified from Ramírez-Tobías et al., 2007a).

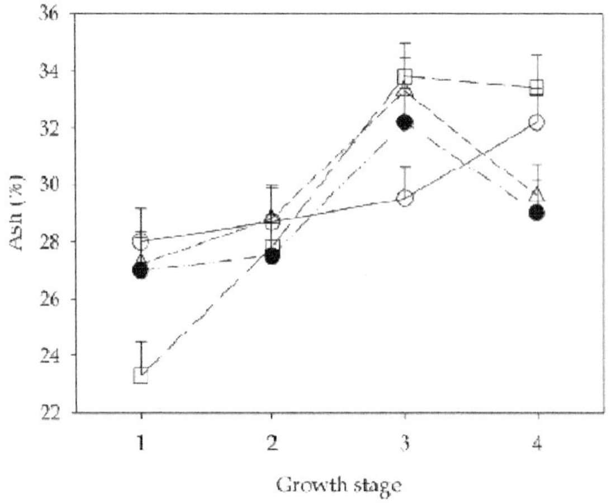

Fig. 10.6. Effect of maturity over the ash content of Opuntia ficus-indica ( ), O. robusta ssp. larreyi ( ), O. undulate x. O. tomentosa ( ) and Nopalea cochenillifera ( ) cladodes produced in hydroponics under greenhouse. Vertical bars indicate the error standard. (From Ramírez-Tobías et al., 2007a).

By the other hand, the augment on ash content on *N. cochenillifera* followed a continuous increase tendency from young nopalito (growth stage one) through mature cladode (growth stage four) (Fig. 10.6). Mineral content in plants depend on diverse factors as species, nutrient availability, age and others (Gárate & Bonilla, 2000). Since hydroponics system under greenhouse provided suitable growth conditions and high nutrient availability, it is possible to infer that different ash content on species over dissimilar growth stage is a genotype expression of different capacity of nutrient absorption and accumulation.

Dry matter percentage of cladodes produced in hydroponics was significantly higher in young nopalito (growth stage one) and decreased as cladode grows until an inflexion point from which it tended to augment (Fig. 10.7). The highest dry matter percentage registered on first growth stage could be explained by a higher rate area/volume observed in young cladodes than in older ones (Ramírez-Tobías et al., unpublished data). As cladode grows, tissues specialized in water store develop and percentage of water content get up, then dry matter percentage get down. Inflexion point of increase in dry matter percentage may be a result of medullary parenchyma development (Rodríguez & Cantwell, 1988) and accumulation of hemicelluloses and lignin associated with maturation. Inflexion point in dry matter percentage of *O. ficus-indica* cv. Tlaconopal and *O. robusta* ssp. *larreyi* was reached at growth stage three and for the hybrid species at growth stage two. Dry matter percentage during maturation of *N. cochenillifera* showed a different pattern from those of *Opuntia* species (Fig. 10.7). Differences in the pattern of dry matter percentage through growth stages on *Opuntia* species, and the highly marked difference observed between *Opuntia* and *Nopalea* genera suggest adaptive differences among genotypes in this respect.

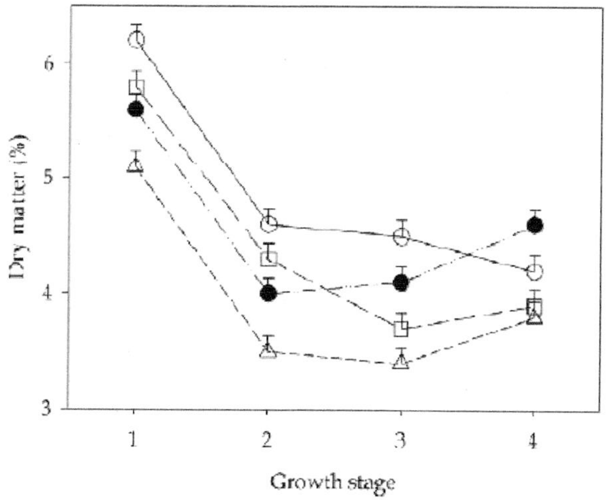

Fig. 10.7. Effect of maturity over the dry matter content of Opuntia ficus-indica cv. Tlaconopal (—□—), O. robusta ssp. larreyi (--△--), O. undulate x. O. tomentosa (- ●-) and Nopalea cochenillifera (—○—) cladodes produced in hydroponics under greenhouse.Vertical bars indicate the error standard. (From Ramírez-Tobías et al., 2007a).

Regardless maturation effect, *O. ficus-indica* cv. Tlaconopal presented the highest insoluble fiber in neutral and acid detergent content and *N. cochenillifera* tend to show the lowest one (Ramírez-Tobías et al., 2007a & Fig. 10.5). High fiber content of *O. ficus-indica* is related to higher resistance to flexion and penetration as regard to other cactus pear species (López- Palacios et al., 2010). Mechanical support is one of the functions of fibers (García & Peña, 1995). Moreover greater fiber content of *O. ficus-indica* cv. Tlaconopal might be an adaptation to support the weight of its cladodes, far bigger than those of *N. cochenillifera* (Ramírez-Tobías et al., 2007a).

Cladodes produced in hydroponics showed divergent dry mater percentage (Fig. 10.7); *N. cochenillifera* had the highest, *O. robusta* ssp. *larreyi* the lowest and *O. ficus-indica* cv. Tlaconopal and the hybrid species medium contents. Since dry matter is the opposite of water content, it could be established that *O. robusta* ssp. *larreyi* have the greatest capability of water store while *N. cochenillifera* have the lowest one. Their greatest capability for water store allows *O. robusta* ssp. *larreyi* to thrive in areas with annual precipitation around 400 mm or lesser, and the smallest capability of *N. cochenillifera* is only sufficient to survive in regions with 1000 mm. Intermediate competence for storing water of *O. ficus-indica* cv. Tlaconopal and the hybrid species require a medium precipitation quantity, 700 mm, as that recorded in places where they are found (García, 2004; Ramírez-Tobías et al., 2007a).

The nutrimental quality of the sprouts produced in hydroponics for animal feed is high. In cladodes produced in hydroponics, crude protein content is four times higher than that of cladodes produced on soil systems (Table 10.5), and it is comparable with that of high quality forages as alfalfa (De Alba, 1971). The high crude protein content in the dry matter of cladodes produced on hydroponics may be explained by the close relationship between the available nutrients in the nutritive solution and the N in the tissues (Gallegos-Vázquez et al., 2000; Nobel, 1983). Also, ash content and insoluble fiber in neutral detergent are higher in cladodes produced in hydroponics than those produced on soil, but insoluble fiber in aciddetergent resulted only slightly superior (Table 10.5). Since ash content and insoluble fiber in neutral detergent of cladodes grown on soil were partially negatively correlated with digestibility (Pinos-Rodríguez et al., 2003), cattle feed basing on cactus pear cladodes has been recommended for short time periods, from two to three months (Gutierrez & Bernal, 2004). Nonetheless, ash of cladodes produced on hydroponics can provide enough Ca, P, K and Zn for maintenance and production of cattle (Mondragón-Jacobo et al., 2001). Moreover, it was stated that the *Opuntia* dry matter of cladodes produced on soil stays for a short time in the rumen (16 to 48 h) and does not affect feed intake (Pinos-Rodríguez et al., 2003), and digestibility of cladodes produced in hydroponics was higher than that of cladodes produced on soil (Mondragón-Jacobo et al., 2001; Pinos-Rodríguez et al., 2003). High ash content in young cladodes (nopalitos) obtained in hydroponics is also relevant in people nutrition since nopalitos could be an important source of minerals (Rodríguez-García et al., 2007). Also richness in fibers may contribute to the control of cholesterol and the prevention of some diseases (Sáenz, 2000).

Form of production	Crude protein	Ash	Insoluble fiber in		Source
			Neutral detergent	Acid detergent	
Soil	4.1	21.6	24.7	16.3	[1]Pinos-Rodriguez et al., 2003
Hydroponics	16.9	31.1	34.5	17.7	[2]Ramírez-Tobías et al., 2007

Table 10.5. Nutrimental attributes of cladodes grown on different systems. 1Data are the average of *O. ficus-indica, O. robusta* and *O. rastrera.* 2Data are the average of *Nopalea cochenillifera, O. ficus-indica, O. robusta* ssp. *larreyi* and *O. undulata* x *O. tomentosa.*

Nopalito Quality

The resistance to the flexion and penetration forces, parameters associated with turgidity and fibrouness, presented differences among species and growth stages. In sprouts produced under hydroponics system, the flexion and penetration resistances increased with the maturation of the cladode (Fig. 10.8) (López-Palacios et al., 2010). In *O. robusta* ssp. *larreyi* from young to development nopalitos (growth stage 1 to 3) had an increased in resistance to longitudinal flexion, whereas nopalitos from other species did not require a greater force to bend. Developing nopalitos (growth stage 3) of the four species required a greater force to bend transversally (Fig. 10.8).

The sprouts of *N. cochenillifera* and *O. ficus-indica* cv. Tlaconopal demanded the higher force to be penetrated, and the hybrid species and *O. robusta* ssp. *larreyi* that required the lower one. During maturity, the nopalitos more developed registered the higher resistance to be penetrated (3 N). Young and middle developed nopalitos (growth stages 1 and 2) of *O. ficus- indica* cv. Tlaconopal and *N. cochenillifera* had higher resistance than those of *O. undulata* x *O. tomentosa* and *O. robusta* ssp. *larreyi*. Nopalitos of *O. ficus-indica* cv. Tlaconopal displayed the highest resistance to penetration, while those of other species did not showed significant differences in penetration resistance (Fig. 10.8).

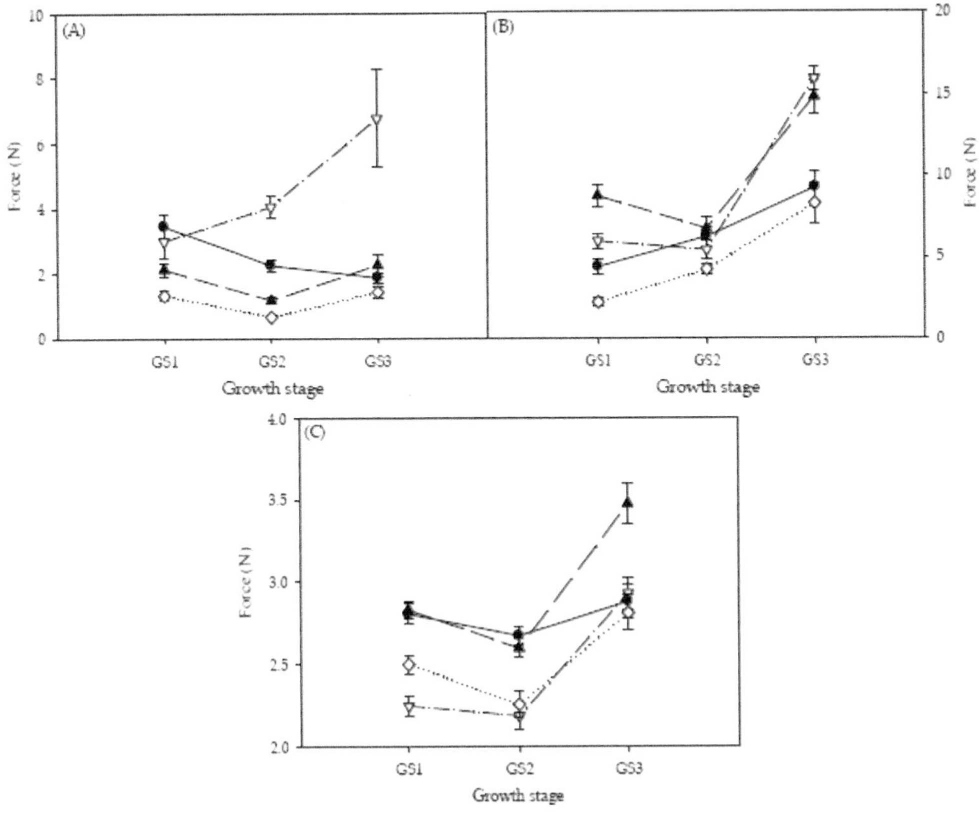

Fig. 10.8. Resistance to the longitudinal (A) and transversal flexion (B) and, (C) penetration force in four species at three growth stages produced in hydroponics system. (•) *N. cochenillifera*, (▲) *O. ficus-indica*, (∇) *O. robusta* ssp. *larreyi* and (◇) *O. tomentosa* x *O. undulata*. (From López-Palacios et al., 2010).

The resistance to flexion and penetration are related to turgidity, which is a parameter defined by Cantwell (1995) as an indicator of good quality in nopalitos. Greater turgidity might result in greater resistance. Physiologically, when there is absence of hydric stress the plant storages the water in the parenchyma cells and produces an increase in the bending force as well as a decrease in penetration resistance. Also the increase in both parameters with development might have derived from an increase in cellulose and lignin content (García & Peña, 1995). Guevara et al. (2003) observed a low resistance to penetration in *O. ficus-indica* cv. Tlaconopal nopalitos associated with a lower crude fiber content, which consist of lignin, mostly cellulose and, in some cases, a small portion of hemicellulose (García & Peña, 1995). Other authors (Rodríguez-Felix & Villegas-Ochoa, 1997) found a higher resistance to penetration in partially dehydrated nopalitos, since stored water causes a reduction in resistance to penetration (Gibson & Nobel, 1986).

The hardness of nopalitos also could be associated with their anatomy (Betancourt- Domínguez et al., 2006). To observe the effect of the cuticle on the resistance to penetration, its thickness was measured in 1 year-old cladodes from each species. The cellulose andlignin distribution was also observed in transversal sections after staining with safranin and fast green respectively. *Nopalea cochenillifera* had the thinnest cuticle and also *O. ficus-indica* the thickest. Hypodermis cell walls displayed the greatest safranin staining and five to six rows of hypodermal cells and calcium oxalate crystals. Likewise, parenchyma cells appeared to have the highest cellulose content. These results seem to indicate that the resistance of penetration depend more on hypodermal cell-wall lignification and calcium oxalate crystals than on the characteristics of the cuticle and epidermis. Nevertheless, the accumulation of calcium oxalate reduces the food quality because these compounds bind essential minerals and form complexes that decrease calcium absorption (Contreras-Padilla et al., 2011). Although we did not make a study about of presence of calcium oxalate during development of sprouts produced under hydroponics, the increased of ash observed above from mid-developed sprouts to mature sprouts is relate to the calcium presence and with the increase in penetration resistance as it was observed by Contreras-Padilla et al. (2011) in cladodes developed on soil.

Mucilage is a slime substance exuded from tissue cutting, characteristic of nopalitos that is part of the soluble dietary fiber. Its content may affect the consumer preference (Calvo- Arriaga et al., 2010), which may vary regionally or according to cultural factors, but no data are available on the optimum mucilage content. In this study, the highest mucilage content was observed in *N. cochenillifera* with 8.27% and the lowest in *O. ficus-indica* cv. Tlaconopal with 5.34%, while the *O. robusta* ssp. *larreyi* and the hybrid species had an intermediate content. Each species studied displayed a different pattern on mucilage content in relation to the stage of growth (Fig. 10.9). Thus, the mucilage percentage of *O. undulata* x *O. tomentosa* nopalitos was directly related with maturity. As for the other species, *O. ficus-indica* cv. Tlaconopal showed the lowest mucilage content during its development. These differences on mucilage content are related with genetics differences among species; thus, the highest mucilage content of *N. cochenillifera* may be related to water balance in the warm and humid environment where it grows. Likewise, the *N. cochenillifera* sprouts showed the highest neutral-detergent insoluble fiber content and an increase in dietary fiber from middledeveloped nopalitos (Ramírez-Tobías et al., 2007a). Some studies have shown that exist an inverse relationship between soluble and insoluble fiber content in nopalitos (López- Palacios et al., 2011; Peña-Valdivia & Sánchez-Urdaneta, 2006), which may explain those differences between *Nopalea* and *Opuntia* genus during development of the nopalitos.

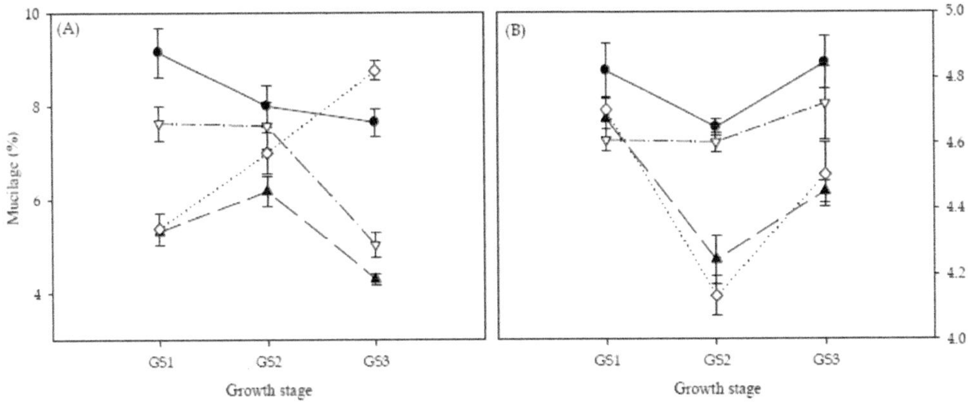

Fig. 10.9. Content of mucilage (A) and acidity as pH (B) in four species of nopalitos at three growth stages produced in a hydroponics system. (•) *N. cochenillifera*, (▲) *O. ficus-indica*, (∇) *O. robusta* ssp. *larreyi* and (◇) *O. tomentosa* x *O. undulata* (Modified from López-Palacios et al., 2010).

The acid flavor of nopalitos affects the consumer preference. The *N. cochenillifera* nopalitos had the highest pH value (4.77), and *O. undulata* x *O. tomentosa* the lowest (4.47). Moreover, the pH values decreased from younger nopalitos (growth stage 1) to middle developed ones (growth stage 2) and then increased from middle developed nopalitos (growth stages 2) to developed ones (growth stage 3) in *N. cochenillifera*, *O. ficus-indica* cv. Tlaconopal and *O. undulata* x *O. tomentosa* sprouts (Fig. 9). These changes observed during the growth were similar to the showed by field-grown nopalitos (Betancourt-Dominguez et al., 2006). In contrast, the pH values in nopalitos of *O. robusta* ssp. *larreyi* remained constant during development. A study registered that the pH decreased in nopalitos as growth progressed (Rodríguez-Felix & Cantwell, 1988); a different pattern was observed in our investigation. Pimienta-Barrios et al. (2005) observed a shift from CAM (crassulacean acid metabolism) to C3 matabolic pathway in nopalitos grown under irrigation which caused an increase of pH. This metabolic shift has already been related to water availability (Cushman, 2001), so what in this shift in the photosynthetic pathway can have happened young and development nopalitos (growth stage 1 and 3, respectively).

The color of nopalitos, as hue angle, was between green and yellow and increased with development while chroma, or intensity of color, decreased. Growth decreased cladode luminosity or brightness, but some species as *O. robusta* ssp. *larreyi* did not change during development (Fig. 10.10). However, the hue angle values were similar among species and production system, which may indicates that this color parameter is common among different nopalito species (Calvo-Arriaga et al., 2010). In this study, none of the species displayed the color attributes considered as indicator of high quality, which had a hue angle near 114° and a chroma and luminosity of 30 and 43, respectively (George et al., 2004). In addition, presence of waxy cuticle and pubescence in species as *O. robusta* ssp. *larreyi* and *O. undulata* x *O. tomentosa* may influence the chroma and lightness values and their acceptatabilty by consumers.

The bending and penetrating resistance and mucilage content of sprouts in the four species studied were lower than those registered on harvested in commercial plantations (Table 10.6). These differences may be due to a continuous water supply and the protective environment of greenhouse. The high bending and penetration resistance forces in nopalitos produced on soil is explained by the fact that, under those conditions, fibers are more robust than those of nopalitos grown in a hydroponic system in a protective greenhouse environment (Rodríguez- Felix & Villegas-Ochoa, 1997).

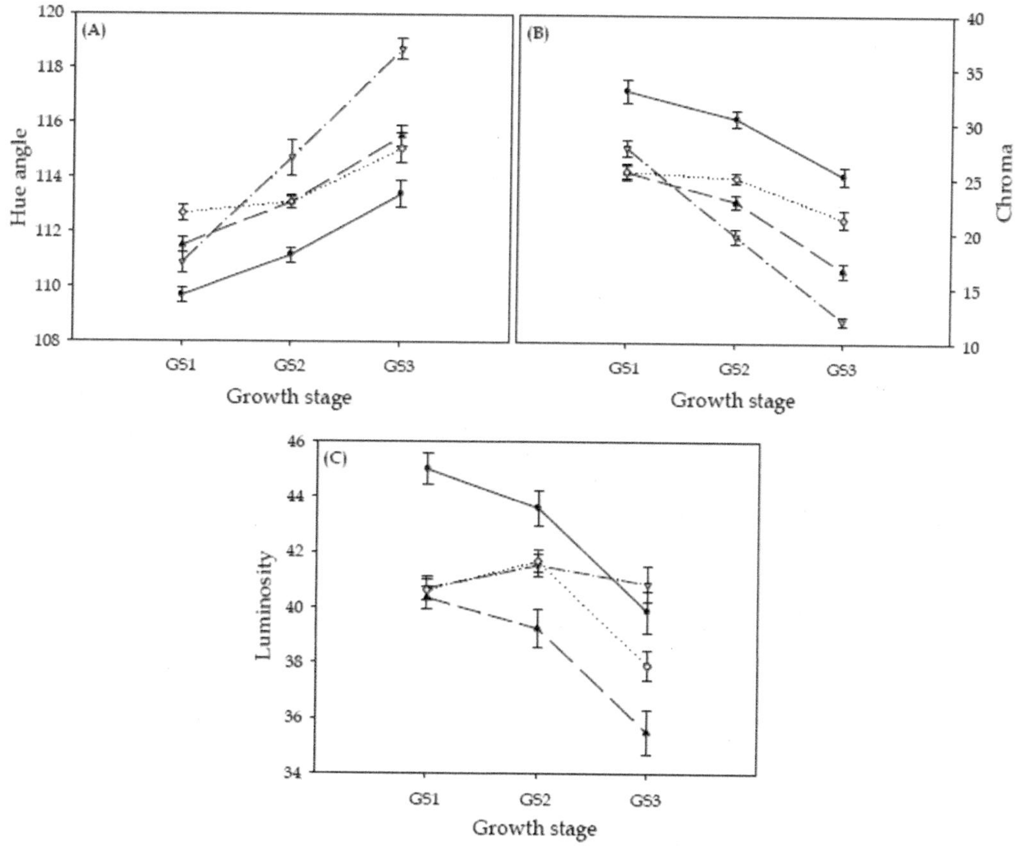

Fig. 10.10. Color parameters as hue angle (a) chroma or intensity of color (b) and luminosity (c) in four species at three growth stages produced in hydroponics system. (•) *N. cochenillifera*, (▲) *O. ficus-indica*, (∇) *O. robusta* ssp. *larreyi* and (◇) *O. tomentosa* x *O. undulata* (Modified from López-Palacios et al., 2010).

Moreover, the lower mucilage content in nopalitos species produced under hydroponic system can be explained by the absence of temperature oscillations and water stress than in an open production systems on soil. Some authors had documented an increase in mucilage production in plants under stress conditions (García- Ruíz, 2007; Goldstein & Nobel, 1991, 1994; Nobel et al., 1992) because the mucilage has an important role on water storage, freeze tolerance, calcium metabolism, and as reserve of carbohydrates. Nopalitos produced under hydroponics system had a lower mucilage higher content due to plenty water availability and to the protective greenhouse environment, thus the dissimilarity observed may be due to genetic differences among species.

Nopalitos grown on hydroponics had similar pH values than nopalitos grown on soil (Table 10.7). This similarity is probably due to that in the field these plants follow the same trend observed on hydroponics system to produce new cladodes. Moreover, the acidity changes as a reaction to environmental conditions also are related to changes in mucilage viscosity (Trachtemberg & Mayer, 1981), which permit to the plant tolerate different stress conditions. Also, none of the nopalitos species produced under hydroponics had similar values in the three color parameters to those produced on soil (Table 10.7). Razo & Sánchez (2002) observed in nopalitos produced on soil differences in color parameters during the day. They associated these changes to modification of pigments; nevertheless, the color of nopalitos also depends on the production system as observe in *O. ficus-indica* (Table 10.7).

Nopalito quality is determinated by a set of anatomical and physiological features that result in perceptible attributes which allow the consumer to develop certain preferences for the various nopalito species available in a particular region (Cantwell, 1995; Figueroa, 1984; López-Palacios et al., 2010). Thus, the yield from nopalitos of different species produced under hydroponics system is an opportunity to supply a market with different preferences. The results obtained in the present study mar serve for future studies on hydroponics nopalito production and their marketing.

Table 10.6. Resistance to the flexion and penetration forces, and mucilage content of nopalitos produced on commercial soil plantations and hydroponic system.

Specie	Production system	Resistance to		Mucilage content (%)
		Flexion (N)	Penetration (N)	
N. cochenillifera [a]	Soil	---	---	15.9
N. cochenillifera [b]	Hydroponics	4.85	2.77	8.27
O. albicarpa [c]	Soil	---	---	6.89
O. ficus-indica [b]	Hydroponics	7.21	2.91	5.34
O. ficus-indica [c]	Soil	---	---	11.72
O. ficus-indica [d]	Soil	---	---	6.90 – 8.57
O. ficus-indica [e]	Soil	21.8 – 30.0	11.0 – 15.0	---
O. ficus-indica [f]	Soil	---	19.0	---
O. hyptiacantha [c]	Soil	---	---	8.47
O. megacantha [c]	Soil	---	---	7.40
O. robusta ssp. *larreyi* [b]	Hydroponics	7.95	2.39	6.76
O. streptacantha [c]	Soil	---	---	7.18
O. undulata x *O. tomentosa* [b]	Hydroponics	3.45	2.52	6.93

[a]Nerd et al., 1997

[b, c]López-Palacios et al., 2010, 2011

[d]Peña-Valdivia & Sánchez-Urdaneta, 2006

[e]Rodríguez-Félix & Villegas-Ochoa, 1997

[f]Guevara et al., 2003

Table 10.7. Color attributes and acidity (pH) of nopalitos produced on commercial soil plantations and hydroponic system.

Species	Production system	Color			pH
		Hue angle	Chroma	Luminosity	
N. cochenillifera [a]	Soil	---	---	---	4.4
N. cochenillifera [b]	Hydroponics	111.14	30.18	43.20	4.8
O. ficus-indica [b]	Hydroponics	112.90	20.46	38.96	4.5
O. ficus-indica [c]	Soil	117.66	16.60	---	---
O. ficus-indica [d]	Soil	116.0 – 120.0	17.0 – 21.0	40.0 – 49.0	---
O. ficus-indica [e]	Soil	---	---	---	4.0 – 6.0
O. ficus-indica [f]	Soil	114.0	29.37	43.03	---
O. megacantha [d]	Soil	---	---	---	5.0 – 6.0
O. robusta ssp. *larreyi* [b]	Hydroponics	114.15	21.13	41.05	4.6
O. undulata x *O. tomentosa* [b]	Hydroponics	113.51	24.26	40.20	4.5

[a]Ochoa et al., 2004

[b]López-Palacios et al., 2010 c Anzures-Santos et al., 2004 d Calvo-Arriaga et al., 2010

[e]Flores-Hernández et al., 2004

[f]George et al., 2004

AT A GLANCE

- Production of cactus pear cladodes in hydroponics represents an opportunity for producing biomass in arid and semiarid regions. Productivity of this system, harvesting nopalito or fodder, overcomes that of soil systems. Besides, quality of sprouts as nopalito or fodder is comparable or better than that recorded on traditional production systems. Nevertheless, establishment of hydroponics require high initial investment. However, economic evaluations of these systems are not documented. In addition, due to the possibility for recycling the water, hydroponics is considered an efficient water use system. Last, arguments that support some investigations and treatise on cactus pear hydroponics, still need to be evaluated and verified.

REFERENCES

Anaya P., M. A. (2001). History of the use of *Opuntia* as forage in México. In: *Cactus (Opuntia spp.) as forage*, Mondragón-Jacobo, C. & Pérez-González, S. (Eds), 3-12, Food and Agriculture Organization of the United Nations, ISBN 92-5-104705-7, Rome, Italy.

Anzures-Santos, G., Corrales-García, J. & Peña-Valdivia, C.B. (2004). Efecto el almacenamiento en algunos atributos de calidad del nopalito desespinado (*Opuntia* sp.) en la variante Atlixco. In: *Memoria del X Congreso Nacional, VIII Internacional sobre el Conocimiento y Aprovechamiento del Nopal y otras Cactaceas de Valor Económico y del Fifth International Congress on Cactus Pear and Cochineal* (CD Edition), Flores V.,

C. (Ed.), Universidad Autónoma Chapingo, Fondo de las Naciones Unidas para la Alimentación y la Agricultura and International Society for Horticultural Science. Chapingo, México.

Betancourt-Domínguez, M.A., Hernández-Pérez, T., García-Saucedo, P., Cruz-Hernández, & Paredez-López, O. (2006). Physico-chemical changes in cladodes (nopalitos) from cultivated and wild cacti (*Opuntia* spp.). *Plant Foods for Human Nutrition*, Vol. 61, No. 3, (September 2006), 115–119, ISSN 0921-9668.

Blanco-Macías, F.R., Valdez-Cepeda, R.D., Ruiz-Garduño, R. & Márquez-Madrid, M. (2004). Producción intensiva de nopalito orgánico durante cuatro años en Zacatecas, In: *Memoria el X Congreso Nacional y VII Congreso Internacional sobre Conocimiento y Aprovechamiento del Nopal y del Fifth International Congress on Cactus Pear*, Flores V., C.A. (Ed.), 2, Chapingo, México.

Cantwell, M. (1992). Aspectos de calidad y manejo postcosecha de nopalitos, In: *Conocimiento y Aprovechamiento del Nopal. V Congreso Nacional y III Internacional. Memoria de Resúmenes*, Salazar, S. & López, D. (Eds.), 110, Universidad Autónoma Chapingo, Chapingo, México.

Calderón, P. N., Estrada L., A. A. & Martínez A., J. L. (1997). Efecto de la salinidad en el crecimiento y absorción nutrimental de plantas micropropagadas de nopal (*Opuntia* spp.). In: *Memoria del Séptimo Congreso Nacional y Quinto Internacional sobre el Conocimiento y Aprovechamiento del Nopal,* Vázquez A., R.E. (Ed.), 165-166, Monterrey, México.

Calvo-Arriaga, A.O., Hernández-Montes, A., Peña-Valdivia, C.B., Corrales-García, J. & Aguirre-Mandujano, E. (2010). Preference mapping and rheological properties of four nopal (*Opuntia* spp.) cultivars. *Journal of the Profesional Assoiation for Cactus Development,* Vol. 12, 127-142, ISSN 1938-6648.

Collins, M. & Fritz, J.O. (2003). Forage quality, In: *Forages*, Barnes, R.F., Nelson, C.J., Collins, M. & Moore, J.K. (Eds.), 363-390, Iowa State Press, Ames, Iowa, USA.

Contreras-Padilla, M., Pérez-Torrero, E., Hernández-Urbiola, M.I., Hernández-Quevedo, G., del Real, A., Rivera-Muñoz, E.M. & Rodríguez-García, M.E. (2011). Evaluation of oxalates and calcium in nopal pads (*Opuntia ficus-indica* var. redonda) at different madurity stages, *Journal of Food Composition and Analysis*, Vol. 24, No. 1, (February 2011), 38-43, ISSN 0889-1575.

Cordeiro dos S., D., M. de Andrade L., F., Marcos D., I., Farias, M.V. & Ferreira dos S., V.F. (2000), Productividade de cultivares de palma forrageira (*Opuntia* e *Nopalea*), In: *II Congresso Nordestino de Produção Animal y VII Simposio Nordestino de Alimentação de Ruminantes,*121-123, Teresina, Brasil.

Cordeiro dos S., D., Lira, M.A., Farias, I., Dias, F.M., Silva, F.G. (2004). Competition of forage cactus varieties under semi-arid conditions in the northeast of Brazil. In: *Memoria del X Congreso Nacional y VII Congreso Internacional sobre Conocimiento y Aprovechamiento del Nopal y del Fifth International Congress on Cactus Pear and Cochineal* (CD Edition), Universidad Autónoma Chapingo, Food and Agriculture Organization e International Society for Horticultural Science, Chapingo, México.

Cushman, J.C. (2001). Crassulacean acid metabolism. A plastic photosynthetic adaptation to arid environments, *Plant Physiology*, Vol. 127, No. 4, (December 2001), 1439-1448, ISSN 0032-0839.

De Alba, J. (1971). *Alimentación del ganado en América Latina* (2nd. Edition), La Prensa Médica Mexicana, México, D.F.

Figueroa H., F. (1984). Estudio de las nopaleras cultivadas y silvestres sujetas a recolección para el mercado en el Altiplano Potosino-Zacatecano, Tesis Profesional, Escuela de Agronomía, Universidad Autónoma de San Luis Potosí, San Luis Potosí, México.

Flores V., C.A. & Aguirre R., J.R. (1979). *El nopal como forraje.* Universidad Autónoma Chapingo, Chapingo, México, 80 p.

Flores V., C.A. & Olvera, J. (1994). *El sistema-producto nopal verdura en México*, Universidad Autónoma Chapingo, Texcoco, México.

Flores-Hernández A., Orona-Castillo I., Murillo-Amador B., Valdez-Cepeda R.D. & García- Hernández J.L. (2004). Producción y calidad de nopalito en la región de la comarca lagunera de México y su relación con el precio en el mercado nacional. *Journal of the Profesional Assoiation for Cactus Development,* Vol. 6, 23-34, ISSN 1938-6648.

Gallegos-Vázquez, C., Olivares-Sáenz, E., Vázquez-Alvarado, R. & Zavala-García, F. (2000). Absorción de nitrato y amonio por plantas de nopal en hidroponía. *Terra,* Vol. 18, 133-139, ISSN 1870-9982.

Gárate, A. & Bonilla, I. (2000). Nutrición mineral y producción vegetal. In: *Fundamentos de Fisiología Vegetal*, Azcón, B.J. & Talón, M. (Eds.), 113-130, Mc Graw Hill, ISBN 9788448151683, Madrid, España.

García de Cortázar, V. & Nobel, P.S. (1992). Biomass and fruit production for the prikly pear cactus, *Opuntia ficus-indica. Journal of the American Society for Horticultural Science*, Vol. 117, No. 4, (July 1992), 558-562, ISSN 0003-1062.

García, E. (2004). *Modificaciones al sistema de clasificación climática de Köppen* (5th edition), Universidad Nacional Autónoma de México, México, DF.

García H., E. & Peña V., C. (1995). *La pared celular. Componente fundamental de las células vegetales*, Universidad Autónoma Chapingo, Chapingo, ISBN 9688843342, Estado de México, México.

García-Ruíz, M.T. (2007). *Procesos fisiológicos y contenido de polisacáridos estructurales en nopalito* (Opuntia *spp.) y su modificación por el potencial de agua del suelo,* Tesis para obtener el grado de Maestro en Ciencias, Colegio de Postgraduados, Montecillo, Texcoco, Estado de México, México.

George, R.S., Corrales G., J. Peña V., C. & Rubio H., D. (2004). Cambios en color, sabor y contenido de mucíalgo en noplito (*Opuntia ficus-indica*) escaldado con tequezquite. In: *Memoria del X Congreso Nacional, VIII Internacional sobre el Conocimiento y Aprovechamiento del Nopal y otras Cactaceas de Valor Económico y del Fifth International Congress on Cactus Pear and Cochineal* (CD Edition), Flores V., C. (Ed.), Universidad Autónoma Chapingo, Fondo de las Naciones Unidas para la Alimentación y la Agricultura and International Society for Horticultural Science. Chapingo, México.

Gibson, A.C. & Nobel, P.S. (1986). *The cactus primer*, Harvard University Press, ISBN 978- 0674089914, Cambrige, Massachusets, USA.

Gonzaga de A., S. & Cordeiro dos S., D. (2005). Palma forrageira. In: *Espécies vegetais exóticas com potencialidades para semi-árido brasilero,* Piedade, L. H. & Assis M., E. (Eds), 91- 127, Empresa Brasileira de Pesquisa Agropecuária, ISBN 8573832878, Brasilia, Brasil.

Goldstein, G.; Nobel, P.S. 1991. Changes in osmotic pressure and mucilage during low- temperature acclimation of *Opuntia ficus-indica*, *Plant Physiology*, Vol. 97, No. 3, (November 1991), 954-961, ISSN 0032-0889.

Goldstein, G.; Nobel, P.S. 1994. Water relations and low-temperature acclimation for cactus species varying in freezing tolerance. *Plant Physiology*, Vol. 104, No. 2, (February, 1994), 675-681, ISSN 0032-0889.

Gregory, R.A. & Felker, P. (1992). Crude protein and phosphorus contents of eight contrasting *Opuntia* forage clones, *Journal of Professional Cactus Development*, Vol. 22, 323-331, ISSN 1938-6648.

Guevara, J.C., Yahia, E.M., Brito, F.E. & Biserka, S.P. (2003). Effects of elevated concentrations of CO2 in modified atmosphere packaging on the quality of prickly pear cactus stems (*Opuntia* spp.), *Postharvest Biology and Technology*, Vol. 29, No. 2, (August 2003), 167-176, ISSN 0925-5214.

Gutiérrez O., E. & Bernal B., H. (2004). Uso del nopal en la nutrición animal, In: *Memoria del X Congreso Nacional y el VII Congreso Internacional sobre Conocimiento y Aprovechamiento del Nopal y del Fisth Intenational Congress on Cactus PEar and Cochineal. Universidad Autónoma Chapingo, Food and Agriculture Organization e International Society for Horticultural Science* (CD Edition), Flores V., C (Ed). Chapingo. Mexico. Edición en DC.

López-García, J.J., Fuentes-Rodríguez, J.M. & Rodríguez-Gámez, A. (2001). Production and use of *Opuntia* as forage in northern of México. In: *Cactus (Opuntia spp.) as forage*, Modragón-Jacobo, C. & Pérez-González, S. (Eds), 29-36, Food and Agriculture Organization of the United Nations, ISBN 92-5-104705-7, Rome, Italy.

López-Palacios, C. (2008). *Evaluación de atributos posiblemente asociados con la calidad del nopalito (*Opuntia *spp. y* Nopalea *sp.)*. Tesis para obtener el grado de Ingeniero, Ingeniería Agroindustrial, Facultad de Ingeniería, Universidad Autónoma de San Luis Potosí, San Luis Potosí, México.

López-Palacios, C., Peña-Valdivia, C.B., Reyes-Agüero, J.A. & Rodríguez-Hernández, A.I. (2011). Effects of domestication on structural polysaccharides and dietary fiber in nopalitos (*Opuntia* spp.), *Genetic Resources and Crop Evolution*, DOI 10.1007/s10722- 011-9740-3, (Online August 2011), ISSN 0925-9864.

López-Palacios, C., Reyes-Agüero, J.A., Ramírez-Tobías, H.M., Juárez-Flores, B.I., Aguirre- Rivera, J.R., Yañez-Espinoza, L. & Ruíz-Cabrera, M.A. (2010). Evaluation of attributes associated with the quality of nopalito (*Opuntia* spp. and *Nopalea* sp.), *Italian Journal of Food Science*, Vol. 22, No., 4, 423-431, ISSN 1120-1770.

MacDougall, D.B. (2002). Colour measurement of food, In: *Colour in Food: Improving quality*, MacDougall, D.B. (Ed.), 33-63, Woodhead Publishing Limited, ISBN 1-85573-590-3, Cambridge, Unit Kingdom.

Mcconn, M.M. & Nakata, P.A. (2004). Oxalate reduces calcium availability in the pads of the prickly pear cactus through formation of calcium oxalate crystals, *Journal of Agricultural Food and Chemistry*, Vol. 52, No. 5, (March 2004), 1371-1374, ISSN 0021-8561.

Mondragón-Jacobo, C, S.J. Méndez G., G. & Olmos O. (2001). Cultivation of *Opuntia* for fodder production: from re-vegetation to hydroponics. In: *Cactus (Opuntia spp.) as forage*, Mondragón-Jacobo, C. & Pérez-González, S. (Eds), 107-122, Food and Agriculture organization of the United States. ISBN 92-5-104705-7, Rome, Italy.

Nefzaoui, A., H. & Ben Salem. (2001). *Opuntia* spp., a strategic fodder and efficient tool to combat desertification in the WANA region. In: *Cactus (Opuntia spp.) as forage,* Modragón-Jacobo, C. & Pérez-González, S. (Eds), 73-90, Food and Agriculture Organization of the United Nations, ISBN 92-5-104705-7, Rome, Italy.

Nerd, A., Dumotier, M. & Mizrahi, Y. (1997). Properties and postharvest behavior of the vegetable cactus *Nopalea cochenillifera*, *Postharvest Biology and Technology*, Vol. 10, Num. 2, (February 1997), 135- 126, ISSN 0925-5214.

Nobel, P.S. (1983). Nutrient levels in cacti-relation to nocturnal acid and growth, *American Journal of Botany*, Vol. 70, 1244-1253, ISSN 0002-9122.

Nobel, P.S. (1988). *Environmental biology of Agaves and Cacti,* Cambridge University press. ISBN 0521543347, New York, United States.

Nobel, P.S., Cavelier, J. & Andrade, J.L. (1992). Mucilage in cacti: its apoplastic capacitance, associated solutes, and influence on tissue water relations. *Journal of Experimental Botany,* Vol. 43, No. 5, (May 1992), 641-648, ISSN 0022-0957.

Ochoa, M.J., Leguizamo, G., Ayrault, G. & Miranda, F.N. (2004). Cold storage and shelf life *Nopalea cochenillifera* behavior: evaluation of quality parameters. In: *Memoria del X Congreso Nacional, VIII Internacional sobre el Conocimiento y Aprovechamiento del Nopal y otras Cactaceas de Valor Económico y del Fifth International Congress on Cactus Pear and Cochineal* (CD Edition), Flores V., C. (Ed.), Universidad Autónoma Chapingo, Fondo de las Naciones Unidas para la Alimentación y la Agricultura and International Society for Horticultural Science. Chapingo, México.

Odum, E.P. (1972). *Ecología* (3rd Edition), Interamericana, México, D.F.

Olmos O., G., Méndez G., S. de J. & Martínez H., J. (1999). Evaluación de 29 cultivares de nopal para producción de forraje en hidroponía. In: *Memoria del Octavo Congreso Nacional y Sexto Congreso Internacional Sobre Conocimiento y Aprovechamiento del Nopal*, Aguirre R., J. R. & Reyes A., J. A. (eds), 105-106, San Luis Potosí, México.

Paiz, R.C., Juárez F., B.I., Aguirre R., J.R., Cárdenas O., N.C., Reyes-Agüero, J.A., García C.,

E. & Álvarez F., G. (2010). Glucose-lowering effect of xoconostle (Opuntia joconostle Web., Cactaceae) in diabetic rats. Journal of Medicinal Plants Research, Vol. 4, No. 22, (November 2010), 2326-2333, ISSN 1996-0875.

Peña-Valdivia, C.B. & Sánchez-Urdaneta, B.A. (2006). Nopalito and cactus pear (Opuntia spp.) polysaccharides: mucilage and pectin. Acta Horticulturae, Vol. 728, No. 1, (December 2006), 241-248, ISSN 0567-7572.

Pimienta-Barrios, E., Zañudo-Hernández, J., Rosas-Espinosa, V.C., Vañenzuela-Tapia, A. & Nobel, P.S. (2005). Young daughter cladodes affect CO2 uptake by mother cladodes of Opuntia ficus-indica, Annals of Botany, Vol. 95, No. 2, (January 2005), 363-369, ISSN 0305-7364.

Pinos-Rodríguez, J.M., Duque B., R, Reyes-Agüero, J.A., Aguirre-Rivera, J.R. & González, S.S. (2003). Contenido de nutrientes en tres especies de nopal forrajero. In: *Memoria del IX Congreso Nacional y VII Congreso Internacional sobre Conocimiento y Aprovechamiento del Nopal*, Esparza F., G., Salas, M.A., Mena C., J. & Valdez C., R.D. (Eds.), 60-63, Universidad Autónoma Chapingo, Universidad Autónoma de Zacatecas e Instituto Nacional de Investigaciones Forestales Agrícolas y Pecuarias. Zacatecas, México.

Pinos-Rodríguez, J.M., Duque B., R., Reyes-Agüero, J.A., Aguirre-Rivera, J.R., García-López, J.C. & González-Muñoz, S. (2006). Effect of species and age on nutrient content and *in vitro* digestibility of *Opuntia* spp., *Journal of Applied animal research*, Vol. 30, 13-17, ISSN 0971-2119.

Ramírez-Tobías, H.M. (2006). *Productividad primaria y calidad nutrimental de nopal (*Opuntia *spp. y* Nopalea *sp.) en condiciones intensivas,* Tesis para obtener el grado de Maestro en Ciencias, Programa Multidisciplinario de Posgrado en Ciencias Ambientales, Universidad Autónoma de San Luis Potosí, México.

Ramírez-Tobías, H.M., Aguirre-Rivera, J.R., Pinos-Rodríguez, J.M. & Reyes-Agüero, J.A. (2010). Nopalito and forage productivity of *Opuntia* spp. and *Nopalea* sp. (Cactaceae) growing under greenhouse hydroponics system, *Journal of Food, Agriculture & Environment*, Vol. 8, No. 3 & 4, (July-October 2010), 660-665, ISSN1459-0255.

Ramírez-Tobías, H.M., Reyes-Agüero, J.A., Pinos-Rodríguez, J.M. & Aguirre R., J.R. (2007a). Effect of the species and maturity over the nutrient content of cactus pear cladodes. *Agrociencia*, Vol. 41, No. 6, (August 2007), 619-626, ISSN 1405-3195.

Ramírez-Tobías, H.M., Reyes-Agüero, J.A., Aguirre R., J.R. (2007b). *Construcción, establecimiento y manejo de un módulo hidropónico para producir nopal*, UASLP-Fundación produce San Luis Potosí, AC-CONAZA, ISBN 970-705-069-1, San Luis Potosí, S.L.P. México.

Razo M., Y. & Sánchez H., M. (2002). *Acidez de 10 variantes de nopalito (*Opuntia *spp.) y su efecto en las propiedades químicas y sensoriales,* Tesis de Licenciatura, Departamento de Ingeniería Agroindustrial, Universidad Autónoma Chapingo, México.

Rodríguez-Félix A. & Cantwell M. (1988). Developmental changes in composition and quality of prickly pear cactus cladodes (nopalitos). *Plant Foods for Human Nutrition*, Vol. 38, No. 1, (March 1988), 83-93, ISSN 0921-9668.

Rodríguez-Felix, A. & Villegas-Ochoa, M.A. (1997). Quality of cactus stems (*Opuntia ficus- indica*) during low-temperature storage. *Journal of Professional Cactus Development,* Vol. 2, 142-152, ISSN 1938-6648.

Rodríguez-García, M.E., de Lira, C., Hernández-Becerra, E., Cornejo-Villegas, M.A., Palacios-Fonseca, A.J., Rojas-Molina, I., Reynoso, R., Quintero, L.C. Del Real, A., Zepeda, T.A. & Muñoz-Torres, C. (2008). Physicochemical characterization of nopal pads (*Opuntia ficus-indica*) and dry vacuum nopal powders as a function of the maturation, *Plant Foods for Human Nutrition,* Vol. 62, No. 3, (September 2008), 107-112, ISSN 0921-9668.

Sáenz, C. (2000). Processing technologies: An alternative for cactus pear (*Opuntia* spp.) fruits and cladodes, *Journal of Arid Environments*, Vol. 46, No. 3, (November 2000), 209-225, ISSN 0140-1963.

Sáenz, C., Sepúlveda, E. & Matsuhiro, B. (2004). *Opuntia* spp. mucilage's: A functional component with industrial perspectives. *Journal of Arid Environments*, Vol. 57, No. 2, (July 2004), 275-290, ISSN 0140-1963.

Sánchez del C., F. & Escalante R., E. R. (1983). *Hidroponía,* Universidad Autónoma Chapingo, Chapingo, México.

Sánchez del C., F. & Escalante R., E.R. (1988). *Hidroponía* (3rd. Edition), Universidad Autónoma Chapingo, Chapingo, Estado de México, México.

Sánchez-Venegas, G. (1995). Estimación del área fotosintética caulinar de *Nopalea cochenillifera* (L.) Salm-Dick. In: *Memoria del VI Congreso Nacional y IV Congreso Internacional sobre Conocimiento y Aprovechamiento del Nopal.* Pimienta B., E., Neri L., C., Muñoz U., A. & Huerta, M. (Eds.), 103-106, Universidad de Guadalajara, Guadalajara, Jalisco, México.

SAS, (1990). *User's Guide: Statistics*. SAS Inst. Inc., Cary, North Caroline, USA.

Stintzing F.C. & Carle, R. (2005). Cactus stems (*Opuntia* spp.): A review on their chemistry, technology, and uses. *Molecular Nutrition & Food Research,* Vol. 49, No. 2, (February 2005), 175-194, ISSN 1613-4133.

Tegegnea, F., Kijorab, C. & Petersb, K.J. (2007). Study on the optimal level of cactus pear (*Opuntia ficus-indica*) supplementation to sheep and its contribution as source of water. *Small Ruminant Research,* Vol. 72, No. 2-3, (October 2007), 157-164, ISSN 0921-4488.

Trachtenberg, S. & Mayer, A.M. (1981). Composition and properties of *Opuntia ficus-indica*mucilage, *Phytochemistry*, Vol. 20, No. 12, (December 1981), 2665-2668, ISSN 0031-9422.

Van Soest, P., Robertson, J. & Lewis, B. (1991). Methods for dietary fiber, neutral detergent fiber, and nonstarch polysaccharides in relation to animal nutrition, *Journal of Dairy Science*, Vol. 75, No. 10, (October 1991), 3583-3597, ISSN 0022-0302.

CHAPTER-11

HYDROPONIC CULTIVATION OF SEEDLINGS FOR FRUIT TREES

INTRODUCTION: AN OVERVIEW

Hydroponic cultivation has emerged as a game-changing method for the production of fruit tree seedlings. This innovative approach, which involves growing these young trees in a soil-less, nutrient-rich water medium, offers numerous advantages. It allows for precise control over the growth conditions, ensuring optimal nutrient intake and environmental parameters, which can lead to healthier and more robust seedlings. Hydroponics also mitigates the risk of soil-borne diseases, providing a clean and disease-free environment for seedling development. With a reduced environmental footprint, increased growth rates, and enhanced consistency in seedling quality, hydroponic cultivation is redefining the future of fruit tree seedling production, offering a promising solution to address the growing demand for high-quality, resilient trees in orchards worldwide.

Fruit production is an important socio-economic activity in Brazil. Data from IBGE (2007) has shown 2 million and 260 thousand hectares and a production of approximately 41 million tons shipped with tropical, subtropical and temperate fruit. In this scenario orange, banana and coconut-the-bay productions deserve to be highlighted with 821,575 ha, 519,187 ha and 283,930 ha planted, respectively.

Hydroponics is the name given to all forms of cultivation in nutrient solution without using soil. The word hydroponics comes from two Greek words: hydro, water and ponos (from Greek), which means work. The combination of words means "work with the water," and implicitly, means the use of solutions and chemical fertilizers for growing plants in the absence of soil (Catellane & Araújo, 1994).

Hydroponic cultivation of plants is an ancient technique of cultivation. Plant growth in water is reported in hieroglyphic files dating hundreds of years before Christ, which describes the cultivation of plants in the River Nile. It is believed that the first use of hydroponic cultivation as a tool was in ancient Babylon, in the famous hanging gardens, known as one of the seven wonders of the ancient world (Prieto Martinez, 2006).

Woodward, in 1699, probably conducted the first experiments testing growing plants in liquid medium without the use of solid substrates. In 1804, Saussure made one of the first attempts to analyze the factors involved in growing plants in nutrient media, establishing the requirement to provide nitrogen in the form of nitrate to the solution of cultivation. In the nineteenth century intensive research were performed involving nutrient solutions and plant growth. Researchers like Sachs, Boussingault and Knop performed experiments with nutrient solution that helped to determine the essentiality of certain chemical elements for plant growth (Nachtigall & Dechen, 2006). Several formulations of culture solutions were developed, mainly from the elaboration growing solution of Hoagland & Arnon (1950).

The creation of the polymer polyethylene in 1930 and the technique of hydroponic cultivation known as Nutrient Film Technique (NFT) in 1965, created by the British Allen Cooper, enabled the use of hydroponics as a commercial scale (Prieto Martinez, 2006). Ever since, the commercial cultivation of vegetables, fruit, culinary medicinal and ornamental plants, using hydroponic techniques, have greatly expanded, especially near large urban centers.

Hydroponics is becoming a very interesting alternative compared to traditional farming cultivating on soil. It can be used in regions where there is limited availability of arable land and in regions where there was an excessive use of the soil, causing imbalance of chemical and biological characteristics, and high infestation of plant pathogens, frequent problems in protected cultivation. Thus, even in tropical countries with abundant land, hydroponics has been used quite successfully. In addition to the high capacity of production, independent of climate and soil conditions, hydroponics also offers high quality products and reduced use of pesticides when compared to the traditional cultivation in soil (Castellane & Araújo, 1995).

Each day, the natural resources like soil and water become scarce requiring new ways to rationalize their use. Nowadays, opening new agricultural frontiers is not feasible due to deforestation and concern for the environment. It is becoming increasingly necessary to enhance the productivity of different species of plants by breeding and/or by other techniques such as hydroponics that guarantees the preservation of natural resources like water and soil and increases the productivity.

The hydroponic cultivation has as main advantages the rational use of water and nutrients supplied to plants, so plants may present a further development in short time intervals. The rationalization of water use, perhaps is the most important feature, because population growth has been directly confronted with the limited water available for agricultural production systems (Zanini et al., 2002). The main advantages of hydroponics in addition to the water savings can be cited as: lower labor demand due to the higher automation system; elimination of operations with agricultural machinery such as plows and fences that harm the soil; has no need for crop rotation; increased productivity by up to three times compared to cultivation in the soil; reduced need for pest control due to reduced wetting of the leaf area; plant uniformity in the development and early production; better ergonomics for workers; and better utilization of nutrients by avoiding waste and leaching.

On the other hand, there are doubts about the effectiveness of hydroponics in order to have high initial cost and initial labor in setting up the structure; greater risk of loss in case of power outages in automated systems; require more knowledge and training of employees; greater risk of loss contamination with pathogens. These drawbacks can be easily circumvented by diluting the initial costs over the production cycles, simple staff training, installation of electrical generator, appropriate pathogen control, cleaning system with sodium hypochlorite and weekly monitoring of the structure.

The fruit trees are increasingly being investigated in hydroponics in order to produce quality fruit with low cost and preserving the environment. The use of hydroponic systems in these types of plants has been taken for some time to attend the consumer market by offering fresh fruit. However, when it comes to hydroponic systems for the production of fruit seedlings, the information is still very scarce. Many researchers have developed their research, but they are still not released due to patent applications, as can be evidenced in Faquin & Chalfun (2008) and Medeiros et al. (2000).

There are few published studies about producing fruit seedlings in hydroponics. The proposal for a hydroponic system to produce these plants has been studied by many researchers, since this is a method that saves water, labor, reduces the application of pesticides and allows the production of seedlings in less time compared to traditional systems in the field and screened nurseries.

Several studies have demonstrated the possibility of growing fruits in hydroponics as cited by Macedo et al. (2003) about pineapple cultivation, Costa & Leal (2008) about strawberry, Dechen & Albuquerque (2000) about grape and Villela et al. (2003) about melon and others crops.

TYPES OF HYDROPONICS: CONSTRUCTION AND MANAGEMENT OF HYDROPONICS

Types of Hydroponics

According to Furlani et al. (1999), the most common types of hydroponic system are:

1. NFT (nutrient film technique) or technique of laminar flow of nutrients. This system forms a layer of cultivation along the canals where plants, especially vegetables, keeps its roots. The technique allows wide range of adaptations and can be performed in rigid or flexible tubes with different sections, diameters and lengths (Figure 11.1 A);
2. DFT (deep technique film) or growing in water or floating or pool. In this system the nutrient solution establish a layer of 50 to 20 cm where the roots are submerged, with no channels of cultivation, but a flat table (Figure 11.1 D);
3. With substrate. This system utilizes pots, tubes, filled with masonry or other inert material such as sand, stones of diverse types, vermiculite, perlite, rockwool, phenolic foam and others compounds to support the plant. The nutrient solution is percolated through these materials and subsequently drained to the bottom of the container, returning to the solution tank for recirculation (Figures 11.1 B and C).

The NFT system if compared to other hydroponic systems, presents as main advantages the lowest cost of install, easiness of operation and equipment sterilization between cultivation, saving water and nutrients due to the closed system for recycling and possible reduced environmental contamination with nutrients and pesticides from the effluent.

The hydroponic systems that employ the subirrigation can use substrates to sustain the plants or replace it with potted plants (Ebb and Flow System). In the most systems with substrates (Figures 11.1 B, C and D) is used sand, gravel, vermiculite and others (Table 11.1), since they have little or no chemical activity, so that the nutrition of plants depends entirely on the provision of a properly balanced nutrient solution (Prieto Martinez, 2006). The interest in systems that employ substrates has increased in recent years in order to reduce the number of fertigation and, consequently, the consumption of electricity (Andriolo et al., 2009).

Among the main advantages of subirrigation can be quoted to have uniform nutrition, good ventilation, easy anchoring of plants and more time for repairs in case of system failures. The disadvantages have been the largest cost of facilities and maintenance when compared with NFT.

The supply of nutrient solution by dripping is recommended for cultivation in substrates such as sand, perlite, sawdust, rice hulls, ground volcanic rock and mineral wool (Figure 11.1 B).

Fig. 11.1. Main types of hydroponics: A) NFT: one of the most used for growing hardwoods. Thisprototype uses fixed hardware and cultivation channels are double-sided facilitating the disinfection of the system; B) and C) Substrate: a system that allows the use of substrate with the nutrient solution percolation. It is also suitable for the production of fruit tree seedlings, D) DFT or Floating: a model suitable for production of seedlings of fruit trees.

This example refers to the production of seedlings "Jenipapo" (*Genipa americana* L.) by using seeds. Source: Figures 1A, 1B e 1 C: Profa. Sheila Isabel do Carmo Pinto, Figure 1 D: IFMG Bambuí, Prof. Ricardo M. Corrêa.

Cultivation can be conducted in pots, mineral wool or plastic bags. Formerly these systems were open, however, the loss of water and nutrients, and the risk of environmental contamination with effluents led to the adoption of the circulating system. When passing through the substrate that composes the cultivation area, the nutrient solution composition has changed, beyond the incorporation of suspended solids; therefore the lixiviated should be filtered, disinfected and returned to a closed system (Prieto Martinez, 2006).

Table 11.1. Materials that can be used for hydroponic cultivation on substrate. Source: Adapted from Bliska Junior, 2008 & Andriolo, 1999.

Mineral origin	Desirable characteristics	Difficulty	Finality	Reference
Vermiculite (mineral 2:1)	Free of pathogens	High cost, suffers breakdown along the crop cycles by reducing the aeration	Seedlings of peach, pear and tangerine Ponkan	Menezes 2010; Souza 2010
Perlite	Good drainage	-	Farming in general	No citation found for fruit
Rock wool	Good water retention, inert and easy to handle	Distribution of air and water disuniform	Depends on regional availability	No citation found for fruit
Phenolic foam	Easy acquisition and low cost	-	Production of vegetable seedlings	No citation found for fruit
Sand	Easy acquisition and generally inert	May contain high levels of calcium requiring care in the neutralization, for being heavy handling difficult	Absorption of macronutrients in grapevine rootstock; tolerance of grapevine rootstocks in saline	Albuquerque & Dechen (1997); Viana et al. 2001
Organic origin				
Cotton	Easy handling and lightweight	High cost	Farming in general	No citation found for fruit
Rail rice	Inert	High C / N ratio	Evaluation on strawberry varieties	Costa & Leal 2008; Fernandes Júnior et al. 2001
Pine bark / sawdust	Porosity and easy to purchase	Presence of phytotoxic substances when new, high C / N ratio	Farming in general	No citation found for fruit
Coconut fiber	Encourage germination, light, porous, easy to use and has low electrical conductivity	Depends on regional availability	Farming in general	No citation found for fruit
Peat	High water holding capacity and low density	High acquisition cost	Farming in general	No citation found for fruit
Foam castor	Not harmful for environment	Depends on regional availability	Farming in general	No citation found for fruit

Property	Substrate							
	Sand	Gravel	Expanded clay	Vermiculite	Mineral wool	Coconut fiber	Polyurethane foam	Phenolic foam
TVP(1) (%)	38-44	42	69-72	-	95	95-96	>95	> 95
Water retention capacity	Moderate/High	Low	Low	High	High	Low	High	High
Soil aeration	Low/Moderate	Moderate	Moderate/High	Moderate	Moderate/High	High	High	Moderate
Diameter (mm)	0,2-2,0	2,0-20,0	4-20	0,75-8,0	-	0,5-2,0	-	-
Density (kg m3-)	High (1500)	High (1530)	Moderate (500-600)	Low 96-160	Low <100	Low (56-75)	Low (55)	Low (10-25)
Capillary action	Moderate	Low	Low	High	High	-	-	-
Water loss by evaporation	Moderate	Moderate	Moderate	High	High	-	-	-
Loss of structure	Low	Low	Low	Moderate	Moderate	Low	No	High
Reusability	Good	Good	Good	Good	Unusual	Unusual	No	No
pH	4,0-8,0	6,9	5-7	5,5-9,0	7,0-8,5	4,9-5,6	6,0-9,0	6,0-7,5
CEC(2) (cmolc dm3-)	Low (0,3-0,5)	Low (0)	Low (0-0,2)	High (5,0)	Low 0-0,1	-	Low	Low

(1)TVP: total pore volume; (2)CEC: cation exchange capacity.

Table 11.2. Main properties of the substrates used in hydroponic systems. Adapted from Martinez Prieto (2006).

Among the advantages of farming systems by dripping can be cited the greater lateral movement of the nutrient solution and greater retention of moisture. However, the system has a higher cost of deployment and maintenance, mainly of the closed systems, and the difficulty of disinfecting after the cycles of cultivation and the possibility of obstruction of emitters (Prieto Martinez, 2006).

There are several options of substrates to be used in hydroponics, considering fruit crops were related vermiculite, sand and rice hulls (Table 11.1). However, there are possibilities of using other materials such as peat, cane bagasse, pruning waste, coconut fiber; it depends on availability, cost of these products in the region and properties (Table 11.2).

In the DFT-type systems a bench containing a blade of 5 to 10 cm of water is used, where trays or tubes are displaced in direct contact with the nutrient solution (Fig. 11.1 D), and irrigation done by capillary . The reservoir level is usually flushed where the excess solution flows through a pipe at a lower level being conducted into a reservoir where it is recirculated. However, this system requires a lot of water and good aeration system. In the seedlings production of woody species as the most fruit trees, DFT or hydroponics substrates seems to be more efficient because these plants are difficult for large-scale management and staking in NFT system. Most of the data about fruit trees seedlings has used the DFT or substrates as cited by Menezes (2010) & Souza (2010).

Each substrate has its own characteristics that must be known (Table 11.1), evaluating their suitability for the crop system and to the culture to be produced. In the choice of substrate should be considered: cost, availability, stability over time and absence of toxins and / or pathogens. The main chemical and physico-chemical characteristics of the substrate that should be assessed are: decomposition rate, pH, buffer capacity, cation exchange capacity (CEC), electrical conductivity, sodium concentration, density and water retention (Prieto Martinez, 2006). In Table 11.2 the main characteristics of the substrates can be observed.

The literature cites several hydroponic systems listing many advantages and disadvantages. According to Menezes (2010) most systems are dynamic, and there is forced circulation of water or air to oxygenate the solution. It is observed that there is a tendency to use the NFT system, Mitchell & Furlani (1999) report that this trend is due to factors such as more effective control of nutrition, reduced cost and easiness in the renewal of crop fields. However, it was observed in the present review, the tendency to use floating systems and substrates for fruit crops.

CONSTRUCTION AND MANAGEMENT OF HYDROPONICS

To consider a hydroponic system efficient it should combine low cost, high production of plants or seedlings, suitable nutrient solution for the species cultivated, as well as hydraulic structure appropriate for the proper functioning of the system.

The seedlings of fruit bearing herb may be performed using NFT systems, sub-irrigation in bed with substrate or Ebb and Flow system or drip. The cultivation of seedlings for shrub fruit (citrus, guava, peach, grape, etc.) is better suited to the cultivation system by sub- irrigation Ebb and Flow System, since due to the need for better anchoring of these plants should be grown potted and is more common to use tubes. The use of tubes in the production of seedlings of fruits makes transplanting them into the field, as well as possible damage to the root system during handling (Figure 11.1 D).

To build a hydroponic system the producer can use simple materials and even materials for recycling. Usually small producers that own little capital can acquire low-cost materials such as treated wood, vases, bottles, scrap wood and others. The reservoir of nutrient solution, depending on the number of seedlings being produced, can be water-tower of 250, 500, 1000 L or more, but should not exceed 5.000 L (due the complicated management of the solution). The use of asbestos water tank should be avoided due to release of chemical compounds in the nutrient solution.

In contrast, currently there are numerous opportunities in the market of companies that lead all construction projects of greenhouses and hydroponic systems. There is a huge variety of equipment that enables the system to automate the most productivity.

To calculate the reservoir volume of nutrient solution should be considerate the number of plants that intended to grow, ranging from 0.1 L up to 5 L. In the case of production of fruit plants that are of larger size is recommended volumes around 3-5 L^{-1} plant. Whether the goal is to produce 500 seedlings, for example, should be planned a reservoir of 2,500 L. It is noteworthy that large reservoirs (bigger than 5,000 L) complicate the management of the solution and it is recommended to mount systems in series with several smaller reservoirs. If happens contamination of the solution and loss of seedlings, the damage is minor.

The place where will be accommodate the motor-pump set must be as fresh as possible to avoid heating the solution. In general the reservoir can be buried in the soil to or build masonry to keep the tank and electrical system that support the pump and timer. In tropical locations the temperature in solution can be very high reaching 104º F, which enables the cultivation. In this sense, especially in warm regions the accommodation of the nutrient solution reservoir should be correctly kept (Figure 11.2 A).

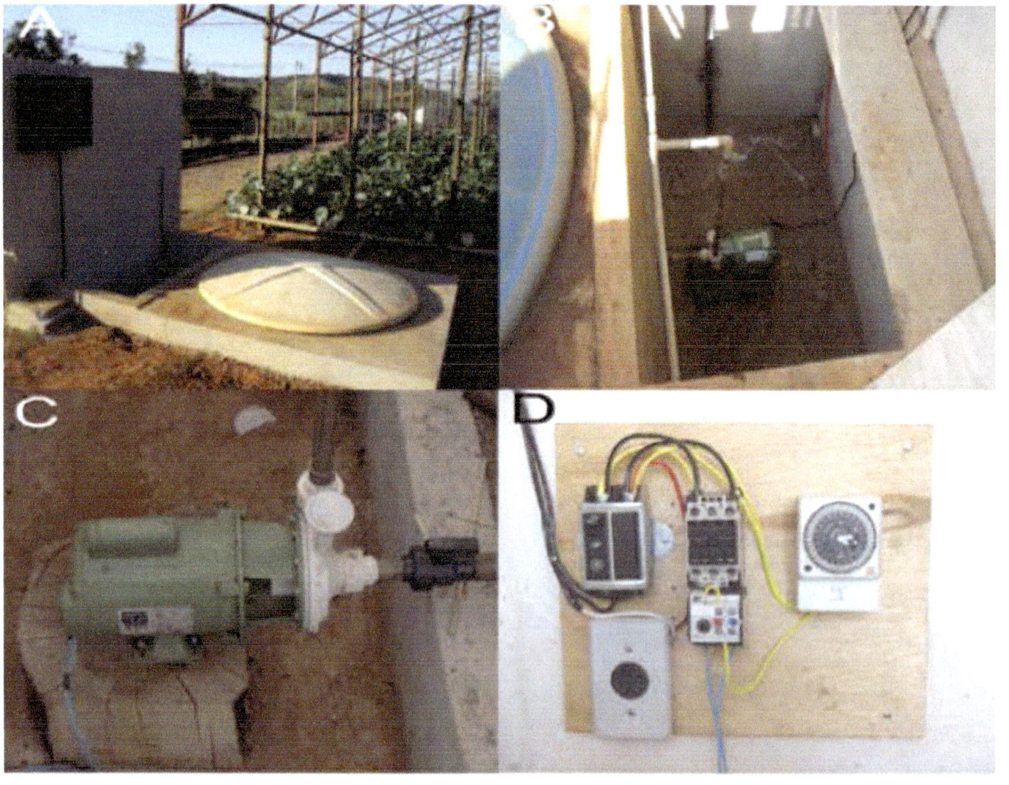

Fig. 11.2. Detail of the pump house (under construction) illustrating: A) Hydroponics being built by focusing the detail of water tank buried in the soil to avoid heating, B) and C) Set motor pump installed in a primed (below the level of reservoir), D) Panel containing outlet, timer relay and nutritious species. Photos: IFMG Bambuí, Professor Ricardo M. Corrêa.

The circulation of the nutrient solution is usually done in the NFT system from 15 to 15 minutes during the day and during the night this interval may be increased to 30 to 60 minutes due to lower evapotranspiration. This caution must be taken and the systemmonitored throughout the day especially when the temperatures are very high, since the lack of water can cause death of plants.

Electrical conductivity is another point that should be well monitored, because it measures the amount of salt added to solution. The solution must be renewed periodically to avoid problems in plant growth. Furlani (1997) recommends renew the solution every month. According to this author, the renewal avoids unnecessary accumulation of components presents in the water not absorbed by plants, and the excess of organic material from decomposed algae and roots, which contribute to the development of microorganisms harmful to plants. In general the conductivity varies between species, 1.4 to 3.0 mS cm-1 depending on the plant size, nutritional requirements and types of drains as fruits, tubers and other.

After each cycle of cultivation it is important to clean the solution reservoir with hypochlorite to reduce algae growth. After cleaning it follows by the renewal of water and dilution of the salts according to the recommendation of the species of nutrient solution.

PROPAGATION AND FRUIT SEEDLINGS PRODUCTION IN HYDROPONIC

Fruit Propagation

The propagation methods can be grouped into two types: sexual propagation, which is based on the use of seeds and asexual propagation, based on the use of vegetative structures. Fundamentally, the difference between the two forms of propagation is the occurrence of mitosis and meiosis. While asexual propagation the cell division involves the simple multiplication (mitosis), keeping unchanged the number of chromosomes, in sexual propagation meiosis provides a reduction in the number of chromosomes (Fachinello et al. 2005). In field or in greenhouse plants can be propagated asexually by cuttings, grafting, layering, through other structures such as stolons, bushes and saplings. The use of seeds in fruit cultivation is more restricted to the formation of rootstocks and breeding, with the exception of papaya and coconut that still rely on seeds to produce seedlings. The current trend in the production of seedlings of fruit trees is to work with asexual propagation in order to maintain the characteristics of the genotype, reduce the period of growing the seedling in the nursery and a consequent cost reduction as well as reduce the juvenile period and size.

Tissue culture is a biotechnological tool that allows obtaining large number of plants in limited time and with high quality plant. However, micropropagation protocols are more developed for herbaceous species such as strawberry, banana and pineapple, while occur more difficulty to growth in vitro woody fruit.

In an in vitro culture one of the techniques researched that still had little advance was the micrografting which consists of micrografts under aseptic conditions, a stem apex, containing two to three leaf primordia, excised from a mother plant on a rootstock established in vitro (Adapted from Peace & Pasqual, 1998). However, due to the difficulty of growing woody species in vitro and the process of micrografting be cumbersome, this technique still goes in slow steps.

Among the methods of asexual propagation, the most used are cuttings, layering and grafting (Simão, 1998), and for some fruit like strawberries and bananas are used morespecific methods such as stolons and division of clumps, respectively. The layering is a process that can be divided into soil layering and air layering (or layering). According to Gomes (2007), the soil layering is rarely used for fruit trees propagation. However, air layering is more applicable in the production of fruit mainly lyche and jaboticaba.

Hydroponics Cultivation of Fruit

Usually most fruit trees are grown in the field due to the need for large areas of cultivation, a soil support to maintain the plant, water, nutrients and also space for canopy growth. Over the years, researchers began to notice that not only leafy vegetables can be grown in hydroponics, but also species such as fruit vegetables as pepper, paprika, cucumber and tomato (Furlani and Morais, 1999, Rocha et al., 2010) , seed potatoes (Medeiros et al., 2002, Correa et al., 2009, Correa et al., 2008), fruits such as strawberries and melon (Furlani & Morais, 1999; Andriolo et al. 2009; Fernandes Júnior et al ., 2001), pineapple (Macedo et al. 2003); Vilela Junior et al., 2003), coffee (Tomaz et al., 2003), eucalyptus and pine (Wendling et al. 2003; Loewe & Gonzalez, 2003) among others.

The hydroponic culture requires special care in installation and conduction of the culture. The plants growth and formation with commercial quality depend on the production of good seedlings. For this, some factors should be considered, such as variety to be cultivated, seed source, substrate to be used, place of germination, seedling growth and management of the nursery (Paulus et al., 2005).

Obtaining seedlings begins with selecting the seeds, which should have been properly collected, processed, stored, packaged, and free of pathogens and pelleted (Prieto Martinez, 2006). Cultivation of plants such as strawberry, melon, watermelon, pineapple and other is held by producers to attend the trade. The seedlings to sustain these crops are made in the field or greenhouse conditions with generally high levels of pests and diseases.

Strawberry and melon plants have growth habits similar to the vegetables and, in this way, they are commonly cultivated in hydroponics where the management is facilitated due the small size of plants. However, considering fruit trees and woody plants such as orange, peach, apple, avocado, there are other restrictions and difficulties of cultivation for commercial production due to the size and weight. But hydroponics can assist in the production of woody seedlings species since the short period of time to be taken to the field.

The production of seedlings to support commercial crops is usually done in the field or greenhouse. The greenhouse crops are being most preferred due to the high phytosanitary control mainly in tropical regions that provide greater proliferation of pests and diseases. The fruit seedlings production in hydroponic cultivation is an alternative to soil cultivation and can reduce the number of applications of pesticides, prevent spread of pests and diseases, increase the efficiency of water use, reduce waste, nutrients, enabling early harvests and reduce time of seedling production.

Fachinello & Bianchi (2006) report the importance of seedlings quality influencing productivity of orchards. These same authors state that to compete in today's market fruit, it is necessary to produce with quality and competitive price. According to those authors, the productivity of orchards is seriously compromised by infection of plants by viruses and similar organisms.

Currently many producers have installed a screen against aphid and countertops held to prevent entry of insects and disease proliferation. In this sense, the hydroponic systems can be deployed in those screened maximizing the quality of producing seedlings in less time.

Souza (2010) reports that the production of pear and peach seedlings in hydroponics is considered unprecedented, becoming a new way to produce seedlings of fruit trees of temperate climates. According to him the production of seedlings in hydroponics has been used in a pioneering way, due to its early production and absence of pathogens, especially related with the soil.

Table 11.3. Seedlings of fruit species grown in hydroponics

Fruit bowl	Type of hydroponics	Nutrient solution	Propagation	Reference
Pear (*Pyrus calleryana* Decne	Floating	Not disclosed (patent in process)	Grafting	Souza (2010)
Peach (*Prunnus persica* L. cv. Okinawa)	Floating	Not disclosed (patent in process)	Grafting	Souza (2010)
Tangerine (*Citrus reticulata*)	Floating	Not disclosed (patent in process)	Grafting	Menezes (2010)
Pyrus communis cv. 'Triunfo', 'Tenra' e 'Cascatense'	Floating	Not disclosed (patent in process)	Grafting	Souza et al. (2010)
Pineapple (*Annanas comosus* L. Merrill var. Perola	Floating	1/5 of Hoagland and Arnon solution (1950)	Micropropagation and termination of the seedlings in hydroponics	Macedo et al. 2003
Guava (*Psidium guajava* L. cv. Paluma and cv. Século XXI)	Aeroponic	Test solutions of Hoagland & Arnon (1950); Sarruge (1975); Castellane & Araujo (1995); Furlani et al. (1999)	Propagation of seedlings of guava	Franco and Prado (2006)

Hydroponics emerges with a viable alternative for the production of fruit seedlings, because this culture system stimulates the production of high number of seedlings per m^2, beyond plant seedlings of high quality, attending an increasingly demanding market. Thus, the use of hydroponics for the production of citrus plants, using conventional methods of propagation, may be a promising activity (Mehta, 2010).

Among different studies developed, the propagation method initially more used in hydroponics was grafted due to be one of the most widely used in fruit growing. Table 11.3 shows some references about some fruit species such as pear, peach, mandarin Ponkan, pineapple and guava. Some of these studies are under patent and details of the hydroponic nutrient solution were not disclosed.

NUTRIENT SOLUTIONS, SUBSTRATE AND ITS MANAGEMENT

A nutrient solution can be defined as a homogeneous system where the nutrients needed by plants are scattered, usually in ionic form and proportion. Besides nutrients, it is assumed that the nutrient solution containing oxygen and proper temperature to the absorption of nutrients by the plants. However, it should be noted that in any system of soilless culture, two important factors on productivity should be observed: the environment, determined by the type of plant protection, especially the cover with transparent plastic films and fabrics for shading, and nutrient solution, which can be free or dispersed in a substrate (Cometti et al., 2006).

The secret of success in hydroponics is not the sophistication of the materials used in the construction of greenhouses and the system itself. The success in hydroponics is reached when it combines low cost assembly, market demand for the product to be produced, type of hydroponic system used, place of location the hydroponic system, nutrient solution and its management and prevention the system against crashes, besides skill labor.

Among the recommendations mentioned above, the nutrient solution is noteworthy since the food that plants need are the nutrients that depend on the amount and availability of these plants. According to Malavolta (2006) there are several aspects to consider about the mineral requirements of crops, which generally apply to all species: (1) total requirements,(2) amounts in harvested product, (3) amounts for the production unit, (4) requirement in the cycle, (5) needs in the agricultural year, (6) reserves mobilization, (7) accumulation in the fruit, (8) cycling. In this sense, the energy that will be demanded during the propagation process, such as emission of roots (in the case of cutting and layering) and growth of shoots (in the case of grafting) depend on the availability of mineral nutrients that must be available at the right time and quantities required. Thus, correct nutrient solution, balanced, correct pH and well managed contributes to the production of seedlings be successful.

The composition of the nutrient solution has been studied for many years, with reports dating to 1865, as the Knopp solution. However, only after 1933 there were concerns about the preparation of a solution containing micronutrients. In 1938, Hoagland & Arnon showed a complete and balanced nutrient solution for tomato, based on the composition of plants grown in pots with nutrient solution

(Hoagland & Arnon, 1950). In 1957, this solution was slightly adapted with respect to NO_{3-}: NH4 $^{+}$ by Johnson et al. (1957), to keep pH close to five. From the solution of Hoagland and Arnon, many others have been developed, such as Clark (1975), but the traditional solution of «Hoagland» remains the most used (Cometti et al., 2006).

Initially a lot of research with fruit adopt nutrient solutions already developed for other species, which are already standardized (Table 11.4). Dechen & Albuquerque (2000) studied the absorption of macronutrients by rootstocks and grapevine cultivars in hydroponics. These authors based the solution of macronutrients on Furlani (1995) and micronutrients on Hoagland & Arnon (1950). It was observed high vigor of rootstocks 'Jales' (IAC 572), 'Tropical' (IAC 313) and 'Campinas' (IAC 766) with high correlation between amount of biomass and nutrients accumulated.

It must be admitted, however, that there is not an ideal nutrient solution to cultivate all crops, since there is a variation depending on various factors such as species, plantdevelopmental stage, time of year, environmental factors, among others. In theory any plant that grows naturally in the soil can be grown in hydroponics, i.e. small species, shrubs, herbaceous plants such as vegetables, ornamental, medicinal and others (Crocomo, 1986). For the production of fruit seedlings like shrubs, the nutrient solutions used are still in research stage and are not published in scientific journals, since the production technology of these plants are going through the process of establishing patent.

In this sense, Franco and Prado (2006) found difficult to compare the results of their research to the literature due to lack of information about the nutrient solution for fruit, especially guava. These authors stated that there are no studies indicating an ideal solution for growing seedlings of guava, and the comparison of results with the literature is impaired.

Macedo et al. (2003) succeeded in terminating pineapple plants derived from in vitro propagation. These authors used a solution of Hoagland & Arnon (1950) diluted in 1 / 5 and concluded that the seedlings produced in the laboratory were more developing ex vitro, using the floating hydroponics. This is another applicability of hydroponics in the acclimation of seedlings, since the micropropagated plants instead of being in acclimatized on conventional nursery with waste water and nutrients can be adopted as the floating hydroponic systems or substrate for acclimatization. All plants from tissue culture must be acclimatized before going to the field and hydroponics can be an alternative to traditional nurseries.

The management of nutrient solution should be very careful done, considering that the absorption of nutrients and water occurs in different proportions, which is a challenge toproper nutrient replenishment and water. Water used to prepare the nutrient solution must be quality (Table 11.5), free of contaminants and excess chlorine.

Table 11.4. Some of the major nutrient solutions used in hydroponic fruit.

Nutrient solution	Composition (mg L-1)	Fruiter	Finality	Reference
Hogland & Arnon (1950)	210,1 (N); 31,0 (P); 234,6 (K); 200,4 (Ca); 48,6 (Mg); 64,1 (S); 0,5 (B); 0,02 (Cu); 0,65 (Cl); 5,02 (Fe); 0,5 (Mn); 0,01 (Mo); 0,05 (Zn)	Pineapple, Guava	Acclimatization of micropropagated plants, production of seedlings	Macedo et al., 2003; Franco & Prado, 2006
Castelane & Araújo (1994)	200 (N); 40 (P); 150 (Ca); 133 (Mg); 100 (S); 0,3 (B), 2,2 (Fe); 0,6 (Mn); 0,3 (Zn); 0,05 (cu) e 0,05 (Mo)	Melon	Production of melon	Costa et al., 2004
Bernardes Júnior et al. (2002)	102,62 (N); 40 (P); 116 (K); 36,16 (S); 76 (Ca); 27 (Mg); 1,89 (Fe); 0,55 (Mn); 0,32 (B); 0,20 (Zn); 0,08 (Cu); 0,02 (Mo)			
Furlani et al. (1999)	202,0 (N); 31,5 (P); 193,4 (K); 142,5 (Ca); 39,4 (Mg); 52,3 (S); 0,26 (B); 0,04 (Cu); 1,8 (Fe); 0,37 (Mn); 0,06 (Mo); 0,11 (Zn)	Guava	Seedling production	Franco & Prado, 2006

(1) EC: electrical conductivity. Adapted of Prieto Martinez (2006) and Böhme (1993).

Martinez Prieto (2006) points out that maintaining a favorable environment for plant growth depends on choice, preparation and maintenance or adjustment of the solution as the plants grow. It is essential the continuous monitoring of nutrient solution, correcting, where necessary, the volume of water, pH and nutrients concentration.

Table 11.5. Quality indices used for water used in hydroponic systems.

Index	Good	Acceptable	Maximum
EC(1) mS cm^{-1}	< 0,75	0,75 - 1,50	2,0
pH	6,50	6,80	7,50
Na^{+} mmol L^{-1}	0,87	1,30	2,61
Ca^{2+} mmol L^{-1}	6,5	10,00	14,00
Cl^{-} mmol L^{-1}	1,14	1,71	2,86
SO_4^{2-} mmol L^{-1}	0,83	1,26	2,08
Fe µmol L^{-1}	-	-	0,08
Mn µmol L^{-1}	-	-	0,04
Zn µmol L^{-1}	-	-	0,02
B µmol L^{-1}	-	-	0,03

The concentration of nutrients can be monitored by measuring the electrical conductivity of the nutrient solution, being used as an indicator of the need for replacement or exchange it. However, it should be noted that the electrical conductivity of nutrient solution is not a quantitative measure of the nutrients present in the solution, but only the concentration of ions in the medium and this may vary due to changes in temperature of the solution.

The pH of the nutrient solution should be adjusted daily, as this varies depending on the differential absorption of cations and anions by plants. High concentrations of H^+ in the nutrient solution can destabilize the cell membranes, causing loss of ions and death of cells of the root. The plants can withstand a pH between 4.5 and 7.5 without major physiological effects. However, indirect effects, such as a reduction in nutrient availability, may seriously compromise the growth of plants, since changes in pH may promote the formation of ionic species that are not readily transported to the cells, impairing the absorption of nutrients (Cometti et al., 2006).

The choice of substrate depends on their physical and chemical characteristics and requirements of the species used for rooting (Verdonck et al., 1981). The substrate affects not only the quality of roots formed, as well as in the percentage of rooting of cuttings (Couvillon, 1988), having also the function of fixing them and keep the environment on the basis of the them, wet, dark and with adequate aeration (Fachinello et al., 1994). The physical relationship between volume of water and air present in the substrate influences the morphology of adventitious roots formed and its branches (Wilson, 1983). The techniques used to produce seedlings in hydroponics adopt different substrates for rooting cuttings. Depending on the type of substrate used, the rooted cuttings may show non- uniformity of adventitious roots, reflecting the fixation and plant development (Paulus et al., 2005). You should opt for those that do not convey pathogens, which are uniform, lightweight, low cost, easy to disinfect to allow re-use, high moisture retention and good aeration, and in the case of transplanted bare root, easily release from the roots. Substrates that adhere to the roots and are difficult to remove lead to a greater stress in transplanting, delaying recovery and resumption of plant growth, and its permanence in the system could promote blockages (Prieto Martinez, 2006).

The temperature is another factor to be considered, since high temperatures of the greenhouses can impair the development of seedlings, causing a reduction in growth rate and areas of necrotic tissue in the leaves and stem. These symptoms occur due to starvation, in the other words, breathing more than photosynthesis, and it becomes worse when the humidity is excessive, because the needs of oxygen for root respiration and an accumulation of toxic products such as ethanol or acetaldehyde (Martinez Prieto, 2006).

It should be emphasized the importance of providing adequate amounts of light and nutrients in the formation of seedlings. Insufficient environmental heatstroke can result in etiolated seedlings with low rate between root / shoot and the excess can cause damage to seedlings. The nutrition of the seedlings must be balanced to provide nutrients in appropriate quantities and proportions, since the substrates normally used in hydroponic systems, have low or no chemical activity.

Barboza et al. (1997) studying pear rootstocks "Taiwan Nascher-C" in field conditions in plastic bags with 5 L of soil observed that the time for 65% of the rootstocks reached the point of grafting was 240 days of growth. However, Souza (2010) could produce these same rootstock to 77 days, 163 days in anticipation of using the hydroponic substrate. That is, in hydroponics were spent only 32% less time in relation to the field for rootstocks in grafting point. Consequently, there is significant cost reduction in producing pear seedlings.

For peach Raseira & Medeiros (1998) studied under field conditions rootstocks 'Okinawa' in 5 L plastic bags containing soil and obtained plants able to the graft 240 days after sowing, in hydroponic conditions. Souza (2010) has reduced for 69 days the formation time of the same rootstock. These data showed that the formation of peach rootstock in hydroponics takes approximately 29% less time spent than in field conditions.

Franco et al. (2007) studied the growth and mineral absorption of nutrients for seedlings of guava. These authors used a solution of Castellane and Araujo (1995) that according to Franco and Prado (2006) was the solution recommended for the guava tree. Franco et al. (2007) concluded that there is an accumulation of dry matter of guava plants along the time of growing and cultivating and the seedlings of Century XXI have greater demand for macronutrients than seedlings of the cultivar Paluma.

At IFMG Bambuí Campus are being driving and testing the system of hydroponics where DFT pools were built measuring 1.2 m wide by 2.2 m in length. These DFT pools contain media with tubes of various kinds of substrates such as bagasse cane, vermiculite and pine bark, shredded pruning waste, among others. These tubes are inserted into various materials such as seeds and propagating cuttings of several species such as guava (*Psidium guajava* L.), loquat (*Eriobotrya japonica* Lindl.), Surinam cherry (*Eugenia florida* DC.) and genipap (*G. americana* L.) (data not published). The adaptability of these species have been observed in pre-tests and further tests will be conducted to study different substrates and methods of propagation of fruit trees (Figure 11.3).

Fig. 11.3. Some possibilities for the production of fruit tree seedlings in hydroponics: A) Guava (*P. guajava* L.), B) Jenipapo (*G. americana* L.), C) Loquat (*E. japonica* Lindl.), D) Pitanga (*E. florida* DC.) E) Peach (*P. persica* L.), and F) Grape (*V. vinifera* L.). Source: Figures 1 A, 1 B, 1 C and 1 D: Professor Ricardo M. Correa. Figures 1 E and F: Courtesy of Prof. Josinaldo Lopes Araujo (Federal University of Paraíba, Brazil).

Therefore, hydroponics depends on skilled labor since any change in plants or in the system can compromise the production. The professionals who deal with hydroponics must know technical details about the species to be produced, plant health problems and how to fix them, symptoms of nutrient deficiency and toxicity, management of nutrient solution, anticipation of possible power outages and the consequent lack of water circulation in the channels, among other skills. In this sense, it is necessary to qualify the employee responsible of conducting hydroponics.

PROPOSAL FOR A HYDROPONIC SYSTEM: PRODUCTION OF GRAPE SEEDLINGS

A simple system is proposed in Figure 11.4 to obtain grape seedlings in plastic tubes using the DFT system. Initially should be prepared a wooden box of 2.2 m long by 1.2 wide and 25 cm high and then cover it with plastics double-sided with the white side facing up so it will waterproofs them. Later trusses are posted on this other side of the plastic. The following is drilling holes along the surface of the plastic tubes for accommodation of the tubes. These tubes are usually of average size of

6 cm wide and 19 cm in height comprising 280 cm^3 of substrate. Smaller tubes are not recommended because they have lower volume of substrate and commitment of the root system. The rootstocks are rooted in advance before being placed in this hydroponic system. After accommodation of grape rootstocks in the pool, the process of grafting is executed. After this process it is observed the success of the graft and subsequent growth of the seedling.

Fig. 11.4. Sequence assembly of hydroponic system to produce seedlings of fruit trees. (A) Support of wood that will hold the nutrient solution and the seedlings of fruit trees, (B) Truss to support the wooden support and sustain the tubes with the seedlings, (C) Support of wood completely covered with plastic to receive the nutrient solution and prevent the entry of light in the nutrient solution to avoid the proliferation of algae, (D) seedlings grown in hydroponic vine, (E) vine seedlings produced hydroponically presenting precocious fruit production, (F) seedlings of different species of fruit produced in hydroponic systems. Photos: Courtesy of Prof. Josinaldo Lopes Araujo (Federal University of Paraíba - Brazil).

AT A GLANCE

- Hydroponics is promising to produce fruit tree seedlings, but still needs research that guide to a better production system and for hydroponic nutrient solution that attend the needs of each species. Results from literature are scarce and discuss briefly the production of seedlings of woody plants such as orange, guava, grape, mango among other important.

- The nutrient solutions used by the researchers are still based on the others developed for different kinds of vegetables such as lettuce, tomatoes and potatoes. However there is no record of a specific nutrient solution developed specifically for fruit, but it is under a patent process.
- Future research should be conducted trying to identify a substrate, a hydroponic nutrient solution and an ideal hydroponic system for each fruit species combining low cost and profitability.

REFERENCES

Albuquerque, T.C.S.; Dechen, A. R. Absorção de macronutrientes por porta-enxertos e cultivares de videira em hidroponia. 2.000. Scientia Agricola. v. 57, n.1. mai/2000. pp. 56-63. ISSN: 0103-9016.

Andriolo, J.L. 1999. Fisiologia das culturas protegidas. ISBN: 85-7391-012-7. Santa Maria/RS. Andriolo, J.L.; Jänisch, D.I.; Schmitt, O.J.; Vaz, M.A.B.; Cardoso, F.L.; Erpen, L. 2009. Concentração da solução nutritiva no crescimento da planta, na produtividade e na qualidade de frutas do morangueiro. Ciência Rural, v.39, n.3, Maio-Junho/2009.pp.684-690. ISSN: 0103-8478.

Barboza, W.; Dall'orto, F.Ac.C.; Ojima, M.; Novo, M.C.S.S.; Betti, J.A.; Martins, F.P. Conservação e germinação de sementes e desenvolvimento de plantas da pereira porta-enxerto "Taiwan Naschi-C". 1997. Scientia Agricola. v.54. n.3. Janeiro/1997. pp. 147-151. ISSN: 0103-9016.

Bliska Júnior, A. Sistemas de cultivo hidropônico: muitas opções diferentes. 2008. Revista Plasticultura. n.4. pp.8-11. 2008.

Castellane, P.D.; Araújo, J.A.C. 1995. Cultivo sem solo: hidroponia. Jaboticabal/SP.

Clark, 1975 R.B. Clark, Characterization of phosphates in intact maize roots. 1975. *Journal of Agriculture Food Chemistry*, v. 23. n. 2, May/June. 1075. pp. 458–460. ISSN: 0021-8591.

Cometti, N.N.; Furlani, P.R.; Ruiz, H.A.; Fernandes Filho, E.I. 2006. Soluções nutritivas: formulações e aplicações. In: Nutrição mineral de plantas. Fernandes, N.S. ed..pp.89-114. Sociedade Brasileira de Ciência do Solo. ISBN: 85-86504-02-5. Viçosa/MG.

Couvilon, G.A. Rooting responses to different treatments. 1988. Acta Horticulturae, v.227, n. 23. Agosto/1988. pp.187-196. ISSN: 0567-7572.

Corrêa, R.M.; Pinto, J.E.B.P.; Pinto, C.A.B.P.; Faquin, V.; Reis, E.S.; Monteiro, A.B.; Dyer, W.E. A comparison of potato seed tuber yields in beds, pots and hydroponic systems. 2008. Scientia Horticulturae. v. 116. n. 4. Setembro/2008. pp. 17-20. ISSN: 0304-4238.

Corrêa, R.M.; Pinto, J.E.B.P.; Faquin, V.; Pinto, C.A.B.P.; Reis, E.S. The production of seed potatoes by hydroponic methods in Brazil. 2009. In: Fruit, Vegetable and Cereal Science and Biotechnology. SILVA, J.T. Ed.. pp. 133-139. Global Science Books. Japão. ISBN: 9784903313269.

Costa, C.C.; Cecílio Filho, A.B.; Cavarianni, R.L.; Barbosa, J.C. Produção do melão rendilhado em função da concentração de potássio na solução nutritiva e do número de frutos por planta. 2004. Horticultura Brasileira. v. 22. n. 1. Maio-Junho/2004. pp. 75-83. ISSN: 0102-0536.

Costa, E.; Leal, P. M. Avaliação de cultivares de morangueiro em sistemas hidropônicos sob casa de vegetação. 2008. Revista Brasileira de Fruticultura. v. 30, n. 2. Agosto/2008. pp-425-430. ISSN: 0100-2945.

Crocomo, O.J. Cultivo fora do solo: hidroponia. 1986. In: Grande manual globo de agricultura, pecuária e receituário industrial. Magalhães, A.; Bordini, M.E. (Ed.). p.209-220. Editora UFSM. Porto Alegre/RS.

Dechen, A.R.; Nachtigall, G.R. Elementos requeridos à nutrição de plantas. 2006. Fertilidade do solo. In: Novais, R.F.; Alvarez V., V.H.; Barros, N.F.; Fontes, R.L.F.; Cantarutti, R.B.; Neves, J.C.L. eds.. pp. 91-132. Sociedade Brasileira de Ciência do Solo. ISBN: 978-85-86504-08-2. Viçosa/MG.

Fachinello, J.C.; Hoffmann, A.; Nachtgal, J.C. 1994. Propagação de plantas frutíferas de clima temperado. UFPEL. Pelotas/RS.

Fachinello, J.C.; Hoffmann, A.; Nachtigal, J.C. 2005. Propagação de plantas frutíferas. Embrapa Uva e Vinho. ISBN: 85-7383-300-9. Bento Gonçalves/RS.

Faquin, V.; Chalfun, N.N.J. 2008. Hidromudas: processo de produção de porta-enxerto de mudas frutíferas, florestais e ornamentais enxertadas em hidroponia. In: Instituto Nacional de Propriedade Intelectual. (BRN.PI 0802792-7). Acesso em: 09 de agosto de 2011. Disponível em <http: www.inpi.gov.br/meu-superior/pesquisas>.

Fernandes Júnior, F.; Furlani, P.R.; Ribeiro, I.J.A.; Carvalho, C.R.L. 2001. Produção de frutos e estolhos do morangueiro em diferentes sistemas de cultivo em ambiente protegido. Bragantia. v. 61. n.1. Agosto/2001. pp.25-34. ISSN: 0006-8705.

Franco, C. F.; Prado, R. M. Uso de soluções nutritivas no desenvolvimento e no estado nutricional de mudas de goiabeira: macronutrientes. 2006. Acta Scientiarum. v. 28. n.2. Maio/2006. pp. 199-205. ISSN: 1679-9275.

Furlani, P.R. Cultivo de alface pela técnica de hidroponia NFT. Campinas: Instituto Agronômico, 18p. 1995. (Documentos IAC, 55).

Furlani, P.R. Instruções para o cultivo de hortaliças de folhas pela técnica de hidroponia NFT. Campinas: Instituto Agronômico, 30p. 1997 (Documentos IAC, 168).

Furlani, P.R.; Silveira, L.C.P.; Bolonhezi, D.; Faquin, V. 1999. Estruturas para o cultivo hidropônico. In: Informe Agropecuário: Cultivo protegido de hortaliças em solo e hidroponia. Silveira, L.C.P.; Bolonhezi, D. (Eds.) ; v. 20. n. 200/201. Maio/1999. pp. 25-40. Epamig. ISSN: 0102-0536. Belo Horizonte/MG.

Gomes, P. Fruticultura Brasileira. 2007. 13º edição. ISBN: 85-213-0126-x. São Paulo/ SP. Hogland, D.R.; Arnon, D.I. The water culture method for growing plants without soil. IBGE Califórnia: The Colllege of Agriculture, 32p. (Circular, 347). 1950. (Instituto Brasileiro de Geografia e Estatística). Agricultura. Acesso em: 22 de agosto de 2011. Disponível em: <http://www.sidra.ibge.gov.br>.

Johnson, C.M.; Stout, P.R.; Broyer, T.C.; Carlton, A.B. 1957. Comparative chlorine requirement of different plant species. *Plant and Soil*, v.8, n.3, maio/Junho 1957.pp.337-353. ISSN: 0032-079X.

Loewe, V.M.; Gonzalez, M.O. Análisis preliminar de La compatibilidad inter e intraespecífica de algumas especies nativas y exóticas en cultivo hidropónico. 2003. Bosque. v.24. n.3. Setembro/2003. pp.65-74. ISSN: 0304-8799.

Macêdo, C.E.C.; Silva, G.M.; Nóbrega, F.S.; Martins, C.P.; Barroso, P.A.V.; Alloufa, M.A.I. 2003. Concentrações de ANA e BAP na micropropagação de abacaxizeiro L. Merril (Ananas comosus) e no cultivo hidropônico das plântulas obtidas in vitro. Revista Brasileira de Fruticultura. v. 25. n. 3. Setembro/2003. pp. 45-53. ISSN: ISSN: 0100-2945.

Malavolta, E. 2006. Manual de nutrição mineral de plantas. Editora Agronômica Ceres. ISBN: 85-318-0047-1. Piracicaba/SP.

Medeiros, C. A. A. B; Raseira, M. C. B. A cultura do pessegueiro. Brasilia: Embrapa-SPI; Pelotas: Embrapa-CPACT, 1998.

Medeiros, C.A.B.; Daniels, J.; Pereira, A.S. 2000. Sistema para cultivo em hidroponia de plantas, tubérculos e bulbos. In: Instituto Nacional de Propriedade Intelectual (BRN.PI 0005711-8 B1). Acesso em: 10 de agosto de 2011. Disponível em <http: www.inpi.gov.br/meu-superior/pesquisas>.

Medeiros, C.A.B. Produção de sementes pré-básicas de batata em sistemas hidropônicos. Horticultura Brasileira. v.20, n.1. Maio/2002. pp.110-114. ISSN: 0567-7572.

Menezes, T.P. 2010. Crescimento de porta-enxertos cítricos em sistema hidropônico. MSc. thesis, Lavras/MG.

Morais, C.A.G.; Furlani, P.R. 1999. Cultivo de hortaliças de frutos em hidroponia em ambiente protegido. Informe Agropecuário: Cultivo protegido de hortaliças em solo e hidroponia. v.20. n.200/201. Maio/1999. pp. 105-113. Epamig. ISBN: 0102- 0536. Belo Horizonte/MG.

Paulus, D.; Medeiros, S.L.P.; Santos, O.S.; Riffel, C.; Fabbrin, E.G.; Paulus, E. 2005. Substratos na produção hidropônica de mudas de hortelã. Horticultura Brasileira. v. 23, n.1, mai/2005. pp .48-50. 2005. ISSN: 0567-7572.

Paz, O.P.; Pasqual, M. 1998. Microenxertia. In: Cultura de Tecidos e Transformação Genética de Plantas. Torres, A.C.; Caldas, L.S.; Buso, J.A. (Eds). v. 1. pp. 45-63. Embrapa. ISBN: 85-7383-044-1.Brasília/DF.

Prieto Martinez, H.E. 2006. Manual prático de hidroponia. Editora Aprenda Fácil. ISBN: 85- 7630-022-2. Viçosa/MG..

Rocha, M.Q.; Peil, R.M.N.; Cogo, C.M. 2010. Rendimento do tomate cereja em função do cacho floral e da concentração de nutrientes em hidroponia. Horticultura Brasileira, v.28. n.4. Outubro/2010. pp. .466-471. ISSN: 0567-7572.

Simão, S. 1998. Tratado de Fruticultura. FEALQ. ISBN: 85-7133-002-6. Piracicaba.

Souza, A.G. Produção de mudas enxertadas de pereira e pessegueiro em sistema hidropônico. 2010. MSc. thesis, Lavras/MG.

Souza, A.G.; Chalfun, N.N.J.; Souza, A.A.; Faquin, V.; Emrich, E.B.; Morales, R. G. F. 2010. Produção de matéria seca e acumulo de nutrientes em mudas de pereira. Proceedlings of XIX Congresso de Pós-Graduação da UFLA. Lavras/ MG. Outubro/2010.

Tomaz, M.A.; Silva, S.R.; Sakiyama, N.S.; Martinez, H.E.P. 2003. Eficiência de absorção, translocação e uso de cálcio, magnésio e enxofre por mudas enxertadas de *Coffea arábica*. Revista Brasileira de Ciência do Solo. v. 27. n.2. Outubro/2003. pp. 888-892. ISSN: 0100-0683.

Verdonck, O.; Vleeschauwer, D.; Boodt, M. 1981. The influence of the substrate to plant growth. Acta Horticulturae, v.126, n. 5. pp.251-258, Setembro/1981. ISSN: 0567- 7572.

Viana, A.P.; Bruckner, C.H.; Martinez, H.E.P.; Huaman, M.Y.; Mosquim, P.R. 2001. Características fisiológicas de porta-enxertos de videira em solução salina. Scientia Agrícola. v. 58. n.4. Agosto/2001. pp.65-72. ISSN: 0103-9016.

Vilela Junior, L.V.; Araújo, J.A.C.; Factor, T.L. 2003. Comportamento do meloeiro em cultivo sem solo com a utilização de biofertilizante. Horticultura Brasileira. v.21, n.2, pp.153-157, Abril/Junho. ISSN: 0567-7572.

Wedling, I.; Xavier, A.; Paiva, H.N. 2003. Influência da miniestaquia seriada no vigor de minicepas de clones de Eucalyptus grandis. Revista Árvore. v. 27. n.5. Janeiro/Fevereiro/2003. pp. 611-618. ISSN: 0100-6762.

Wilson, G.C.S. Use of vermiculite as a growth medium for tomatoes. Acta Horticulturae, v.150, n.2. Abril/Maio/1983. pp.283-288. ISSN: 0567-7572.

Zanini, J.R.; Bôas, R.L.V.; Feitosa Filho, J.C. 2002. Uso e manejo da fertirrigação e hidroponia.

INDEX